D0583481

Field Emission in Vacuum Microelectronics

MICRODEVICES
Physics and Fabrication Technologies

Series Editors: Ivor Brodie, *Consultant*
Paul Schwoebel, *SRI International and University of New Mexico*

Recent volumes in this series:

COMPOUND AND JOSEPHSON HIGH-SPEED DEVICES
Edited by Takahiko Misugi and Akihiro

ELECTRON AND ION OPTICS
Miklos Svilagyi

ELECTRON BEAM TESTING TECHNOLOGY
Edited by John T.L. Thong

FIELD EMISSION IN VACUUM MICROELECTRONICS
George Fursey

ORIENTED CRYSTALLIZATION ON AMORPHOUS SUBSTRATES
E.I. Givargizov

PHYSICS OF HIGH-SPEED TRANSISTORS
Edited by Juras Pozela

THE PHYSICS OF MICRO/NANO-FABRICATION
Ivor Brodie and Julius J. Murray

PHYSICS OF SUBMICRON DEVICES
Ivor Brodie and Julius J. Murray

THE PHYSICS OF SUBMICRON LITHOGRAPHY
Kamil A. Valiev

RAPID THERMAL PROCESSING OF SEMICONDUCTORS
Victor E. Borisenko and Peter J. Hesketh

SEMICONDUCTOR ALLOYS: PHYSICS AND MATERIAL ENGINEERING
An-Ben Chen and Arden Sher

SEMICONDUCTOR DEVICE PHYSICS AND SIMULATION
J.S. Yuan and Peter Rossi

SEMICONDUCTOR LITHOGRAPHY
Wayne M. Moreau

SEMICONDUCTOR MATERIALS
An Introduction to Basic Principles
B.G. Yacobi

SEMICONDUCTOR PHYSICAL ELECTRONICS
Sheng S. Li

A Continuation Order Plan is available for this series. A continuation order will bring delivery of each new volume immediately upon publication. Volumes are billed only upon actual shipment. For further information please contact the publisher.

Field Emission in Vacuum Microelectronics

George Fursey
Vice-Chairman, Russian Academy of Natural Sciences
Moscow, Russia

Edited by

Ivor Brodie
Consultant
Palo Alto, California

Paul Shwoebel
Senior Research Engineer
SRI International
Menlo Park, California
and
Research Professor of Physics
University of New Mexico
Albuquerque, New Mexico

Kluwer Academic / Plenum Publishers
New York, Boston, Dordrecht, London, Moscow

Library of Congress Cataloging-in-Publication Data

Fursey, George.
Field emission in vacuum microelectronics / George Fursey.
p. cm. — (Microdevices)
Includes bibliographical references and index.
ISBN 0-306-47450-6
1. Vacuum microelectronics. 2. Electromagnetic fields.
I. Title. II. Series.

TK7874.9.F87 2003
621.381—dc21 2002034120

ISBN: 0-306-47450-6

233 Spring Street, New York, New York 10013

http://www.wkap.nl/

10 9 8 7 6 5 4 3 2 1

A C.I.P. record for this book is available from the Library of Congress

Printed in the United States of America

PREFACE

The field electron emission (FEE) is a unique quantum-mechanical effect of electrons tunneling from a condensed matter (solid or liquid) into vacuum. The efficiency of this emission process is tens of millions of times higher than in other known emission processes. The extremely high current density in FEE and the fact that no energy is consumed by the emission process proper afford exceptionally wide possibilities for practical application of this effect.

Currently, the FEE is being infused with new life due to emergence of *vacuum microelectronics*, a principally new branch of the micro- and nanoelectronics.

Point field emission cathodes (FECs) find novel applications in the *fine-resolution electron microscopy*, Auger spectroscopy, atomic-resolution electron holography, and other areas of superfine diagnostics of atomic surfaces.

The FEE is also a process capable of initiating and sustaining the generation of high-power and superhigh-power electron beams (thousands and millions of amperes) through a phenomenon called the *explosion electron emission*, which is the basis of the modern high-current electronics.[485–488]

This monograph is devoted to the phenomenon of FEE per se, its governing relationships, attainable parameters of the field emission process, and its application to problems of vacuum microelectronics.

The book is intended for researches in various branches of physics, postgraduate and undergraduate students, specialists in physical electronics (emission electronics, micro- and nanoelectronics), engineers, and technologists working in the field of vacuum microelectronics, specialists interested in electron microscopy and electron-probe systems of high resolution.

The research in FEE has a long history and a number of excellent books and reviews had been written on this subject. Among them there is a review by W. P. Dyke and W. W. Dolan (1956),[27] monographs by M. I. Elinson and G. F. Vasil'ev (1958),[25] a book by R. Gomer (1961),[23] a review by R. Fischer and H. Neumann (1966),[149] monographs by A. G. J. Van Oostrum (1966)[17] and by L. W. Swanson and A. E. Bell (1973),[19] a volume of review articles on Non-Incandescent Cathodes edited by Elinson,[28, 65, 66, 179] a book by A. Modinos (1984),[29] a topical review on vacuum microelectronics by I. Brodie and C. A. Spindt (1992),[87] a review by I. Brodie and P. R. Schwoebel (1994),[428] a review by G. N. Fursey (1996),[152] and others.

Lately, new data on the FEE process have been uncovered and further areas of research took shape. Besides, stimulated by the needs of the vacuum microelectronics, a strong interest has developed in some features of the field emission process under specific conditions: in microwave fields, at extreme current densities, and when FECs operate at very short pulse durations. New aspects emerged in the problem of stability and formation of the surface of FECs. Very little attention is given to studies of FEE from nanoscale objects (emitters of just this size are currently used in the vacuum microelectronics). Renewed interest is seen toward FEE from semiconductors. In the earlier reviews, this information is either lacking or insufficient. This monograph has the purpose to fill this gap.

Very important information on the properties of field electron emitters is provided by field emission microscopy and, accordingly, the data accumulated in field emission imaging experiments. The field emission patterns represent an immediate source of data on variation of the submicron geometry of an emitting surface, its work function and many other characteristics. Of great importance is the unique possibility, afforded by the field emission microscopy, of studying in situ the processes on the surface taking place in superhigh electric fields of $\sim 10^8$ V/cm. Much effort has been invested in order to present in this book a wide range of images for different materials from metals, semiconductors, and carbon nanoclasters. Such a range of original field emission images of different objects in a wide range of electric fields is published for the first time.

Special emphasis is made on the limiting properties of the field emission process (Chapter 3). Distinctive features of the FEE in microwave fields are considered in Chapter 4. In Chapter 5, detailed consideration is given to the FEE from semiconductors, and in Chapter 6, data on FEE statistics are presented. Great attention is paid to analysis of new ideas in the theory of FEE (Chapters 1 and 3) and to fundamental, crucial experiments providing deeper insight into the mechanism of the effect under extremely high electric field (Chapter 3).

In Chapter 8, the effect of FEE is treated as a possible basis for designing various devices of the vacuum microelectronics. Here we attempted to clarify with all possible detail the problems of a new class of low-voltage displays (Chapter 8) and the use of point FECs in electron probe systems (Chapter 7). Special attention is given to the possibility of application in the vacuum microelectronics of self-organizing structures based on carbon clusters, which have a very low emission threshold, such as diamond-like films, fullerenes, and nanotubes.

Initially, the book was supposed to cover also studies of the field emission from the superconducting state and polarization effects in emission from surfaces coated with ferromagnetic films. However, because of insufficient amount of data on these problems, such a review was considered premature.

The absence of data on the energy distribution of field emission electrons in this book can be puzzling. This subject has been excellently dealt with in an original work by R. D. Young (1959),[15] in a paper by R. D. Young and E. W. Muller (1959),[16] in reviews by L. W. Swanson and A. E. Bell (1973)[19] and A. G. J. Van Oostrum (1966),[17] and in a book by A. Modinos (1984).[29] I could not convince myself to just repeat their account of the problem and refer the reader to the above publications.

This book would not have been written without the help from my colleagues, associates, and friends, whose assistance in preparing this monograph cannot be overestimated. First of all, I would like to express my deepest gratitude to the outstanding specialists in the field of vacuum electronics, Henry Gray and Charles Spindt, for many discussions, hints, and

the access to detailed information concerning their published works. My great appreciation to professor Francis Charbonnier for his interest in the book and the overall support.
Many thanks are due to Prof. Aivor Brodie and Dr. Paul Schwoebel for taking part in the work on this book, in particular, the valuable additions they made to the book, editing work and very useful comments on the manner of presentation of the material.
I am very thankful to Prof. E. I. Givargizov and Prof. A. J. Melmed for making available to me excellent pictures of field ion images of some semiconductors, and Prof. V. N. Shredink for the original photographs of field emission pictures of nanoscale emitters.
I am also grateful to Prof. L. M. Baskin, Dr. D. V. Glazanov for the fruitful discussion and, A. D. Andreev, A. N. Saveliev, V. M. Oichenko, and D. V. Novikov for the technical assistance in preparing the book for publication and to B. N. Kalinin for translating it into English. My sincere thanks go to S. I. Martynov for help with artistic design of the book cover layout.
And, finally, I express my most heartfelt thanks to my wife Ludmila Fursey for her patience and forbearance, for the work she assumed of listening to and correcting many fragments of this writing.

HISTORICAL OVERVIEW

The phenomenon of the electron field emission was discovered by R. W. Wood in 1897.[1] Initial theoretical insight into this process was provided by W. Schottky (1923),[2] who assumed that the electrons are emitted over a potential barrier at the surface lowered by the applied electric field. It was subsequently shown experimentally that fields capable of initiating emission are in fact 10–50 times lower than this value.

R. A. Millikan and C. F. Eyring[3] and B. S. Gossling[4] discovered (1926) that emission currents are not affected by temperatures up to 1500 K. Soon afterwards R. A. Millikan, and C. C. Lauritsen[5] found that the emission current varied exponentially with the applied electric potential. In 1928, R. H. Fowler and L. W. Nordheim developed a theory of field emission based on quantum-mechanical tunneling of electrons through the surface potential barrier.[6,7] This theory accurately described the dependence of the emission current on the electric field and the work function. It also followed from this theory that no external excitation is required for the initiation of this process (as distinct from thermal and photo emission). Clear evidence to this was obtained by J. E. Henderson et al. in their studies of the energy distribution of electrons[8–10] and measurements of the calorimetric effect.[11,12] A more rigorous study of the energy distribution was later undertaken by E. W. Müller,[13,14] R. D. Young,[16] A. G. J. Van Oostrum[17] and L. W. Swanson.[18,19] An important development in the study of field emission was the invention of the field emission microscope by E. W. Müller in 1936.[20,21] It was, in fact, at about this time that the systematic accumulation of data began on the surface properties of field emitters. With the field emission microscope, it was possible to investigate many phenomena, such as factors causing instabilities of the field emission process and the way in which the emitter tip is affected by the applied electric field, temperature, adsorption of foreign atoms, and electron and ion bombardment. The high magnification (10^5–10^6 times) and resolution (10–30 Å) of the field emission microscope, in combination with the possibility of actively influencing the object under study *in situ*, made it an indispensable tool for studying adsorption, desorption, epitaxy, surface diffusion, phase transitions, etc. (see reviews).[14,19,22–29]

One of the most important results of the quantum-mechanical theory was the prediction of extremely high field emission current densities, far in excess of those possible with thermal electron emission. In 1940, R. Haefer[30] made use of the transmission electron microscope in a quantitative study of emitter's shape and emitting area and experimentally proved the feasibility of achieving high current densities, by obtaining stationary current densities of $j \sim 10^6$ A/cm^2.

The beginning of basic research in the high current range and the first achievements in the practical use of field emission are associated with research by W. P. Dyke and his group.[31–36] Using high-vacuum equipment, pulsed techniques, and state-of-the-art technologies, these researchers were able to study electron emission in higher fields and current densities than those explored previously. Current densities of 10^7–10^8 A/cm^2 were achieved under pulsed and steady-state conditions, respectively.[31, 32] W. P. Dyke and J. K. Trolan[31] observed appreciable deviations from the Fowler–Nordheim theory in the high-field region. These deviations manifested themselves as a subexponential dependence of the field emission current on the applied potential. They ascribed this observation to space charge induced lowering of the electric field strength near the emitter surface. J. P. Barbour et al.[34] provided a more detailed theoretical and experimental validation of this interpretation. T. J. Lewis[37] suggested that the observed departure from the theory could be related to the fact that the shape of potential barrier differs from that assumed in the image force theory. This difference becomes greater in the high-field region where barrier dimensions are of the same order as the interatomic separations, an idea that has been elaborated upon further.[38]

A number of important results obtained by the Dyke's group relate to the causes of instabilities and degradation of field emitters.[31–33] It appeared that the main cause of the emitter degradation was Joule heating of the tip by the emission current.[32, 33] In the case of clean and smooth emitter surfaces, the instability develops throughout the emitting tip. This heating can also occur locally when isolated micro-nonuniformities develop on the emitter as a result of, for example, ion bombardment of the cathode. By eliminating factors leading to the formation of micro-nonuniformities, stable operation of the field emission cathode in the continuous mode for more than 7500 hr was demonstrated.[31–33] In these experiments, current densities were close to the maximum achievable values of $\sim 10^7$ A/cm^2).

M. I. Elinson and co-workers[40–44] found that the maximum current values were dependent on emitter geometry and showed that by increasing the tip cone angle, the current density could be increased by about an order of magnitude without emitter tip damage. Researchers in this group outlined a program to search for suitable materials for the field emission cathodes and studied materials based on metal-like and semiconductor compounds, such as LaB_6[42–44] and ZrC.[43, 44] A number of investigations in the high current density region have also been performed by G. N. Shuppe's group.[45, 46]

Further progress in research on field emission at extremely high current densities has been conducted by G. N. Fursey et al. Improvement in the experimental techniques employed made it possible to increase sensitivity of the pulsed measurements by 5–7 orders of magnitude.[47, 48] This allowed for the first measurements to be made of the maximum current densities from localized emission areas on the emitter tip crystal. The pulsed measurements were extended to the range of 10^{-9}–10^{-3} sec.[51–54] Experiments under quasi-steady-state conditions were performed in the range of 10^{-2}–10 sec. In direct current experiments, thermal effects due to field emission at high current densities were demonstrated.[48, 50] A new type of instability caused by the spontaneous change of the cathode surface microgeometry near the thermal destruction threshold[51, 52] was discovered. In studies by G. A. Mesyats and G. N. Fursey, current densities of 10^9 A/cm^2 were observed with nanosecond range pulse lengths.[53, 54] Current densities up to 5×10^9 A/cm^2 were demonstrated for field emission localized to nanometer-scale emitting areas.[55] In experiments by V. N. Shrednik et al., current densities up to $(10^9–10^{10})$ A/cm^2 are recorded from nanometer-sized tips under steady-state conditions.[56] Recently, G. N. Fursey and D. V. Galazanov, using tips with an apex radius of ~10 Å, were able to reach current densities of $10^{10}–10^{11}$ A/cm^2.[57] These current densities are close to the theoretical supply limit of a metal's conduction band when the electron tunneling probability is unity.

Experimental and theoretical studies have been conducted in order to determine a method of increasing the stability of the field emission current and preventing ion bombardment of the cathode.[25, 58, 59] In fairly recent studies by our group,[60, 61] it was shown that applied microwave fields could reduce the intensity of cathode ion bombardment by several orders of magnitude due to a repulsive potential for ions near the surface.

The practical application of field emission began with the founding of the Field Emission Corporation by W. P. Dyke in 1959. This company produced unique commercial instruments such as pulsed X-ray apparatus for recording high-speed processes[62, 63] and compact X-ray sources for medicine.[39] Subsequently, it was discovered[66] that the phenomenon taking place in these instruments was not only conventional field emission, but a related phenomena referred to as explosive emission.[67–70]

In the 1960s, A. V. Crewe and co-workers[71–74] demonstrated the possibility of using field emitters as an electron source for atomic-scale resolution electron microscopy. The field emitter became a promising electron source for Auger spectroscopy,[75, 76] X-ray microscopes,[77] and semiconductor lithography.[78, 79] Quite recently, reports have been published on the atomic-scale electron holography using field emitters.[80–86]

Of particular interest is the use of field emitters in microwave devices.[61, 87, 88] A series of unique microwave devices has been proposed based on field emission.[58, 89–101, 111]

K. R. Shoulders[102] suggested using arrays of field emission cathodes in various microelectronic components and devices. Research along this avenue was initiated in the United States by C. A. Spindt.[87, 103] Interest in this area has grown rapidly over the last 10 years in the USA,[104] France,[105] Japan,[443] and Russia.[107] The most striking example of the application of field emission cathodes to an area of technological interest is in flat-panel displays[112–118] that are remarkable in their potential for high brightness, resolution, and low price when compared to existing displays.

In recent years, new ideas and experimental techniques and approaches have been developed to address observations that could not be reconciled with previous theoretical models. Examples are the energy spectra of the field emitted electrons, which show broadening of the energy spectrum at high current densities,[120] the presence of high- and low-energy tails,[121, 122] and additional peaks thought to be due to the bulk band structure and surface states (see p. 223).[19] In some studies,[123] additional peaks have been observed using very small, atomically sharp emitters.

As a consequence of the potential impact of vacuum electronics in the field of nanoelectronics, there is a need for insight into the physics of field emission when emitter dimensions are equal to or smaller than the width of the surface potential barrier.[124–127] Again, the inadequacy of the one-particle approximation of the field emission is apparent in this case.[121, 122, 128] Several related studies have been conducted on a variety of topics. F. I. Itskovich[130, 131] indicated the possible effect of the Fermi surface structure on field emission. Also,[61, 132–135] adiabaticity has been considered with electron tunneling in high-frequency alternating fields. For field emission from metallic electron emitters coated with ferromagnetic films, polarized electrons have been detected (19: pp. 256–258, 29:

pp. 155–164).[136–144, 146] A number of researchers have also studied statistical processes in field emission[147] and found that in some cases many-particle tunneling occurred with oxidized surfaces and from high-temperature super conducting materials.[148]
Lastly considerable advances have been made in research on field emission from semiconductors (Chapter 8, pp. 149, 150–152).[29] Recently, interest in field emission from diamond and diamond-like films and fullerenes has grown dramatically.[153–158, 427, 446, 525]

CONTENTS

1 Field Electron Emission from Metals **1**

1.1 Fowler–Nordheim theory 1
1.2 Thermal-field emission 4
1.3 Extending the theory of field electron emission from metals 6
1.3.1 Deviations from the Fowler–Nordheim Theory and Peculiarities of Field Electron Emission from Small-Scale Objects 6
1.3.2 The Effect of Fermi Surface Structure 12
1.3.3 Many-Particle Effects 13
1.4 Resume 17

2 Characteristics of Field Emission at Very High Current Densities **19**

2.1 Deviations from the Fowler–Nordheim theory in very high electric fields 19
2.2 Space charge effects in field emission 19
2.3 Space charge with relativistic electrons 22
2.4 The shape of the potential barrier in high electric fields 26
2.5 Resume 30

3 Maximum Field Emission Current Densities **31**

3.1 The theoretical limit of field emission current 31
3.2 Effects preceding field emitter explosion 32
3.2.1 Dyke's Experiments 33
3.2.2 Pre-Breakdown Phenomena 34
3.2.3 Time-Dependent Observations 35
3.2.4 Quantifying Local Effects 36
3.3 Heating as the cause of field emission cathode instabilities 36
3.3.1 Experimental Demonstration of Field Emitter Heating 37
3.3.2 Analysis of Thermal Processes 38
3.3.3 Three-Dimensional Analysis of Emitter Heating 44
3.4 Build-up of the field emitter surface at high current densities: thermal field surface self-diffusion 47
3.5 The highest field emission current densities achieved experimentally 49
3.5.1 Experimental Current Density Values 50

3.5.2 The Limiting Current Densities Attained Cooling 50
3.5.3 Current Densities from Nanometer-Scale Field Emitters 54
3.6 Resume 56

4 Field Emission in Microwave Field **57**
4.1 Introduction 57
4.2 The adiabatic condition—tunneling time 58
4.3 Experimental verification of Fowler–Nordheim theory 60
4.4 Maximum field emission current densities 62
4.5 Ion bombardment of the cathode 63
4.6 Electron energy spectra: Transit time 64
4.7 Field emission from liquid surfaces 67
4.8 Resume 69

5 Field Emission from Semiconductors **71**
5.1 Introduction 71
5.2 Emitter surface cleaning 72
5.3 Current–voltage characteristics 75
5.4 On preserving the initial surface properties of a field emitter 77
5.5 Voltage drop across the sample and the field distribution in the emitting area 78
5.6 Theory of the field electron emission from semiconductors 81
5.6.1 Stratton's Theory 82
5.6.2 Modeling a Semiconductor Emitter: Statement of the Problem 83
5.6.3 Zero Current Approximation 87
5.6.4 The "Nonzero Current" Approximation with p-Type Semiconductors 88
5.6.5 The Effect of High Internal Fields 90
5.6.6 Field Emission from n-Type Semiconductors 90
5.6.7 Results of Numerical Calculations 91
5.7 Transition processes in field emission from semiconductors 97
5.7.1 Time Variation of the Field Electron Emission 98
5.8 The stable semiconductor field emission cathode 99
5.9 Adsorption on semiconductor surfaces 101
5.9.1 Oxygen Adsorption on Ge 101
5.9.2 Electroadsorption 101
5.10 Resume 103

6 Statistics of Field Electron Emission **105**
6.1 Formulation of the problem 105
6.2 Method of investigation 106
6.3 Field emission statistics from metals 108
6.4 Field emission statistics at cryogenic temperatures 109
6.5 Multielectron field emission from high temperature superconducting ceramics 111
6.6 Investigations of field emission statistics for highly transparent barriers 113
6.7 Resume 114

7 The Use of Field Emission Cathodes in Electron Optical Systems: Emission Localization to Small Solid Angles **115**
7.1 Introduction ... 115
7.2 The optimum crystallographic orientation of the field emission cathode ... 116
7.3 Field emission localization by thermal-field surface self-diffusion ... 117
7.3.1 Localization by Build-Up Processes ... 117
7.3.2 Localization by Micro-Protrusion Growth ... 118
7.3.3 Large Micro-Protrusions ... 119
7.3.4 Fine Micro-Protrusions ... 122
7.4 Field emission localization by a local work function decrease ... 124
7.4.1 Basic Principles of Localization by Decreasing the Work Function ... 124
7.4.2 Zr on W ... 125
7.4.3 The Influence of the Vacuum Conditions ... 125
7.4.4 The Dispenser Cathode Approach ... 128
7.4.5 Stabilization by Oxygen Treatment ... 128
7.4.6 Shaping of the Zr/W Emission Sites in an Applied Electric Field ... 129
7.5 Field emission from atomically sharp protuberances ... 131
7.6 Applications of field emission cathodes in electron optical devices ... 131
7.7 Resume ... 136

8 Advances in Applications **137**
8.1 Introduction ... 137
8.2 Short historical review and main development stages ... 138
8.3 Field electron microscopy ... 140
8.4 Field emission displays ... 144
8.4.1 Requirements to the Tips and Lifetime of Matrix Field Emission Cathodes ... 148
8.4.2 Comparative Characteristics of the Display Types: Advantages of Field Emission Displays ... 150
8.5 Other applications of field emission ... 152
8.5.1 Miniature Field Emission Triodes and Amplifiers Based on Field Emission Arrays (FEAs) ... 153
8.5.2 Microwave Devices for Millimeter Wave Amplification ... 155
8.5.3 Nanolithography ... 158
8.5.4 Vacuum Magnetic Sensors ... 158
8.5.5 External Pressure Sensors ... 159
8.5.6 Mass-Spectrometers with Field Emission Cathodes ... 159
8.5.7 Use of Field Emission in Gas Lasers ... 160
8.6 Arrays made of carbon nanoclusters ... 161
8.7 Resume ... 169

References **171**

List of Main Notation **193**

Index **197**

1

FIELD ELECTRON EMISSION FROM METALS

1.1. FOWLER–NORDHEIM THEORY

The field emission (FE) process is a unique type of electron emission as it is due exclusively to quantum-mechanical effects—tunneling of electrons into vacuum. This phenomenon occurs in high electric fields. This phenomenon occurs in high electric fields 10^7–10^8 V/cm. In order to produce such high fields using reasonable potentials the emitter is usually formed into a tip with the apex radius of curvature ranging from tens of angstroms to several microns. The high electric field narrows the potential barrier at the metal–vacuum interface sufficiently for the electrons to have a significant probability of tunneling from the solid into the vacuum.

The quantitative description of the process, as given by the Fowler–Nordheim (FN) theory,[6, 7, 159] provides an adequate picture. Quantifying the FE process involves calculating the FE current density as a function of the electric field. To do this the probability for the electron to tunnel through the potential barrier must be determined, that is, the barrier transparency, D, and the flow, N, of electrons incident on the barrier from within the metal must be calculated and then integrated over the electron energy on which both D and N are dependent. Usually this is the portion of the electron energy whose velocity component is normal to the barrier.

FN theory is based on the following main assumptions:

1. The metal is assumed to obey the Sommerfeld free electron model with Fermi–Dirac statistics.
2. The metal surface is taken to be planar, that is, the one-dimensional problem is considered. This assumption is accurate because in most cases the thickness of the potential barrier in fields of 10^7–10^8 V/cm is several orders of magnitude less than the emitter radius. Thus, the external field can be taken to be uniform along the surface.
3. The potential $U_1(x)$ within the metal is considered constant ($U_1(x) = \text{const} = -U_0$). Outside the metal the potential barrier is regarded as entirely due to the image forces $U_x = -e^2/4x$ with the externally applied electric field having no effect on the electron states inside the metal.
4. The calculation is performed for the temperature $T = 0$ K.

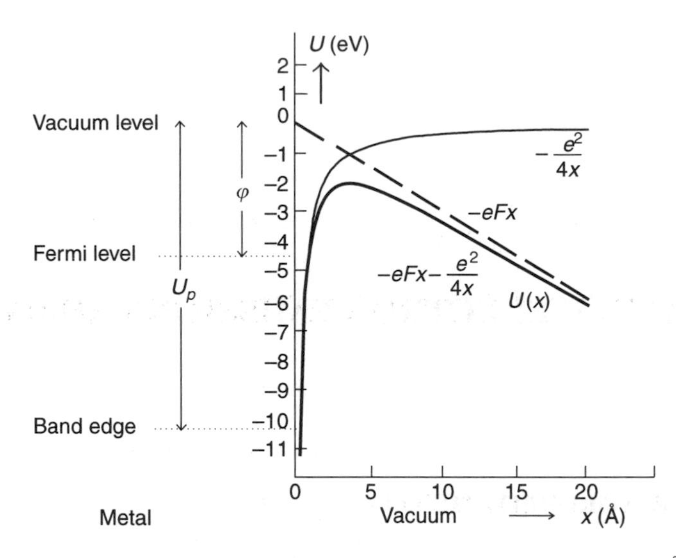

Figure 1.1. The potential energy of electron $U(x)$ (in eV) as function of the distance x (in Å) from the metal surface. $-e^2/4x$ is the image force potential; $-eFx$ is external applied potential; $U(x)$ is total potential; U_p is total potential well depth in the metal; φ is work function; F is electric field strength.

Under these assumptions, the current density is given by the equation

$$j = e\int_0^\infty n(E_x)D(E_x, F)\,dE_x, \tag{1.1}$$

where e is the electron charge, $n(E_x)$ is the number of electrons per second having energies between E_x and $E_x + dE_x$, incident on 1 cm^2 of the barrier surface from within the metal; $E_x = p_x^2/2m$ is the part of the electron kinetic energy carried by the momentum component p_x normal to the surface, m is free electron rest mass, and F is applied electric field.

The barrier transparency is calculated using the semiclassical method of Wentzel–Kramers–Brillouin (WKB) approximation.[160, 161] With an applied electric field F the following potential function

$$U(x) = -\frac{e^2}{4x} - eFx \tag{1.2}$$

describes the barrier (Fig. 1.1).

The calculations show that for such a potential barrier the transparency is given by

$$D(Ex, F) = \exp\left[-\frac{8\pi(2m)^{1/2}}{3he}\right]\frac{|E_x|^{3/2}}{F}\vartheta(y), \tag{1.3}$$

where $\vartheta(y)$ is the Nordheim function.

$$\vartheta(y) = 2^{-1/2}[1 + (1 - y^2)^{1/2}]^{1/2} \cdot [E(k) - \{1 - (1 - y^2)^{1/2}\}]K(k), \tag{1.4}$$

having for an argument

$$y = \frac{(e^3 F)^{1/2}}{\varphi}, \tag{1.5}$$

and

$$E(k) = \int_0^{\pi/2} \frac{d\alpha}{(1 - k^2 \sin^2 \alpha)^{1/2}}, \quad K(k) = \int_0^{\pi/2} (1 - k^2 \sin^2 \alpha)^{1/2}\, d\alpha \tag{1.6}$$

are complete elliptic integrals of the first and second kinds, with

$$k^2 = \frac{2(1 - y^2)^{1/2}}{1 + (1 - y^2)^{1/2}}.$$

Using (1.3), the FE current density at $T = 0$ follows the classic FN formula

$$j = \frac{e^3}{8\pi h} \frac{F^2}{t^2(y)\varphi} \exp\left[-6.83 \cdot 10^7 \frac{\varphi^{3/2}}{F} \vartheta(y)\right], \tag{1.7}$$

where φ is the work function. Substituting the values of constants and expressing φ in eV, F in V/cm, and j in A/cm^2 we have,

$$j = 1.54 \cdot 10^{-6} \frac{F^2}{t^2(y)\varphi} \left[-6.83 \cdot 10^7 \frac{\varphi^{2/3}}{F} \vartheta(y)\right], \tag{1.8}$$

where $y = 3.79 \cdot 10^{-4} \cdot \sqrt{F}/\varphi$, $t(y) = \vartheta(y) - (2y/3)(d\vartheta(y)/dy)$.

$\vartheta(y)$ and $t(y)$ had been tabulated[106] and can be found in a number of works.[25, 162] $t(y)$ in the preexponential factor is close to unity and varies weakly with the argument. In many cases it is justifiably set to unity The Nordheim function $\vartheta(y)$ varies significantly with y and, correspondingly, so does F (Fig. 1.2).

Formulas (1.7) and (1.8) give an excellent description of the experimentally observed exponential dependence of the emission current on field strength F and work function φ. In the so-called FN coordinates the functional dependence $\ln j = f(1/F)$ or, correspondingly,

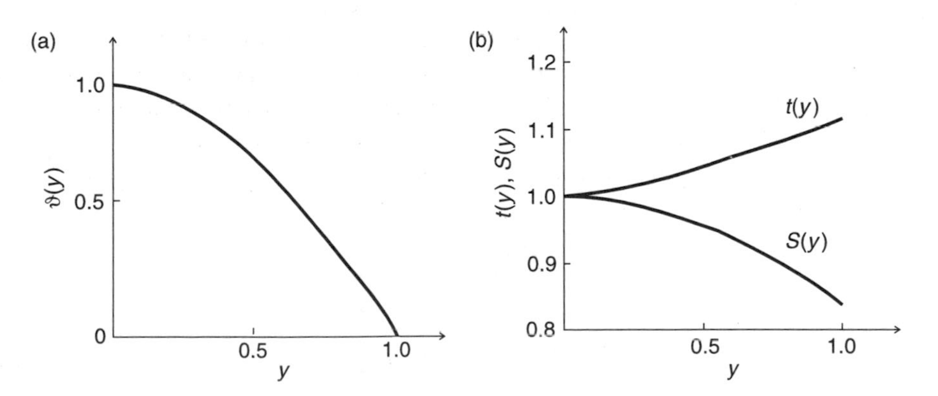

Figure 1.2. The behavior of ϑ, t, and S as a function of y.

$\ln I = f(1/U)$ is a straight line over a wide range of emission current values. Here, $I = j \cdot S$ is the emission current, with S as the emitting area and $F = \beta U$, where β is a geometric quotient determined by the geometry of the vacuum gap.

The Nordheim function $\vartheta(y)$, as noted earlier, is strongly dependent on y (Fig. 1.2), however it does not significantly affect the linear behavior of the current–voltage characteristic because $\vartheta(y)$ is very close to a parabolic curve of the form $\vartheta = 1 - by^2$ (b = const). In this approximation the additional dependence on F through $\vartheta(y)$ is transferred to the preexponential factor and has no effect on $d(\ln j)/d(1/F)$. The correction to the slope of the so-called FN characteristic $\ln(j/F^2) = f(1/F)$ is given by the expression

$$S(y) = \vartheta(y) - \frac{y}{2} \cdot \frac{d\vartheta(y)}{dy}, \tag{1.9}$$

which, as with $t(y)$, is very close to unity (Fig. 1.2). Then,

$$\frac{d\left[\lg(j/F^2)\right]}{d(1/F)} = -2.98 \cdot 10^7 \varphi^{3/2} S(y) \tag{1.10}$$

with φ in eV, F in V/cm, j in A/cm^2.

1.2. THERMAL-FIELD EMISSION

At nonzero temperatures the energy spectrum of electrons in a metal will contain electrons at energies above the Fermi level. These electrons begin to contribute appreciably into the emission current. Emission of this sort is referred to as *thermal-field (T-F) emission*. In this case the limit for the Fermi functions $f(E_x, 0) = 1$ in formula (1.1) can no longer be used and expression (1.1) should be integrated using the general Fermi function

$$f(E_x, T) = \frac{1}{\exp((E_x - E_F)/kT) + 1} \tag{1.11}$$

where E_F is the Fermi energy and k the Boltzmann constant.

Calculations of the T-F emission current density have been made by a number of authors.[163–166] A fairly simple analytic expression can be obtained only for rather low temperatures,[164] at $T \leq 1000$ K. At higher temperatures the analytic solution is possible only over limited temperature intervals and the expressions are cumbersome.

The ratio of $j(F, T)$ to that at zero temperature, $j(F, 0)$ may be expressed as[164]

$$\frac{j(F, T)}{j(F, 0)} = \frac{\pi\omega}{\sin \pi\omega}, \tag{1.12}$$

where

$$\omega = \frac{4\pi\sqrt{2m}k\sqrt{\varphi}t(y)}{he}\frac{T}{F} \cong 9.22 \cdot 10^3 \sqrt{\varphi}\frac{T}{F} \tag{1.13}$$

Expression (1.12) holds as long as $\omega < 0.7$,[165] such that $j(F, T)/j(F, 0) \leq 5$.

Numerical calculations for different temperatures, work function values, and field strengths have been made by Dolan and Dyke[167] using the WKB method.[160, 161] More rigorous numerical calculations[168] used a more accurate expression for the barrier transparency,[169] which is applicable throughout the energy range of interest, both above and below the potential barrier:

$$D(E_x, F) = \left[1 + \exp\left(-2ih^{-1}\int_{x_1}^{x_2} p(x)\,dx\right)\right]^{-1}, \tag{1.14}$$

where

$$p(x) = \left\{2m\left[E_x + \frac{e^2}{4x} + eF\right]\right\}^{1/2}$$

and x_1 and x_2 are zeros of the radical defined such that $x_1 < x_2$. The results of numerical calculations[168] are shown in Fig. 1.3. These curves allow for an estimation of T-F emission current at various temperatures, work functions, and field strengths. The effect of the temperature is most pronounced for low fields and low work function values. For example, at 1000 K with a work function value $\varphi = 4.5$ eV, $j(F, T)/j(F, 0) \approx 5$ at $F = 2 \cdot 10^7$ V/cm where as at $F = 3 \cdot 10^7$ V/cm we have $j(F, T)/j(F, 0) \approx 1.7$. For higher field strengths the temperature contribution is even less.

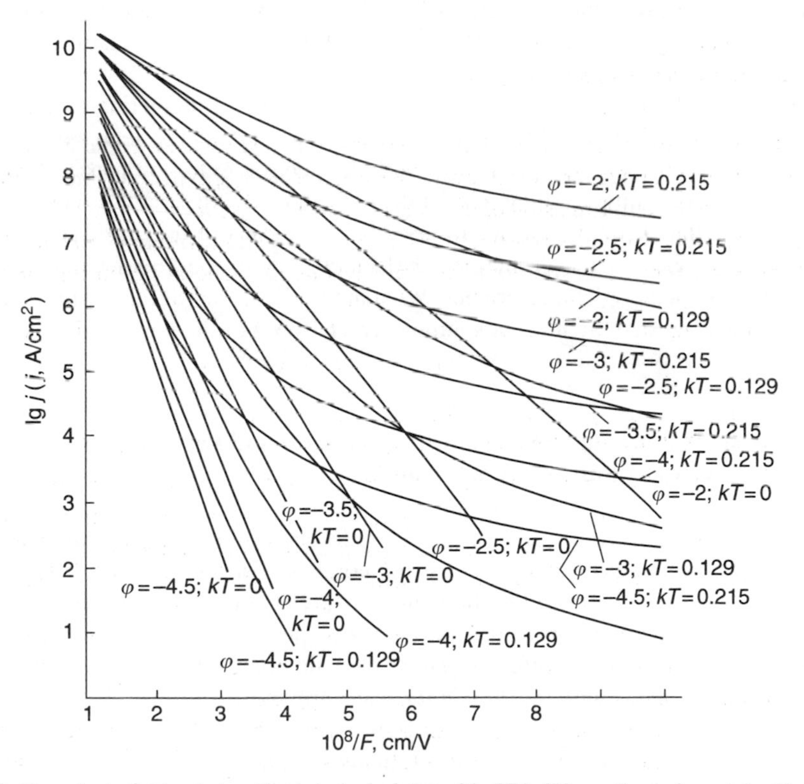

Figure 1.3. Thermionic-field emission. Numerical calculation of the FE in FN coordinates for metals with different work function values ($\varphi = 2, \ldots, 4.5$ eV) in the temperature interval 0–3000 K.

At higher temperatures, $\omega > 0.7$, the emission process moves into the regime of Schottky emission (thermal emission through a barrier lowered by an electric field) and at still higher temperatures to the regime of pure thermionic emission.

For a fixed work function value, the regions over which particular types of emission dominate will be determined by the temperature and electric field strength.

1.3. EXTENDING THE THEORY OF FIELD ELECTRON EMISSION FROM METALS

The model of the bulk metal and its surface in FN theory is very simplified. The free electron model, assumed atomic scale surface smoothness, and the one-particle approximation. There is also an inconsistency with the quasi-classical approach of combining the flow of essentially classic particles with the quantum-mechanical phenomenon of tunneling. In addition, FN theory considers the one-dimensional problem with a potential profile that only accounts for image forces. Attempts to improve upon field electron emission theory have been the result of the discovery of new facts; the departure from FN behavior at high fields and high current densities,[27, 31, 47, 48, 50] the presence of high- and low-energy tails in the emitted electron energy spectrum,[121, 122] the detection of fine structure in the FE spectra from certain crystallographic orientations[19] (p. 223), anomalous broadening of the energy spectrum in very high electric fields,[120, 170, 171] and the emergence of additional peaks in the spectra for atomically sharp microtips.[123]

With the development of atomically sharp emitters used in tunneling spectroscopy[172] and electron holography[80, 82, 83] there came a need to understand the degree of localization of the tunneling process. The physics of the field electron emission process from ultra-small-size emitters is now exciting a great deal of interest[173–177, 219, 222] due to the rapid development of the vacuum nanoelectronics and new means of diagnosing surfaces with atomic resolution. However, at the moment we lack a theory that permits accurate calculation of the cathode's operating characteristics. What are available are attractive ideas, tentative estimates, and suggested methods of solution. The purpose of the next section is to review these.

1.3.1. Deviations from the Fowler–Nordheim Theory and Peculiarities of Field Electron Emission from Small-Scale Objects

As shown in a number of works,[17, 27, 178, 179] the one-dimensional approximation gives a fair description of the electron FE process for atomically smooth emitters having a radius greater than $\approx 0.1\,\mu\text{m}$. In this case the width of the potential barrier is significantly less than the cathode's radius of curvature. Atomic-scale surface roughness and the variation of the work function between different emitter faces do not result in a significant deviation from the results obtained with the one-dimensional approximation.[17, 179]

Vacuum microelectronics employs cathodes having radii in the nanometer range (emitters 10–200 Å in radius). Even sharper tips (atomic sharpness) are used as point sources of electrons in ultra-high-resolution electron spectroscopy, electron holography, and tunneling spectroscopy. With these field emitters the radius of curvature is close to or less than the barrier width, so the assumptions of an one-dimensional barrier and field uniformity over the apex of the tip are no longer justified. Strictly speaking, for these nanometer emitters a fundamental revision of the theory is needed. In particular, it is necessary to solve the three-dimensional Schrödinger using an asymmetric potential barrier, and calculate the behavior of the potential near the surface accounting for its variation with radius r_e and polar angle θ As solving such a problem involves formidable difficulties at present only rough calculations are being made.

One approach[124–126] has been to calculate the potential barrier transparency for the simple case of a rectangular axially symmetrical barrier formed inside the scanning tunneling microscope (STM) gap. In this case, the transparency coefficient is given by,

$$D(E) = \frac{\lambda \exp(-2p_0 d_l)}{8d_l \sqrt{\det(\hat{\Omega}_1 - \hat{\Omega}_2 + 2d\Omega_2\Omega_1)}}, \tag{1.15}$$

where d_l is the length of the most probable tunneling path, $p_0 = \sqrt{2(E-U)}$, λ is a constant, and Ω_j is a matrix representing the curvature of surface apex of the STM probe at the point of its intersection with the most probable tunneling path.

Transparency calculations for a spherically symmetrical FE from an STM tip of radius R[124–126] yield

$$D(E) = \frac{\lambda R_{\mathrm{eff}}}{4d} \exp(-2p_0 d_l), \tag{1.16}$$

where

$$R_{\mathrm{eff}} = \left[\left(\frac{1}{R} + \frac{1}{R_1} + \frac{d}{RR_1}\right)\left(\frac{1}{R} + \frac{1}{R_2} + \frac{d_l}{RR_2}\right)\right]^{-1/2}$$

and R_j are the principal radii of curvature of the surface under investigation at their point of intersection with the most probable tunneling path.

Assuming that the emission originates predominantly at a single atom and using the Wigner model of the localized state[145] it was found that

$$D(E) = \frac{\frac{1}{2}\Gamma_1\Gamma_2}{(E - E_0) + \frac{1}{4}(\Gamma_1 + \Gamma_2)^2}, \tag{1.17}$$

where E_0 is the energy of the localized state and Γ_1 and Γ_2 the localized state decay widths in the STM tip and the sample, respectively.

Also, in the simpler case of a rectangular barrier profile and with FE from a single atom

$$D(E) = \frac{\mu\lambda R_{\mathrm{eff}}}{4(d+R)} \exp(-2p_0 d_l), \quad \text{where } \mu \approx 1. \tag{1.18}$$

Tentative estimates show that in the case of the STM gap, accounting for the three-dimensional character of the barrier makes only for a slight difference to the preexponential factor when compared to that of the one-dimensional approximation. With experimental techniques this difference cannot be detected.

Multidimensional tunneling of electrons with application to STM and hence, FE was also investigated by Lucas et al. and Huang et al.[216, 217] In Lucas et al.[216] the scattering theoretic technique of localized Green functions was applied to the calculation of emission current distribution at the planar STM electrode. The obtained results were applied to the explanation of the lateral resolution of STM. The work[216] has been devoted to the general formalism of the WKB approach to the problem of particle tunneling in multidimensional, in particular, nonseparable, potentials.

Dyke,[27] Elinson,[58] and later in the case of nanoscale emitters Rodnevich[127] noted that, a more rigorous calculation of the field near the tip will require taking into account the field inhomogeneity over the surface. For conventional-size tips Dyke et al.[218] made detailed calculations of $F(\theta)$ and compared these with experimental data for emitters with $r_e > 0.1\ \mu$m. The calculation results were found to be in good agreement with the experiment.

Rodnevich[127] and Cutler et al.[175] considered the emission from a sharp micro-roughness on a planar macro-cathode surface. It was shown that in this case the expression for the electric potential and, consequently, the electron potential energy in the Shrodinger equation must be taken in a more complicated form as compared with the usual "plane" description.[127] The usual expression for the potential energy is

$$U_1 = -\frac{e^2}{4x} - eF_0 x + E_F + \varphi \tag{1.19}$$

In the case of a planar cathode with a micro-roughness it is correct only where $F_0 =$ const. Taking into consideration that $F = \beta F_0$, where β is a geometric factor and F_0 the electric field strength at the cathode surface, (1.19) can be written as

$$U_2 = -\frac{e^2}{4x} - \beta e F_0 x + E_F + \varphi \tag{1.20}$$

Rodnevich[127] drew attention to the fact that representing the potential barrier in the form of U_1 or U_2 raises doubts because the potential distribution near a micro-roughness is essentially nonlinear and should be written as (Rodnevich[177])

$$U_3 = -\frac{e^2}{4x} - \Phi(x) + E_F + \varphi \tag{1.21}$$

where $\Phi(x)$ is a nonlinear function generally depending on the shape of the micro-roughness. A suitable form of the function $\Phi(x)$ should be used with each particular microtip shape. As shown by Porotnikov and Rodnevich,[177] the barrier structure, the FE current density, and the energy distribution markedly depend upon a dimensionless parameter $M = \varphi/F_0 a$, where a is the micro-roughness height. For $M \ll 1$ the barrier structure U_3 is close to U_2 and for $M \gg 1$ the barrier practically coincides with $U_1(x)$ although the surface field strength is given by βF_0 instead of F_0 as for the barrier described by $U_1(x)$. Analysis of the potential distributions near a sharp micro-roughness for different tip shapes shows that over a certain range of M values, $3 \cdot 10^{-2} < M < 30$, current–voltage characteristics are nonlinear and appear more intricate than predicted by the FN theory.

Similar ideas, but in a more consistent and rigorous manner, have been discussed by Culter et al. and Jun He et al.[175, 176] In these studies a numerical calculation has also been made for the model situation that corresponds to a small-gap Spindt cathode design. In Cutler et al.[175] the influence of emitter nonplanarity upon the barrier shape was theoretically investigated and current–voltage characteristics were calculated. It was found that taking into account of the field strength variation along the direction away from the emitter (potential function nonlinearity), and, consequently the corrected barrier shape leads to the changes of current–voltage characteristic nonlinearity, that is, to deviation from FN theory predictions. Cutler et al.[175] described this effect as "dramatic increase in the current density relative to the FN result for planar model of the tip with the same voltage applied in each case"[175] (p. 388). We note, that this increase is connected with the "β-factor"—field strength increase on the emitting surface of the tip to be compared with the abovementioned "plane" case. We suppose that more correct comparison of the results must be done for identical values of the field strength at the emitting surface.[219] For identical F values at the surface, the shape of the potential barrier will be changing as the distance from the cathode is increased, as shown in Fig. 1.4. It is evident from the figure that decreasing the field emitter radius causes a marked increase of the barrier width and, consequently, the drop in the barrier transparency and the tunnel current. So, the "β-factor" effect will be correctly included in the interpretation of results in both cases. The effect of this factor can also be seen from Cutler et al.[175] The increase of current density J, plotted as a function of voltage can be compared with the decrease of J when plotted as a function of $F = \beta \mathrm{V}$. Jun He et al.[176] is a theoretical investigation of the interaction of electron and image charges for emitters of various geometries. The obtained results are compared with the planar case, again for the same anode–cathode voltages.

Calculations[222] show that image potential for the small-scale field emitters (10–100 Å) practically is the same as for the planar cathode.

To illustrate specific features of the field electron emission process in small-size emitters, Fursey and Glazanov[219] performed numerical calculations of J–F relationships in usual FN coordinates and the total energy distribution of emitted electrons. Calculation results for the spherical cathode model were compared with the planar model.

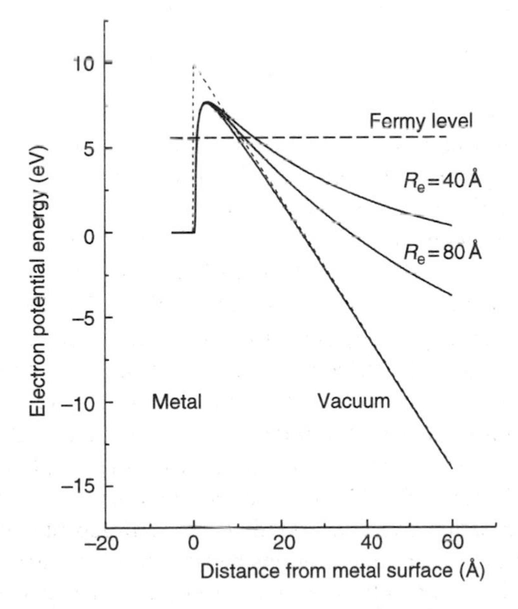

Figure 1.4. Dependence of the barrier shape upon emitter radius R_e for tungsten.[219]

For description of the emission from sharp microtips with the apex curvature radius R_e a spherically symmetric model was chosen. In this model the potential variation with distance x from the emitting surface is given by

$$U(x) = -\frac{e^2}{4x} - eF_0x\left(\frac{R_e}{x + R_e}\right) + E_F + \varphi \tag{1.22}$$

where F_0 is the field strength at the emitting surface.

The results are presented in Fig. 1.5 for $R_e = 80$ and 40 Å, respectively. The deviation from the FN plot (lower current density J for the same values of F_0) is more significant for small R_e values. The reason is that the barrier width is larger, and therefore $D(E_x)$ is less for sharp emitters due to the widening barrier shape (Fig. 1.4).

The barrier can be considered plane if

$$R_e \gg d$$

where d is the barrier width for electrons near the Fermi level. For a plane barrier $d \approx \varphi/F$. Noticeable FE is found for fields 0.3–0.9 V/Å, which corresponds to the range of d values ($\varphi = 4.5$ eV for tungsten) from 15 Å ($F = 0.3$ V/Å) to 5.0 Å ($F = 0.9$ V/Å). So, the

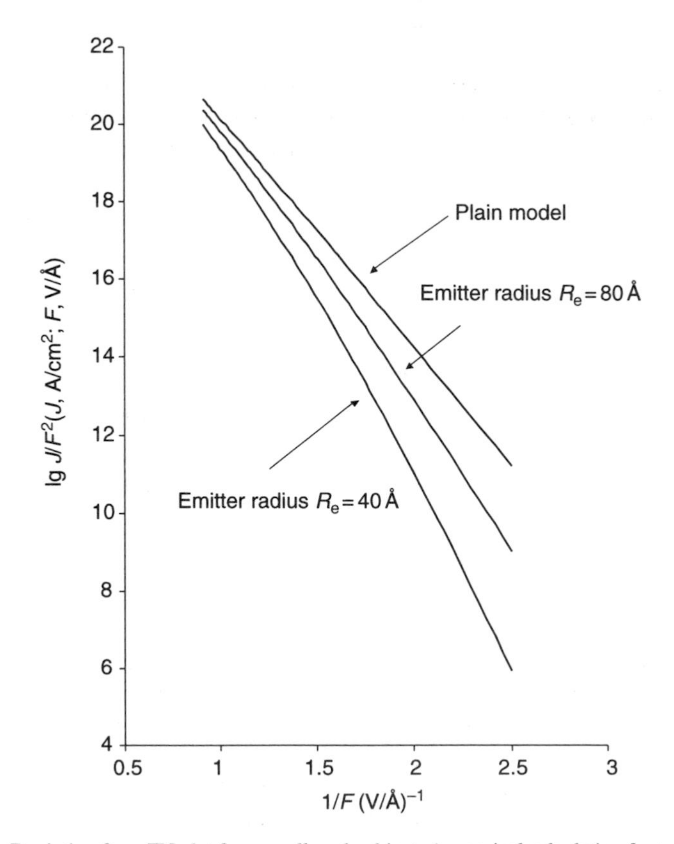

Figure 1.5. Deviation from FN plot for a small-scale objects (numerical calculation for tungsten).[219]

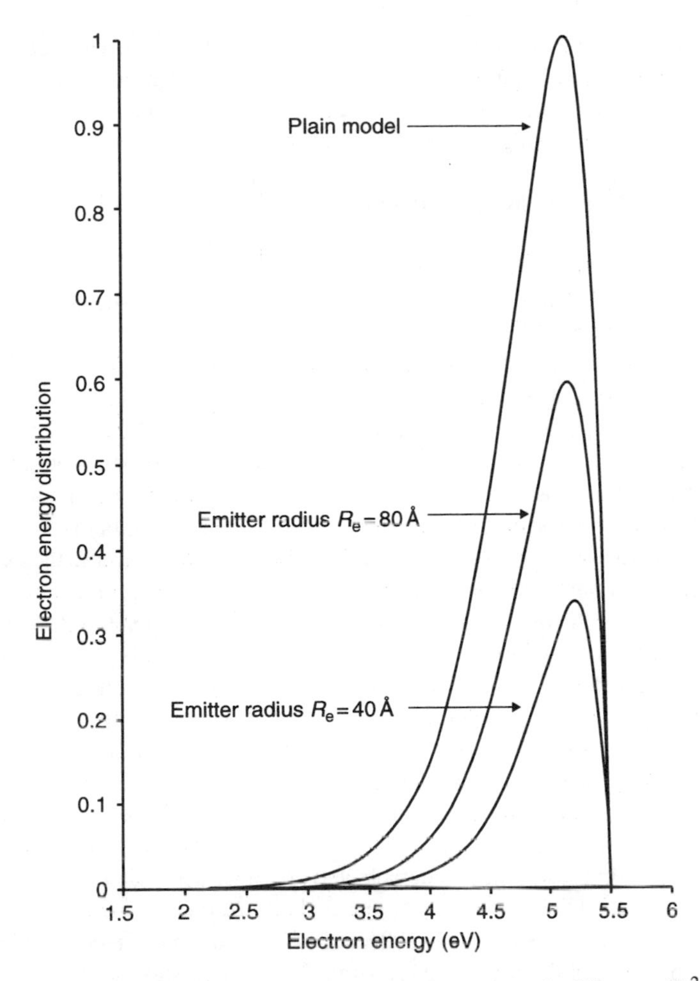

Figure 1.6. Electron energy distribution from tungsten emitters for different radii.[219]

condition $R_e \gg d$, with the emitter radius fixed, is violated in weak fields more significantly than shown in Fig. 1.5.

Energy distributions of emitted electron in high fields are presented in Fig. 1.6. One can see that the area under the curve is greater for the plane case. Also, the width of energy distribution curve is less for sharp emitters ($R_e = 20$, 40 Å), because the barrier width increase is more significant (compared with planar case) for low-energy electrons.

Theoretical and experimental investigations of the field electron emission from nano-emitters are important for technology of point electron sources, particularly those used in vacuum microelectronics devices, and with regard to the fundamental aspects of the emission phenomena involving different areas of physics, such as quantum mechanics, solid state physics, surface physics, etc.

We note also that in the case of nano-emitters traditional theoretical models such as "solid state," "free-electrons," or "band-structure," etc. need to be analyzed more extensively and more thoroughly. It is important to bear in mind that, as shown experimentally, in specially formed nanotips the emitting area includes just a few atoms. (In Fursey et al.[222] the curvature radii of micro-emitters were found to be in the range 10–80 Å.) Also, ideas relating to emission from a single atom at the emitter apex are discussed.[221]

1.3.2. The Effect of Fermi Surface Structure

The FN theory is based on the Sommerfeld's free electron model. This means that the FN equation has been derived using the Fermi–Dirac energy distribution function. The actual situation might prove to be much more complex.

The modern electronic theory of metals is based on the idea that electrons in metals behave as quasi-particles displaying a complex energy dispersion law.[180] One of the first serious attempts to consider the problem of the electron FE with an arbitrary dispersion law is to calculate the transparency coefficient[130, 131] using the Bloch function formalism. In this case, the Fermi surface is traversed by an axis P_z that is perpendicular to the cathode surface, then the FN formula (1.7) based on the free electron theory holds (to within the preexponential factor). In those cases where P_z axis does not traverse the Fermi surface (so-called open Fermi surfaces), then due to the conservation of tangential quasi-momentum for the electron exiting the metal, the work function, φ, in the exponential term in Eq. (1.7) must be replaced with some effective quantity $\varphi^* > \varphi$. It has also been shown that the energy spectrum is different than that obtained using the free electron model.[131]

Theoretically, it was found that the temperature dependence of the electron FE differs in principle from what is predicted by the free electron model, namely, the emission current in this case decreases instead of increasing as in the free electron model.

Despite the significance of these results,[130, 131] the effects have not yet been observed experimentally. This seems to be due to the fact that typically the most strongly emitting areas of the cathode tip crystal (high index planes) are those associated with Fermi surfaces perpendicular to P_z.[179] Thus the anomalies should be looked for in the low-index crystallographic areas.[130, 131] Such features for {100} and {110} faces have been revealed primarily in the energy spectrum of the tunneling electrons[18] as will be shown below.

Note that the shape of the constant energy Fermi surfaces might affect the FE, not so much through the exponential term in the expression for emission current, as through the preexponential factor, which is significantly different than that given by the free electron model. Research performed in our laboratory[181, 182] showed that the experimentally determined constant in the preexponential factor coincides, to within order of magnitude, with the value calculated from the FN theory only in the case of tungsten. For other materials, such as molybdenum, niobium, and tantalum, experimental values might differ from theoretical values by several orders of magnitude.

Unfortunately, due to the mathematical difficulties involved, the formalism developed by Itskovich[131, 132] did not yield the value of the preexponential factor. As he noted, the calculations could be performed only in the case of nearly free electrons. Attempts to calculate the preexponential factor with an arbitrary dispersion relationship were undertaken by Gadzuk[129] and Politzer and Cutler.[183, 184] They showed that for tunneling from the c-band the preexponential factor might reduce the emission by a factor of 100–1000.

Gadzuk[129] also calculated D for a Bloch metal using tight binding wavefunctions and transfer Hamiltonian methods. Gadzuk[129] further postulated a tight binding d-band that is described by a linear dispersion of the form $E = b(k)$, where b is a parameter. The linear $E(k)$ relationship does not greatly alter the functional form of the energy distribution expression from that of a quadratic $E(k)$ relationship; however, the transmission coefficient of d-band tunneling appears to be reduced relative to s-band tunneling by a preexponential factor of 10^{-2}–10^{-3}.

1.3.3. Many-Particle Effects

There are many phenomena, which cannot be described in terms of the one-electron approximation. These problems include the description of the variation of the potential barrier within a small distance from the surface (less than two or three angstroms) and the analysis of correlation effects involving electron–electron and electron–phonon interactions in the near-surface region.

FN theory as well as subsequent theories, including the general theory of electron emission from metals,[185] is essentially an one-electron theory. With the progress in quantum-field methods with statistical physics it became possible to develop a multielectron theory of field electron emission. One of the first attempts to apply this approach was undertaken by Gadzuk.[128] An analysis of the emergence of these ideas was made by Baskin.[186] Gadzuk used a technique employed previously in solving the problem of Josephson tunneling[187] in which the initial state of the anode–cathode system prior to emission is described by two independent Hamiltonians: the cathode Hamiltonian H_{L} (assumed to be located on the left-hand side) and the anode Hamiltonian H_{R} (located on the right-hand side).

The external field applied to the cathode inducing electron emission is equivalent to the introduction of a tunneling Hamiltonian H_{T}, which mixes states on the right and left thereby causing an effective flow of particles from one side to the other. Ultimately, the problem is reduced to the use of a complete Hamiltonian

$$H = H_{\mathrm{L}} + H_{\mathrm{R}} + H_{\mathrm{T}} \tag{1.23}$$

where $H_{\mathrm{L}} = \sum_{k,\sigma} \mathrm{A}_k C_k^+ C_k$ and $H_{\mathrm{R}} = \sum_{q,\sigma} \mathrm{A}_q d_q^+ d_q$ are unperturbed Hamiltonians, and C^+, C, d^+, d are the electron generation and annihilation operators on the left (right) obeying the Fermi anticommutation rules. In the absence of electron pairing the C and d operators commutate; in the general case these are quasi-particle operators.

The tunneling Hamiltonian can be written,[128]

$$H_{\mathrm{T}} = \sum_{k,q,\sigma} (T_{kq} C_{k\sigma}^+ d_{q\sigma} + T_{kq}^* d_{q\sigma}^+ C_{k\sigma}). \tag{1.24}$$

Here, the matrix element T_{kq} was expressed in the form:[188]

$$|T_{kq}|^2 = \frac{1}{4\pi^2} \frac{\delta_{k_t q_t}}{\rho_\perp^{\mathrm{R}} \rho_\perp^{\mathrm{L}}} \exp\left[-2 \int_{x_1}^{x_2} k_\perp(x)\, dx\right] \tag{1.25}$$

The exponent in the Eq. (1.25) can be approximated by[6]

$$\exp\left[-2 \int_{x_1}^{x_2} k_\perp(x)\, dx\right] \approx \exp\left[-c + \frac{E_k^\perp - E_F}{d}\right], \tag{1.26}$$

where as usual

$$C = \frac{4(2m\varphi^3)^{1/2}}{3\hbar e F_s} \vartheta \left[\frac{(e^3 F_s)^{1/2}}{\varphi} \right],$$

$$d = \frac{\hbar e F_s}{2(2m\varphi)^{1/2}} t^{-1} \left[\frac{(e^3 F_s)^{1/2}}{\varphi} \right],$$

E_F is the Fermi energy; $\rho_\perp$ the density of states with the momentum normal to the surface barrier.

The current can be written as

$$j = e\langle \dot{N}_L \rangle, \tag{1.27}$$

where the number of particles operator

$$N_L = \sum_k C_k^+ C_k \tag{1.28}$$

obeys the equation

$$i\dot{N}_L = [N_L, H] = [N_L, H_T] \tag{1.29}$$

The last equation is based on the property that the N operator commutates both with H_L and H_R. Using Eqs. (1.29) and (1.24), the current can be written,

$$j = -ei\langle [N_L, H_T] \rangle = -2e \operatorname{Im} \sum_{k,q} T_{kq} \langle C_k^+ d_q \rangle. \tag{1.30}$$

Expression $\langle C_k^+ d_q \rangle$ in an abridged form will be

$$\langle C_k^+ d_q \rangle = \langle S^{-1}(t) C_k^+(t) d_q(t) S(t) \rangle, \tag{1.31}$$

where $C_k(t) = \exp(iH_L t) C_k \exp(-iH_L t)$ and

$$S(t) = \exp\left[-i \int_\infty^t \exp(\eta t') H_T(t')\, dt' \right].$$

Assuming that the spectral function of the electron gas in the metal is $\sim\delta(E_K^r - E_q^r)$, then calculation of the matrix elements using Green's function method gives[189] the usual expression for the FN function

$$j = (2\pi m c d^2 / h^3) e^{-c} \tag{1.32}$$

Calculations show generally that the magnitude of the emission current depends on the properties of the anode metal due to the assumption that the anode states are the final states of electrons after tunneling. This is true only in those cases where the anode–cathode separation is less than the electron correlation length. Experimentally, as a rule, the opposite is true. As the final state, the electron state in a uniform field should be chosen.

It follows from Eq. (1.32) that the validity criterion for the FN equation is the closeness of the spectral function of electrons in the metal to the delta-function. Because of the fact that this is always the case for the electrons near the Fermi surface[189] the one-electron approximation will be valid when most of the emitted electrons are in states close to the Fermi level.

Gadzuk[128] formulated a more rigorous statement of the problem, which, in principle, can account for many-particle effects. With this formulation the validity of the one-electron approximation appears to be more justified. The Hamiltonian H_L used by Gadzuk does not take into account electron correlation.

Lea and Gomer[121] and, independently, Gadzuk, and Plummer[122] noticed that high- and low-energy tails are observed in the measured total-energy distributions of electrons for a number of faces {111}, {120}, and {112} (Fig. 1.7). Lee and Gomer suggested that the high-energy tails in the electron energy distributions are manifestations of electron–electron scattering in the emitter crystal, making possible the simultaneous emission of electrons in pairs. The emergence of such a pair can occur in the following way. Emission of an electron that resides below the Fermi level at energy $E_F - \Delta$ generates a "hot" hole in the emitter near-surface region (Fig. 1.8). A conduction electron scattered by this hole acquires an energy $E_F + \Delta$ and so is raised above the Fermi level where the tunneling probability P_2 is greater than the tunneling probability P_1 of the first electron. Because of the very short electron–electron tunneling time ($\sim 10^{-13}$ sec) both electrons find their way into the vacuum simultaneously as a pair. The probability of the simultaneous tunneling of the two electrons is proportional to a product of the tunneling probabilities for each electron and may be represented in the form[121]

$$P = P_1 \cdot P_2 \approx \exp\left[\frac{-c(\varphi - \Delta)^{3/2}}{F}\right] \cdot \exp\left[\frac{-c(\varphi - \Delta)^{3/2}}{F}\right] \approx \exp(-2c\varphi^{3/2}/F), \quad (1.33)$$

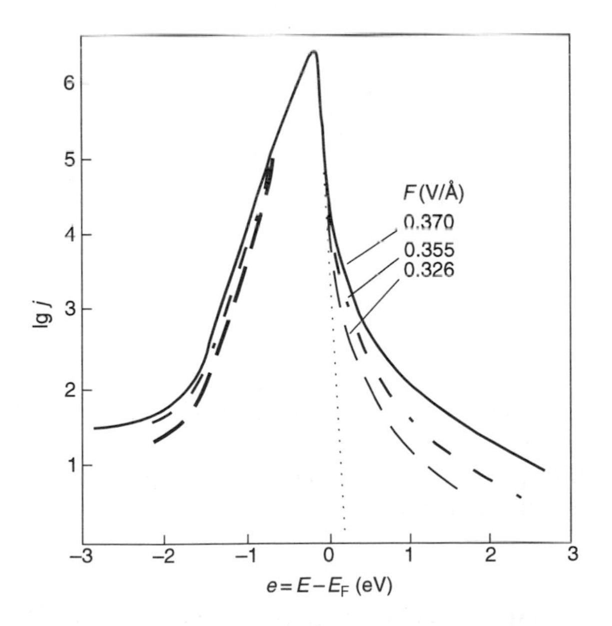

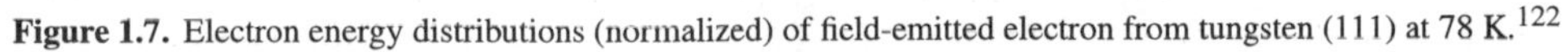
Figure 1.7. Electron energy distributions (normalized) of field-emitted electron from tungsten (111) at 78 K.[122]

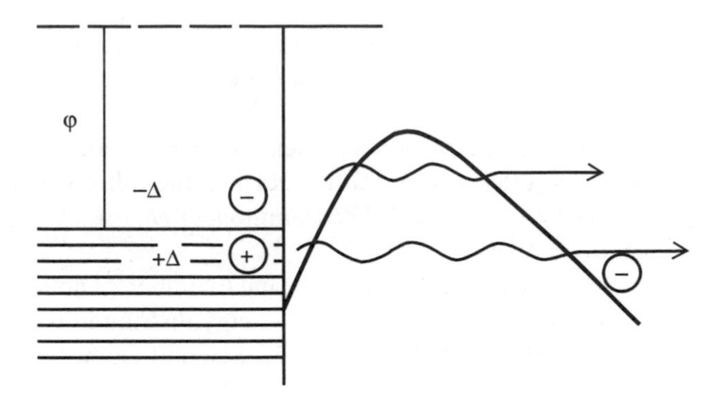

Figure 1.8. Schematic diagram of FE electron pair tunneling.

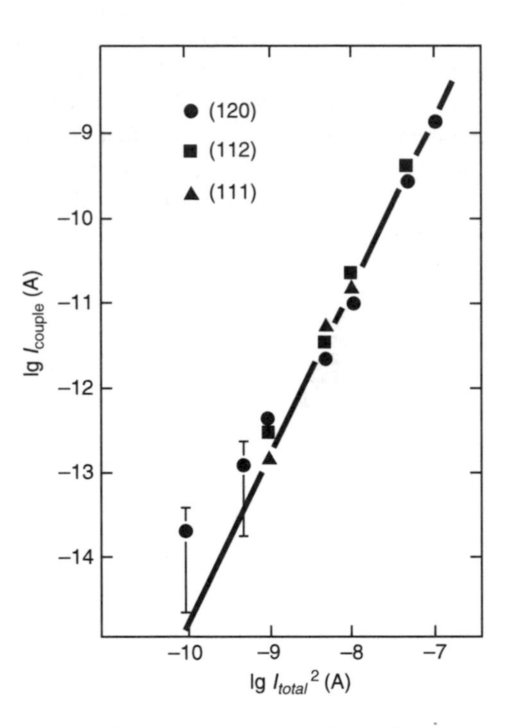

Figure 1.9. Experimental observation of quadratic dependence of the high-energy electron current (so-called pair electrons) versus the total current in the probe.[122]

where φ and F denote, as usual, the work function and the applied field, respectively, and $c = 0.683$ in eV, Å units.

As the total current, by Eq. (1.7), is proportional to $\exp(-2c\varphi^{3/2}/F)$, the current due to paired electrons should be proportional to the square of the total current

$$I_{\text{couple}} \sim I_{\text{total}}^2 \tag{1.34}$$

It has been shown experimentally that I_{couple} is indeed proportional to the square of the total current (Fig. 1.9).[121, 122, 190] It can be seen from Fig. 1.9 that at high current densities the fraction of current due to "paired" electrons can be several percent. The idea of pair tunneling has been proposed to explain an anomalously high work function value found for the {011} face of tungsten.[191, 192]

As noted in the Introduction to this section, a many-particle problem, though of entirely different character than that described above, must be solved to describe the variation in the potential barrier over atomic distances from the surface. An analysis of the barrier configuration near the surface accounting for this variation was attempted by Bardeen[193] and will be considered further in the next chapter.

1.4. RESUME

The Fowler–Nordheim theory gives a fairly adequate account of the basic features of field electron emission and its main deductions have a solid experimental corroboration.

1. The tunneling nature of the emission has been unambiguously proved in direct experimental determinations of the energy distributions of FE electrons by the retarding potential method.
2. The FN theory explains the exponential dependence of j pm F, or, to put it another way, the linear form of the current–voltage characteristics of the field electron emission ($\ln(i) = f(1/u)$). The exponential variation of j with $\varphi^{3/2}$ has been confirmed experimentally.
3. It should be noted that accurate comparison of the theory and experiment is impeded by difficulties encountered in measurements of the electric field strength F. The error in measurements of F under standard experimental conditions is usually not less than 15 percent, posing problems in the use of the theory for accurate absolute measurements.
4. The theory of T-F emission also has a reliable support in experiment. However, the FE emission theory does not provide simple analytical relationships. Straightforward formulas are available only for the region of relatively low temperatures (formulas 1.12 and 1.13). Formulas for high temperatures are cumbersome; therefore, for specific cases numerical calculations are more preferable. Such calculations are illustrated in Fig. 1.3.
5. As to the refinements of the theory (Section 1.3), it should be noted that all of them are of the fundamental nature but as yet could not be reliably confirmed because of insufficient accuracy of measurements.
6. The strongest effect is the deviation from the traditional FN theory of the current–voltage characteristics in the case of nanometer-scale field emitters. Preliminary studies (Subsection 1.3.1) provided consistent explanation of this effect assuming nonlinear variation of the potential near the emitter surface. This case is of particular importance because of the widening use of the nanometer-scale field emitters in various vacuum microelectronics devices. Further thorough studies are required in order to get a clearer insight into the field electron emission of such small-scale emission centers.

2

CHARACTERISTICS OF FIELD EMISSION AT VERY HIGH CURRENT DENSITIES

2.1. DEVIATIONS FROM THE FOWLER–NORDHEIM THEORY IN VERY HIGH ELECTRIC FIELDS

Basic research into Field Emission (FE) at high current densities was initiated by Dyke and his colleagues in the mid-1950s.[31–34] Modern vacuum techniques, pulsed voltage methods, and other advanced technologies allowed their use of higher electric fields and higher emission current densities than were previously possible. Pulsed current densities of 10^8 A/cm^2 with tungsten (W) cathodes and, later, up to 10^7 A/cm^7 in the dc mode[35] were achieved in electric fields 5–8 $\times$ 10^7 V/cm.

Deviations from Fowler–Nordheim (FN) theory (Fig. 2.1) were first detected on W emitters in high fields.[31, 47, 50, 59, 64, 119] Such deviations were also observed for Mo, Ta, Re, W_2C by Fursey with co-workers[194, 195] and by Elinson and Kudintseva[44] for ZrC and LaB_6.

Dyke and Trolan[31] interpreted the deviations from linearity in the FN current–voltage characteristics as due to lowering of the electric field near the cathode due to space charge from the emitted electrons. Lewis[37] suggested a connection between the deviation in the current–voltage characteristics and the dissimilarity between the actual shape of the potential barrier at the metal–vacuum boundary and the approximate shape used in the image force potential theory. This dissimilarity is pronounced in strong electric fields where the barrier thickness approaches interatomic distances. We will discuss these ideas in more detail in the following sections.

2.2. SPACE CHARGE EFFECTS IN FIELD EMISSION

The effect of space charge on FE was first discussed briefly by Stern, Gosling, and Fowler.[196]

It follows from the Poisson equation for a potential distribution between infinite planar electrodes that

$$\frac{d^2U}{dx^2} = -k \cdot j \cdot U^{-1/2}, \tag{2.1}$$

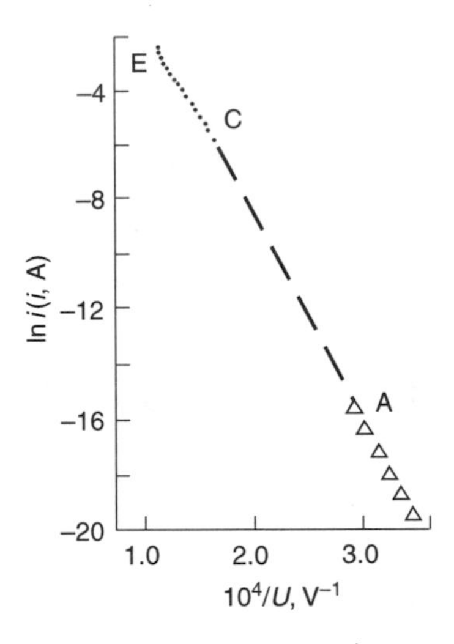

Figure 2.1. Deviations from FN theory at high electric fields: AC—Region of FN current–voltage characteristics; CE—region of anomalous behavior.[32]

where U is the potential, x is the coordinate; $k = 2\pi(2m_e/e)^{1/2}$, and j is the current density. The solution of Eq. (2.1) with the boundary conditions

$$U|_{x=0} = 0 \quad U|_{x=d} = U_a \quad dU/dx|_{x=0} = 0 \tag{2.2}$$

(typical for thermionic emission) leads to a well-known formula of Boguslavskii–Langmuire

$$j = \frac{4}{9kd^2} \cdot U_a^{3/2}, \tag{2.3}$$

where d is the inter-electrode separation.

In the case of FE the electric field strength, F, at the emitter surface is nonzero. Hence, the boundary conditions (2.2) take the form

$$U|_{x=0} = 0 \quad U|_{x=d} = U_a \quad dU/dx|_{x=0} = F. \tag{2.4}$$

The relationship between the electric field, current density, and the potential was found,[196] by integrating Eq. (2.1) using boundary conditions (2.4), as

$$F = \frac{U_a}{d}\left[1 - \frac{16}{3}\pi(m_e/2e)^{1/2} j \cdot \frac{U_a^{1/2}}{F^2}\right]. \tag{2.5}$$

Since $F = U_a/d$, if space charge is ignored, the criterion for neglecting the space charge effect is,

$$\frac{16}{3}\pi(m_e/2e)^{1/2} j \cdot \frac{U_a^{1/2}}{F^2} \ll 1. \tag{2.6}$$

This problem can be analyzed in more detail.[34] The validity of choosing the one-dimensional approximation to represent the actual geometry of the point-cathode diode

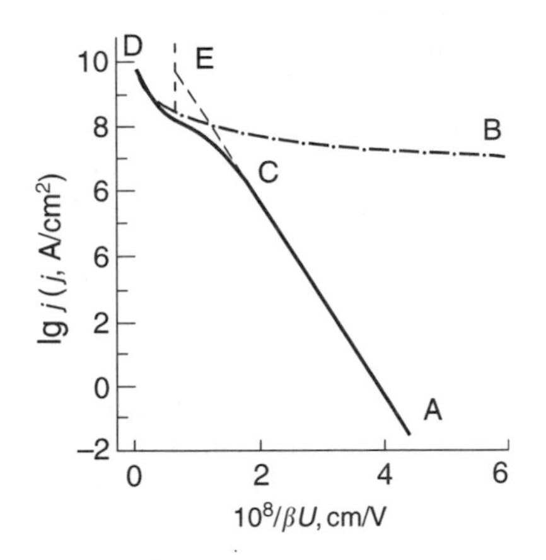

Figure 2.2. Theoretical dependence of the current density on the applied voltage.[34]

was justified by the fact that the space charge affecting the electric field near the emitting surface could be considered as concentrated within a very small distance from the cathode.

By solving Eq. (2.5) in combination with the FN equation,

$$j = AF^2 \exp(-B/F), \tag{2.7}$$

where $A = 1.54 \cdot 10^{-6}/\varphi$; $B = 6.83 \cdot 10^7 \varphi^{3/2} \vartheta(y)$, and eliminate the current density j from Eqs. (2.5) and (2.7), we have an equation linking the field strength F to the potential U

$$4kAU_a^{3/2} \exp(-B/F) - 3U_a = 9k^2 A^2 F^2 d^2 \exp(-2B/F) - 3Fd. \tag{2.8}$$

Equations (2.8) and (2.7) can be used to calculate the current density for fixed values of F as a function of the applied voltage. The usual relationship between the field and the potential at zero current, $F = U_a/d$, follows from Eq. (2.8) at small F. At much larger F, the exponential terms become dominant, and Eq. (2.8) is reduced to Eq. (2.3).

The parameter d was chosen in Barbour et al.[34] to be equal to $1/\beta$, where β is a geometrical factor relating the field strength to the potential at current densities, where the effect of space charge can be neglected.

Figure 2.2 is a plot of the current density versus voltage. The curve ACE was calculated using Eq. (2.7) neglecting the space charge; the curve ACD was calculated using Eqs. (2.7) and (2.8) and allowing for the space charge; the BD curve shows the Boguslavskii–Langmuire law (2.3). The initial part of the curve coincides with the FN straight line at small values of U. As U increases, the curve becomes sublinear and asymptotically approaches the Boguslavskii–Langmuire curve.

The effect of space charge on FE for a spherical diode has also been investigated.[197–200]

The Poisson equation in the case of a spherically symmetric electron flow has the form

$$\frac{d}{dr}\left[r^2 \frac{dU}{dr}\right] = \frac{i}{(U(r) + U_0)^{1/2}} \sqrt{\frac{m_e}{2e}}, \tag{2.9}$$

where U_0 is the initial electron energy, assuming, for simplicity, that at the cathode surface $U = 0$. The exact solution of the equation with boundary conditions like Eq. (2.4) was found by Poplavskii,[201] but his solution is cumbersome and cannot be presented in elementary functions.

Isenberg[197] estimated the additional difference of potentials, ΔU_{m}, required, accounting for space charge, to make the field strength near the cathode equal to the value found in the absence of space charge for the same total current:

$$\Delta U_{\mathrm{m}} = \frac{i}{\sqrt{r_{\mathrm{k}} E_{\mathrm{k}}}} \sqrt{\frac{m_{\mathrm{e}}}{2e}} \left(\ln \frac{4 r_{\mathrm{a}}}{r_{\mathrm{k}}} - 1 \right), \tag{2.10}$$

where $i = j 4\pi r_{\mathrm{k}}^2$ is the total current in a spherical diode, r_{k} is the radius of the cathodes, r_{a} is the radius of the anodes, and $r_{\mathrm{a}} \gg r_{\mathrm{k}}$. Furthermore, the formula is valid only if $|\Delta U / U_{\mathrm{a}}| \ll 1$, and thus can be used for estimating the effect of space charge only when deviations from the FN are small.

The ΔU_{m} values determined from Eq. (2.10) have been compared with exact calculated values from the Poplavskii's formulae.[201] The maximum error in ΔU_{m} calculated from Eq. (2.10) is not greater than a few percent when the emission current is reduced by a factor of two due to space charge.

Kompaneets[199, 200] analyzed the asymptotics of the exact solution of Eq. (2.10) found by Poplavskii[201] and was able to simplify the system of equations and to obtain an estimate of ΔU_{m} having the same order of accuracy as those given by Eq. (2.10).

2.3. SPACE CHARGE WITH RELATIVISTIC ELECTRONS

The preceding discussion of space charge is relevant when the velocities of the electrons are small compared to the velocity of light (accelerating voltages of tens of kV). Relativistic effects influence the electron mass and, therefore, the electron velocity can be a factor in the emission process for electron energies in the range of hundreds of kV.[202, 203] It is obvious that retardation of the electron motion due to the relativistic effects increases the effect of space charge. It has been shown by calculation[202] in that the voltage required to maintain a given current density in the planar diode model is increased several times when relativistic effects are accounted for. In the majority of real cases, however, the field cathode is manufactured as a tip and, hence, the spatial distribution of the beam current density is similar to the distribution in a spherical diode.

Accounting for the nonuniformity of the potential distribution in a real diode by adjusting the inter-electrode distance d in a planar diode, based on the assumption that most of the screening charge is concentrated within a distance approximately equal to the tip radius, is no longer a good approximation. Qualitatively, the reason for this lies in the fact that the total screening charge in the inter-electrode gap, d, at high currents will still be quite large despite the fact that the space charge density decreases rapidly away from the cathode surface. The error involved is especially significant when relativistic effects cause the density of the screening charge ρ_{e} to decrease as $1/r^2$, that is, much slower than in the nonrelativistic approximation ($\rho_0 \sim r^{-2}[U(r)]^{-1/2}$, where $U(r)$ is the electric field potential).

A more rigorous approach is to solve a system of equations involving the Poisson equation and energy conservation and the continuity equation for the current density in a spherical diode model[203]

$$\nabla^2 U = 4\pi \rho_e, \tag{2.11}$$

$$m_e c^2 + eU = m_e c^2 (1 - v^2/c^2)^{-1/2}, \tag{2.12}$$

$$\nabla \cdot \vec{j} = 0, \tag{2.13}$$

with boundary conditions

$$U|_{r=r_k} = 0 \quad dU/dr|_{r=r_k} = F. \tag{2.14}$$

The emission current density $j = j|_{r=r_k}$ is determined by the FN equation (2.7). The approximate solution of Eqs. (2.11–2.14) has the form

$$\Delta U_m = \frac{j}{c} \left\{ \frac{1+\eta}{(\eta(2+\eta))^{1/2}} \ln \left[2(2+\eta) \frac{r_a}{r_k} \right] + \ln \left[\sqrt{\eta(2+\eta)} - (1+\eta) \right] - \frac{1+\eta}{(\eta(2+\eta))^{1/2}} \right\}, \tag{2.15}$$

where ΔU_m, as in Eq. (2.10), is the correction accounting for space charge when relativistic effects are important, and $\eta \equiv eFr_k/m_e c^2$. Equation (2.15), like Eq. (2.10), was obtained under an assumption that $r_a \gg r_k$ and its use is justified only if

$$|\Delta U_m / U_0| \ll 1, \tag{2.16}$$

where U_0 is the potential in a spherical diode in the absence of space charge and is related to the electric field strength F by

$$U_0(r_a) = F r_k [1 - (r_k/r_a)]. \tag{2.17}$$

In accordance with Eqs. (2.7) and (2.15–2.17), a fixed current density j_e and fixed anode voltage U_a correspond to a fixed F value. Using these relationships, one can plot the nonlinear, in $\lg[j(1/U_a)]$ coordinates, current–voltage characteristics of a spherical diode. As before, the approximate character of the solution restricts the validity of these relationships to small deviations from the FN law.

One can derive from Eqs. (2.15–2.17) an explicit formula for the current density at which the deviation occurs by assuming, as the deviation criterion, twice the current value calculated from the FN formula disregarding the space charge and relativistic effects at the same applied voltage.

$$\ln j_{dev} = \ln \left[\frac{\ln(2F^{3/2}(2e/m_e)^{1/2})}{4\pi(2 + B\varphi^{3/2}/F)} \right] - \frac{\ln r_k}{2} - \ln \left[\ln \frac{4r_a}{r_k} - 1 \right]. \tag{2.18}$$

The dependence of the deviation-point current on the emitter radius for two work function values is shown in Fig. 2.3.

The current–voltage characteristics begin to deviate from the straight line at a current value that increases with emitter radius and decreases with increasing work function. The

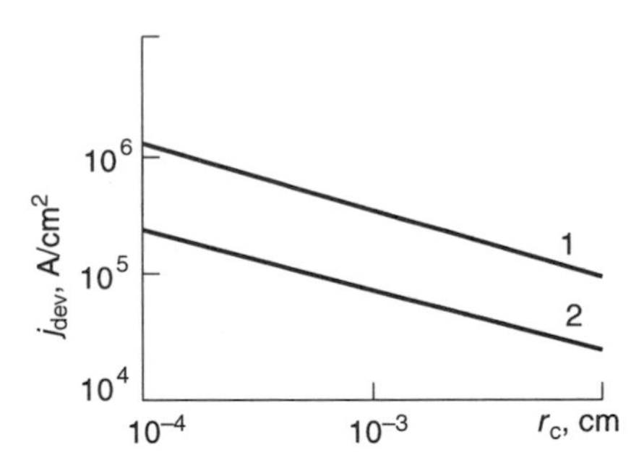

Figure 2.3. Dependence of the current density on the applied voltage when the influence of electron space charge is detected:[203] 1: $\varphi = 4.5$ eV; 2: $\varphi = 2.0$ eV.

calculation predicts that the influence of relativistic effects must be taken into account for emitter radii greater than 10^{-3} cm. At high FE currents in the spherical diode,

$$i \geq 1.7 \cdot 10^4 \text{ A}, \tag{2.19}$$

the system of Eqs. (2.11–2.13) is simplified and an approximate expression can be written for U:

$$U = (i/c)[\ln(r_a/r_k) - 1]. \tag{2.20}$$

It follows from Eq. (2.20) that the value of the current is not dependent on the work function of the cathode material. FE current is limited only by space charge. The same analysis for a cylindrical electrode system leads to

$$U = (4\pi/c)r_a r_k j_e. \tag{2.21}$$

Note that the cylindrical model applies in the practically important case of a sharp edge–plane system (blade cathodes). If the conditions (2.16) and (2.19) are not satisfied, the approximate analysis is irrelevant and one must solve Eqs. (2.11–2.13) numerically. In Fig. 2.4, the current–voltage characteristics of spherical diodes with W cathodes are shown. The plots were made from the results of numerical calculations.[203] The characteristics include a rather extended portion where the increase in current with voltage is slower than in the linear region. (The increased slope of the characteristic at high voltages is related to scaling effects.)

This sublinearity is pronounced for $r_k \geq 10^{-3}$ cm and at current densities $j \geq 10^7$ A/cm^2. The decrease of current due to the space charge in this case is so significant that it becomes very difficult to achieve current densities of 10^8–10^9 A/cm^2, which a FE cathode can easily produce when space charge is not important. Such current densities have been observed experimentally in pulsed regimes on sharp tips ($r_k = 10^{-4}$–10^{-5} cm).[31, 47, 54, 119, 173, 174, 204] For example, to achieve a current density of $j_e = 10^8$ A/cm^2 at $r_k = 10^{-2}$ cm, the voltage must be increased by roughly an order of magnitude over the value obtained when space charge can be neglected.

Distributions of the potential and the field in a spherical diode have also been calculated by numerical methods,[203] see Figs. 2.5 and 2.6. It should be noted that at relativistic electron energies and high current densities, the potential in the diode rises rather monotonically

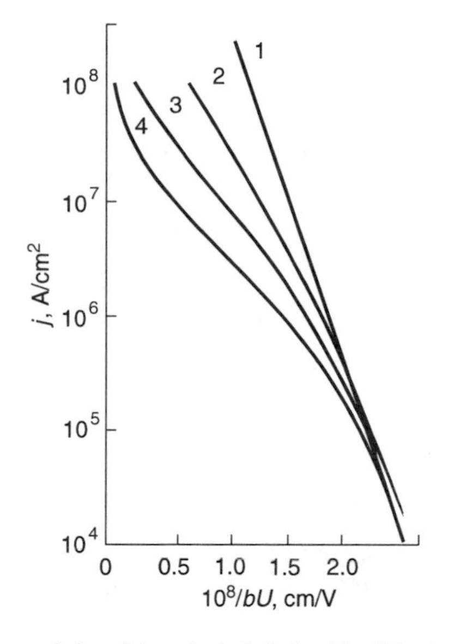

Figure 2.4. Current–voltage characteristics of the spherical diode with a W cathode accounting for space charge and relativistic effects ($\varphi = 4.5$ eV, $\beta = F/U$ at $j = 0$). 1: FN line ignoring the effects of space charge; 2: $r_k = 10^{-4}$ cm, 3: $r_k = 10^{-3}$ cm, 4: $r_k = 10^{-2}$ cm.[203]

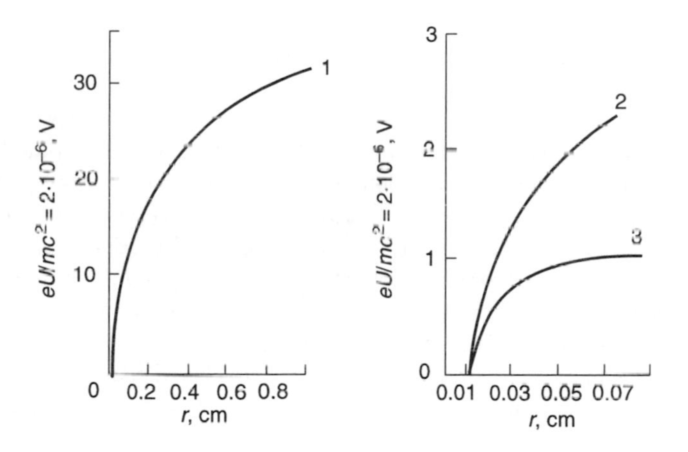

Figure 2.5. The potential distribution in the spherical diode for different emission current densities: 1: $j = 10^8$ A/cm^2; 2: $j = 10^7$ A/cm^2; 3: $j = 10^6$ A/cm^2 ($r_k = 10^{-2}$ cm).[203]

despite the large difference between the cathode and anode radii. The field strength under these conditions is at a maximum some distance from the cathode surface instead of at the cathode surface (Fig. 2.6).

In conclusion, we note that the spherical diode only approximates the point diode potential symmetry. The principal dissimilarity is that in this spherical diode model, the intrinsic magnetic field due to the emission current is not taken into account. Recall that the ratio of magnetic component of the Lorentz force to the force of electrostatic repulsion

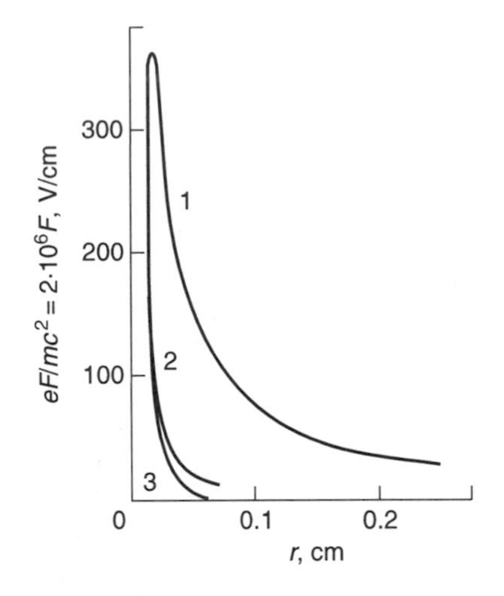

Figure 2.6. The electric field distribution in the spherical diode for different emission current densities: 1: $j = 10^8$ A/cm^2; 2: $j = 10^7$ A/cm^2; 3: $j = 10^6$ A/cm^2 ($r_k = 10^{-2}$ cm).[203]

in an electron beam is of the order of v^2/c^2. If the electron energy in is the MeV range, the electron velocity is then comparable to the velocity of light, and the resulting magnetic field can significantly affect the emitter's current–voltage characteristics.

2.4. THE SHAPE OF THE POTENTIAL BARRIER IN HIGH ELECTRIC FIELDS

FN theory and its corollaries have been well validated with W emitters in fields of $\sim 4 \cdot 10^7$ V/cm. The assumption of a smooth surface and the use of the simplified model of the mirror image forces potential (Fig. 1.1) are quite justified at these field strengths.

In higher electric fields, the dimensions of the potential barrier at the solid–vacuum interface become on the same order as interatomic distances and the radii of close-range interactions. The width of the image force barrier, $\Delta x = |x_1 - x_2|$, where x_1 and x_2 are the classical turning points (Fig. 2.7a), can be easily calculated if the kinetic energy of electrons in the direction of emission is zero at the turning points. The electron energy at the Fermi level is $E_x = -\varphi$. If the potential function is the image potential and that of an external field F, then

$$E_x = -e^2/4x - eFx \tag{2.22}$$

and

$$\Delta x = (\varphi^2 - e^3 F)^{1/2}/eF. \tag{2.23}$$

The distance x_m from the interface of the potential maximum and the barrier height H_m are found from the condition of the potential extremum

$$x_m = 1/2(e/F)^{1/2}, \tag{2.24}$$

$$H_m = \varphi - (e^3 F)^{1/2}. \tag{2.25}$$

In Fig. 2.7, the plots of $\Delta x(F)$, x_m, and $H_m(F)$ are shown.

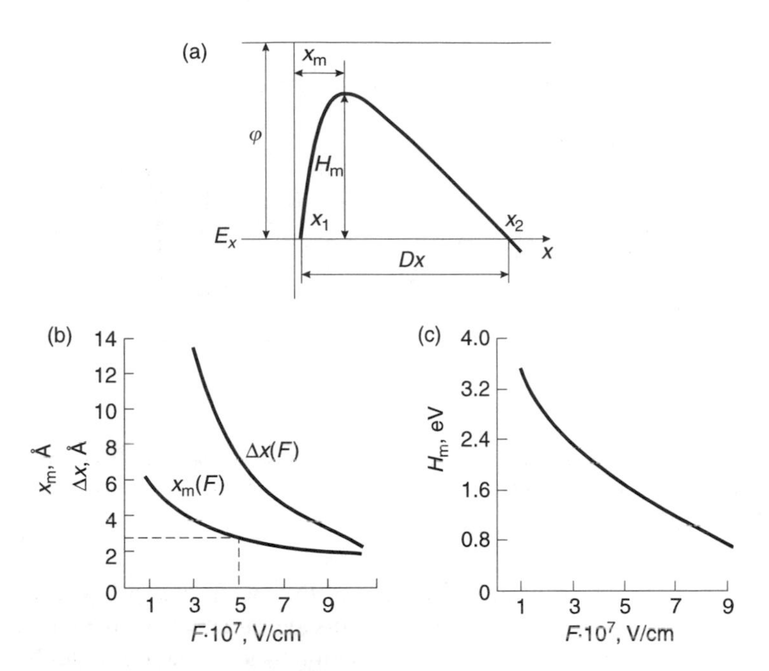

Figure 2.7. The dependence of the barrier width on image forces (a) the position of the maximum in the barrier, (b) the barrier width and the position of its maximum as a function of applied field, (c) the maximum height of the barrier as a function of applied field.

It can be seen that in fields exceeding $5 \cdot 10^7$ V/cm, the barrier width and x_m values are of the same order as the lattice constant for W ($a = 2.37$ Å). It is obvious that the previous assumptions regarding the metal surface in such strong fields are unreliable, and it is necessary to know the exact nature of forces acting on the electron near the surface.

In this case, the surface can no longer be treated as structureless and perfectly smooth. The interaction with surface atoms affects the electron motion and, strictly speaking, to describe the potential one needs to solve a multidimensional, quantum-mechanical many-body problem (see also Chapter 1). This problem has not been solved in the general case. Attempts to solve the problem in the three-dimensional case, even in the one-electron approximation, have failed due to the mathematical difficulties caused by the asymmetry of the forces at the metal–vacuum boundary.

The most comprehensive quantum-mechanical theory describing the potential barrier for electrons near the metal boundary is by Bardeen.[193, 205] The problem was solved by a self-consistent field method in the framework of Hartree–Fock approximation. The important result of Bardeen's research[193] was the demonstration of the fact that the barrier due to the image potential is a good asymptotic approximation to the potential at sufficiently large distances from the surface.

Juritchke[206] found that, in quantum-mechanical terms, one can interpret the classic polarization of a metal in terms of an electron and a "hole" coupled by the exchange interaction and moving in opposite directions.

Bardeen's potential is shown in Fig. 2.8a. The potential function outside the metal is a smooth curve coinciding with the image force law at large values of x and asymptotically

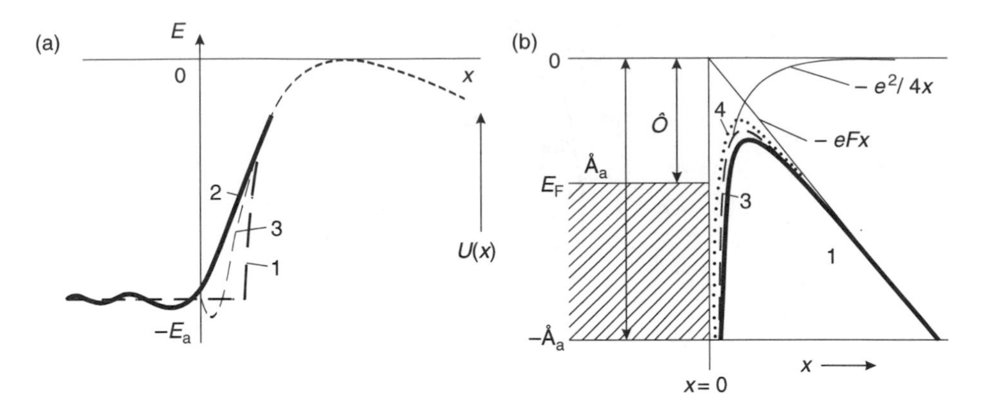

Figure 2.8. Different models of the potential barrier at metal–vacuum interface: (a) no external electric field; (b) with an external electric field F. 1: image potential barrier; 2: Bardeen's barrier; 3: Cutler–Gibbons barrier; 4: Seits–Vasiliev–van Oostrum barrier.

approaching the energy of the lowest state E_a in the metal. The potential curve has a shallow minimum in the near surface region and decays quickly from the surface. Loucks and Cutler[207] have accounted for Coulomb correlation by using Paince–Bohm's formalism and more carefully calculating the exchange portion of the potential. The correction led to only a small increase in the potential barrier height.

It is easily seen from Fig. 2.8 that with Bardeen's potential, the barrier under the applied electric field is wider and higher than that given by the image potential barrier alone. It is also obvious that the dissimilarity should be greater in the high field region where the influence of the near-surface potential is important. Lewis[37, 208] was the first who called attention to the fact. His qualitative analysis has shown that this effect should lead to progressive damping of the emission current with increasing field strength. A more rigorous analysis of the influence of the barrier shape on the current–voltage characteristics in strong electric fields was made by Cutler and Nagi.[38] Since Bardeen's potential, and that of Juretchke and Loucks,[206, 207] have a complicated analytical form and do not readily lend themselves to exact calculation of the transmission coefficient from the Schrödinger equation, Cutler and Nagi[38] used a simplified expression for the potential that included a quantum-mechanical correction obtained by Cutler and Gibbons.[209] The function has the form,

$$U(x) = -\frac{e^2}{4x} + h\frac{e^2}{4x^2} \equiv E(x) + hx^{-1}|E(x)| \quad \text{at } x > x_s > 0$$
$$U(x) = -E_a \quad \text{at } x < x_s \tag{2.26}$$

where h is a parameter depending on the metal and has been calculated to be on the order of 0.069 Å for $E_a = 10$ eV,[38, 209] and 0.9 Å for $E_a = 9.3$ eV.[210] Although this approximation is rough and does not consider details of the barrier structure, it nevertheless reflects the basic features of Bardeen's potential: transition to the image potential at large distances from the metal, the occurrence of the potential minimum near the surface, and the reduction of the force acting on the electron in comparison with the image force alone. FN curves calculated using this modified potential are shown in Fig. 2.9b.

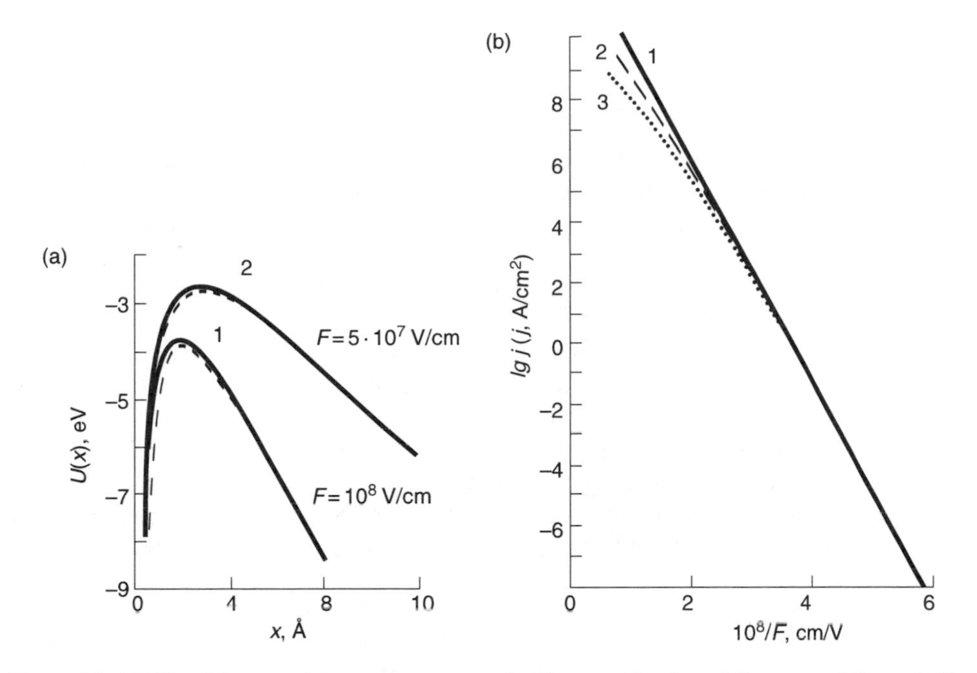

Figure 2.9. (a) Plot of the potential energy near a metal surface as a function of distance at different fields; (b) illustration showing the deviation from linearity of the current–voltage characteristics due to changes in barrier shape by the image force: 1: image potential; 2: simplified potential of Cutler–Gibbons; 3: potential of Seits–Vasiliev–van Oostrum.

Earlier, Vasiliev[211] used the potential function proposed by Seits[212] to analyze the FE current in strong fields (later this potential was used by van Oostrom[17] in his calculations). The image potential has a discontinuity as $x \to 0$. This discontinuity can be eliminated by writing the potential function in the form

$$U(x) = -e^2/(4x + e^2/E_a). \tag{2.27}$$

Vasiliev also used a more general expression derived by Miller and Good[213] for the transmission calculations which is valid throughout the range of the electron energies, both below and above the top of the barrier:

$$D(E_x F) = \left\{1 + \exp \frac{2i}{h} \int_{x_1}^{x_2} [2m_e(E_x + eFx + U(x))]^{1/2}\, dx\right\}^{-1}. \tag{2.28}$$

The current–voltage characteristic calculated with the Seits–Vasiliev–van Oostrom potential is also shown in Fig. 2.9b.

As shown in Fig 2.9, the use of the corrected form of the potential near the surface leads to the deviation of the current–voltage characteristic at lower currents. It is important to note that the values of the current density and field, at which deviation from FN theory occurs, are about the same as for which space charge affects the emission characteristics FE.

2.5. RESUME

Summarizing this chapter, its main results are as follows.

1. In the region of high fields and current densities, a noticeable deviation is observed from the FN law, which is apparent as a slowing down of the current rise with increasing voltage (Figs. 2.1–2.3). This effect is observed not only for tungsten emitters but also for Mo, Ta, Re, W_2C and other materials, where high current densities ($>10^6$ A/cm^2) can be reached. This deviation is stronger at high current densities and in materials with lower work function.
2. Theoretical analyses and experiments indicate that the slowing down of the FE current rise with increasing electric field strength is caused by two factors: build-up of the space charge of emitted electrons and increasing contribution from the potential in the near-surface region because of quantum-mechanical effects. Under conditions usual in a FE experiment and at $r_e = 10^{-5}$ cm, voltages up to 10–20 kV and work function values of $\varphi \approx$ 4–4.5 eV are both effective. At greater r_e and small φ, as well as at very high current densities, the space charge becomes a dominant factor.
3. In blunt emitters ($r_e > 10^{-3}$ cm) at high anode voltages ($u_a > 300$ kV), relativistic effects become clearly apparent in the field electron emission. According to calculations, the space charge of relativistic electrons can change the anode voltage necessary for obtaining high current densities (10^7–10^8 A/cm^2) by more than an order of magnitude. Under ultra-relativistic conditions (at $u_a > 10$ MV for tungsten emitters), the emission current becomes practically independent of the work function and rises linearly with voltage (see Eq. (2.20)).

3

MAXIMUM FIELD EMISSION CURRENT DENSITIES

3.1. THE THEORETICAL LIMIT OF FIELD EMISSION CURRENT

One of the most remarkable results of field emission (FE) quantum theory is the prediction of very high emission current densities. Very high current densities are possible due to two factors: (1) No energy is required for maintaining the emission process if electrons exit the solid by a tunneling mechanism, that is, the emitter does not need to be heated, irradiated, or otherwise excited by some external energy source; and (2) there is a very large reservoir of electrons near the Fermi level of a metal.

As the density of conduction band electrons in metals is of the order of 10^{22}–10^{23} cm^{-3}, it can be assumed that the electrons in the metal flowing toward the metal/vacuum interface can sustain a current density of about 10^{11} A/cm^2.[223] In the limiting case, of a potential barrier transparency of 1 (i.e., the barrier is fully eliminated by the external electric field), the maximum electron current density passing through a metal–vacuum interface can be expressed in the free electron model as,

$$\vec{j} = e \iiint_{(\vec{p},\vec{n})>0} f(\vec{p})\vec{n} \cdot \frac{(\vec{p},\vec{n})}{m} \cdot \frac{d^3\vec{p}}{h^3}, \tag{3.1}$$

where, e and m are the electron charge and mass, respectively, $\vec{p}$ is the electron momentum, $\vec{n}$ is the unit vector normal to the emitting surface, and f is the Fermi function. At $T = 0\,\mathrm{K}$, $f(\vec{p})$ equals unity at the Fermi surface, and $f(\vec{p})$ outside it, and Eq. (3.1) takes a more straightforward form:

$$\left|\vec{j}\right| = \frac{e}{m} \iiint_{(\vec{p},\vec{n})>0} (\vec{p},\vec{n}) \cdot \frac{d^3\vec{p}}{h^3}. \tag{3.2}$$

For a spherical Fermi surface, Eq. (3.2) in spherical coordinates has the form:

$$j = \frac{e}{m} \int_0^{2\pi} d\phi \int_0^{\pi/2} d\theta \cdot \sin\theta \cdot \cos\theta \int_0^{p_F} p^3 \frac{dp}{h^3}, \tag{3.3}$$

where p_F is the momentum of an electron at the Fermi surface. Integration of Eq. (3.3) gives a simple relation for the limiting emission current due to conduction band electrons in the metal

$$j = \frac{\pi \cdot e m_e E_F^2}{h^3} \cong 4.3 \cdot 10^9 E_F^2 \; [\text{A/cm}^2], \tag{3.4}$$

where E_F is the Fermi energy (in eV) measured from the bottom of the conduction band.

The above treatment can be extended to cover the case of elliptic energy bands where the dispersion law has the form:

$$E_F = \frac{p_x^2}{2m_x} + \frac{p_y^2}{2m_y} + \frac{p_z^2}{2m_z}. \tag{3.5}$$

Assuming that x is normal to the emitting surface, and introducing the new variables $p_x^2/m_x = \zeta^2$, $p_y^2/m_y = \vartheta^2$, $p_z^2/m_z = \chi^2$, we have

$$|j| = \frac{e}{m} \iiint\limits_{\substack{(\zeta^2+\vartheta^2+\chi^2)\leqslant 2E_F \\ \zeta>0}} m_x^{1/2} \zeta \cdot d\zeta \cdot d\vartheta \cdot d\chi \cdot J_{tr}, \tag{3.6}$$

where J_{tr} is the Jacobian of the transition from coordinates p_x, p_y, p_z to ζ, ϑ, χ. Integration of Eq. (3.6) gives a formula for the limiting emission current in the case of elliptic Fermi surfaces

$$j_{max} = 4.3 \cdot 10^9 m_x^* (m_y^* m_z^*)^{1/2} E_F^2, \tag{3.7}$$

where $m_x^* = m_x/m$, $m_y^* = m_y/m$, and $m_z^* = m_z/m$ are the ratios of the effective mass components to the free electron mass.

Note that the limiting emission current density j_{max} for the elliptic Fermi surface can be lower than in the case of the spherical Fermi surface for particular ratios of effective masses m_x, m_y, m_z.

The FE current density is usually lower than that predicted by Eqs. (3.4) and (3.7) due to the initiation of a vacuum arc because of the onset of FE instability at high current densities. When an arc is induced, the cathode begins to melt and ceases to be a controllable FE cathode.

3.2. EFFECTS PRECEDING FIELD EMITTER EXPLOSION

A very important case of vacuum breakdown initiation is observed with so-called "conditioned electrodes" in high vacuum. This mechanism involves FE from areas on the electrode surface where the field strength is enhanced due to micro-protrusions such as microtips. High current causes the electrical explosion of these micro-protrusions and the formation of a dense plasma near the cathode surface facilitating breakdown and arc initiation. This phenomenon is of a very general character.

This mechanism not only facilitates vacuum breakdown but also plays an important role in the subsequently ignited vacuum arc. It is possible to develop a model of the pre-breakdown situation at an individual microtip and to study quantitatively emission phenomena and processes at its surface in the time period just preceding the breakdown. These investigations allow one to experimentally study the initiation phase of breakdown and provide a key to understanding the mechanism of vacuum breakdown. This approach was suggested and brilliantly conducted in the well-known studies of Dyke and his collaborators.[31–33] Such studies were continued[47, 48, 50, 194, 195] and further advanced with improved experimental techniques including increased experimental sensitivity (5–6 orders of magnitude), larger pre-breakdown time ranges, and special conditions that enabled the observation of individual stages of the process. Mention should be made of the use of the nanosecond time scale techniques for the excitation and registration of pre-breakdown processes,[53, 54] the use of long current pulses, 10^{-3}–10^{-1} sec, to simulate quasi-stationary conditions,[49] and the extensive utilization of the FE microscopy and electron transmission and high resolution shadow microscopy.

Increased sensitivity enabled studies of emission processes localized to surface areas less than 10^{-6} cm in size[55, 56, 173, 174, 219, 222] at temperatures down to 4.2 K.[224]

3.2.1. Dyke's Experiments

In a set of elegant experiments by Dyke and co-workers,[31–33] it was shown that the destruction of a field cathode is caused by large emission currents. In the case of atomically clean and smooth emitters, instabilities can develop due to the formation of micro-protrusions at the surface as a result of ion bombardment or other reasons.[36, 51, 52] Elimination of the conditions leading to formation of micro-protrusions allows one to achieve current densities of 10^7 A/cm^2 dc[35, 36] or 10^8 A/cm^2 for a pulse duration of 10^{-6} sec.[31, 32]

Using pulsed excitation and registration in their investigations on tungsten (W) emitters, Dyke and co-workers were able to identify and control a number of effects during the transition to vacuum breakdown preceding field emitter destruction.[32] One of these effects is the occurrence of a bright ring in the emission images ("ring" effect)—Fig. 3.1a; the other is a spontaneous increase in the emission current during a voltage pulse ("tilt" effect)—Fig. 3.1b.

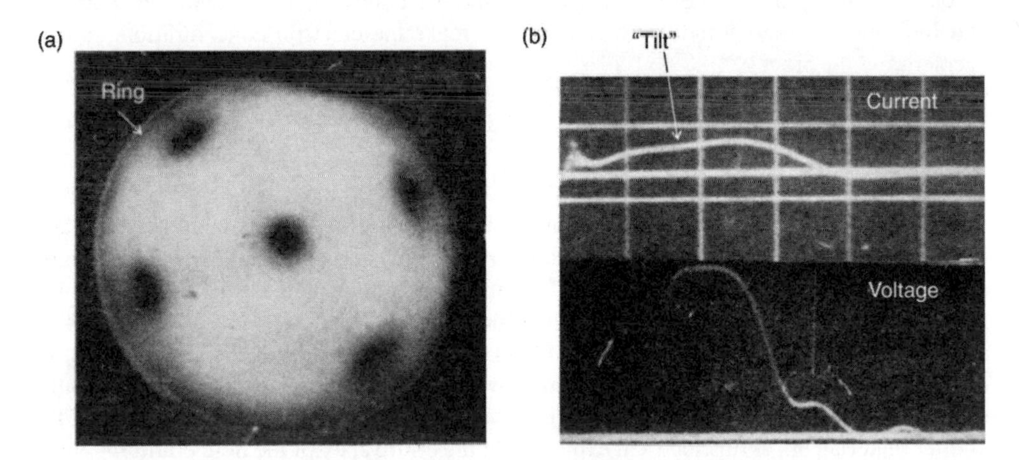

Figure 3.1. Observation of pre-breakdown phenomena [32]: (a) "ring" effect; (b) "tilt" effect—increase in current with time with voltage held constant. (top current oscillogram, $I = 10^{-2}$ A), (bottom voltage oscillogram, $U = 9.12$ kV), pulse duration 0.5 μsec.

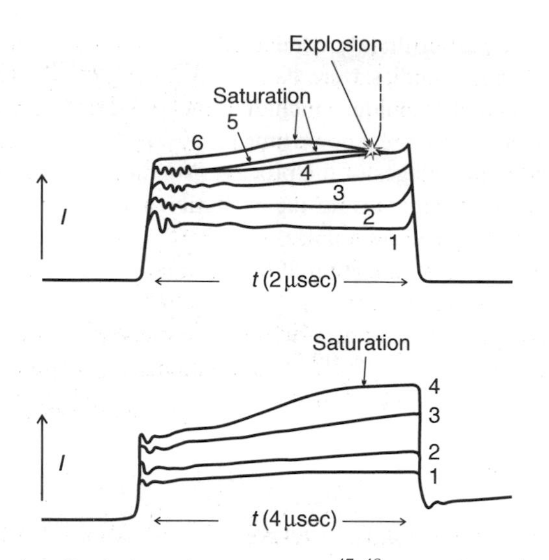

Figure 3.2. Current saturation effect in the pre-breakdown stage.[47, 48] [1–6]: current versus time traces for different values of pulse voltage $U_n > U_m$ for $n > m$.

Unless the emitter is destroyed, the pre-breakdown effects are fully reversible. The Dyke experiments were verified and advanced in investigations of Elinson,[59, 64] Shuppe,[45, 46] and in early works of Fursey's team.[47–50, 194, 195]

Both these effects have been interpreted as being caused by emitter heating due to the electron current, that is, Joule, or resistive, heating. Results of preliminary calculations based on joule heating[33] were consistent with the model. It was suggested that heating of the field cathode tip leads to an increase of the emission current because cold FE gives way to thermal FE. Higher emission current raises the cathode temperature still further and so on until the field cathode is destroyed. At the final stage of the process, evaporation and ionization of the cathode surface atoms increase the electron concentration, further accelerating the process.

3.2.2. Pre-Breakdown Phenomena

New features in pre-breakdown phenomena were observed by utilizing more precise techniques of generating and regulating the pulsed fields and considerably increasing the sensitivity to pre-breakdown currents (6–7 orders of magnitude increase in sensitivity).[47, 48] First, pre-breakdown effects were identified more clearly (Fig. 3.2). It was found that no avalanche-like current increase occurred if high current densities were approached at a slow rate and, on the contrary, a saturation region was observed (Fig. 3.2). This fact led to the suggestion that explosive breakdown occurs not as a result of a gradual evaporation of the emitter material, but as an electrical explosion of the entire apex of the field emitter.[47, 48, 67]

The second notable feature was that the "ring" observed by Dyke was not a single ring, but in fact, a number of concentric rings of oscillating intensity.[48, 50, 194] In a number of cases, these rings overlay the central part of the emission image (Fig. 3.3). The mechanism

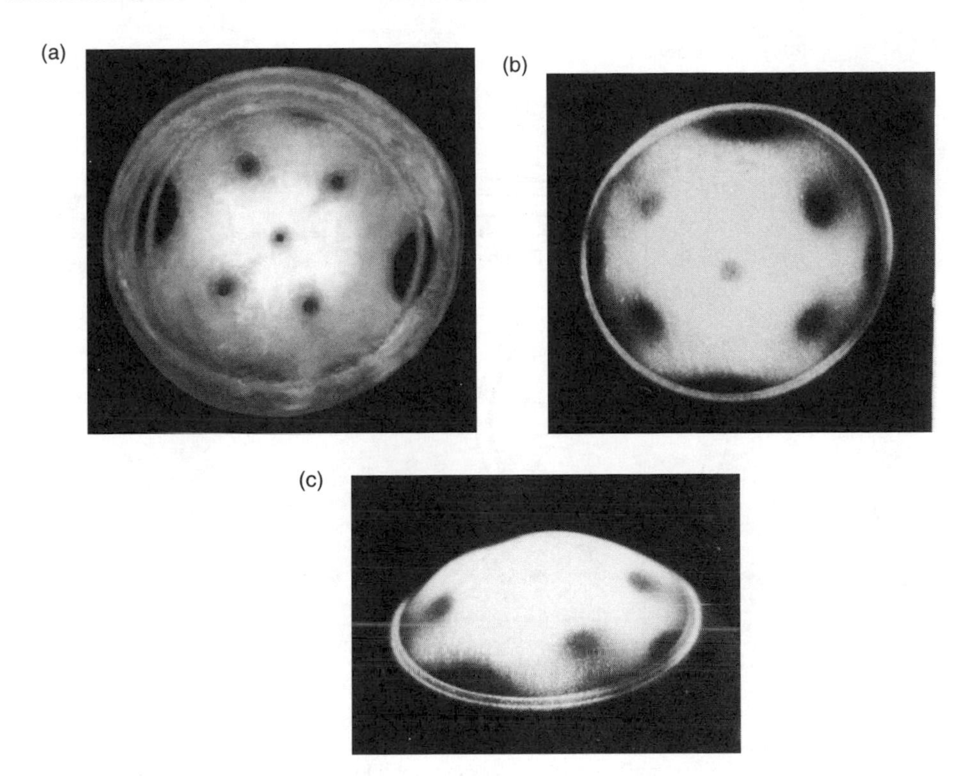

Figure 3.3. Multiple rings in the FE pattern: (a) W;[51] (b) and (c) Ta.[194] Photograph in (c) is taken with a spherical microscope (side view).

of the ring formation is not completely understood at present. Their outward appearance and multiplicity enable us to presume that these rings are due to an electro-optic effect.

The pre-breakdown effects (the "ring" and the "tilt") are observed not only on tungsten but represent general manifestations of the pre-breakdown state for a number of refractory materials. They have been observed with tantalum (Fig. 3.3b,c),[194] molybdenum,[194] rhenium,[195] tungsten carbide,[194] niobium,[51, 52] and systems involving absorbates.[50, 55]

3.2.3. Time-Dependent Observations

To understand pre-breakdown phenomena, we need to quantify the time dependence of the effects that are observed. Time dependent processes were investigated by carefully approaching the critical pre-breakdown state at different field pulse lengths.[47–49, 51, 52]

Experiments have been conducted with pulse lengths from 10 sec in duration[49] down through the millisecond range[51, 52] to the 10^{-7}–10^{-9} sec regime.[53, 54, 204] Current versus time traces in the pre-breakdown phase have similar general features over a wide range of pulse time intervals. However, the mechanism of the phenomenon appears to be different for different pulse durations.

In the nano- and microsecond regions, the current variation is due to a transition from FE to thermal FE by heating of the entire emitter tip. In the millisecond regime[51, 52]

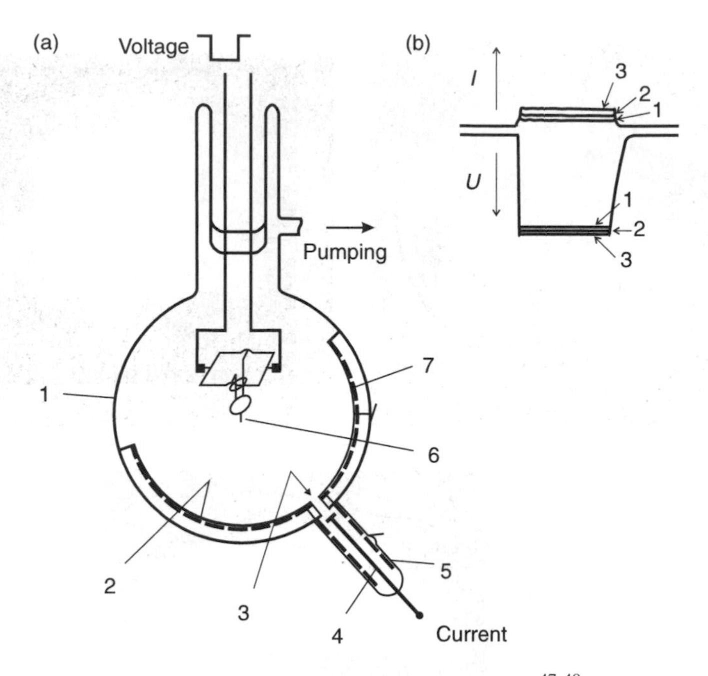

Figure 3.4. Probe experiments with registration of high FE current densities.[47,48] (a) Tube with a rotating field emitter on a gimbal. 1: windows for photoregistration. 2: luminescent screen. 3: probe hole. 4: collector of emitted electrons. 5: collector appendix. 6: FE cathode. 7: conducting film screening the collector from high-frequency interference. (b) Current and voltage oscillograms: 1, 2, 3: current oscillograms for different pulsed voltage values.

and under quasi steady-state conditions,[49] modification of the emitter geometry under the influence of the field and thermally activated surface diffusion play a dominant role.

3.2.4. Quantifying Local Effects

High measurement sensitivity has enabled investigation[47,48] of the FE currents in the pre-breakdown phase from separate areas of the emitting surface using a probe-hole (Fig. 3.4). Figure 3.5 shows current oscillograms taken in the ring and central parts of the FE image. Note that the current in the ring appears with a considerable delay.

Current in the central portion of the image (Fig. 3.5) varies very much like the total current from the tip. The "inertial" behavior of the current in the ring shows that it may be influenced by emitter tip Joule heating.

3.3. HEATING AS THE CAUSE OF FIELD EMISSION CATHODE INSTABILITIES

At present the hypothesis that the FE current is limited by thermal destruction of the emitter has been demonstrated experimentally and confirmed by rigorous theoretical calculations.

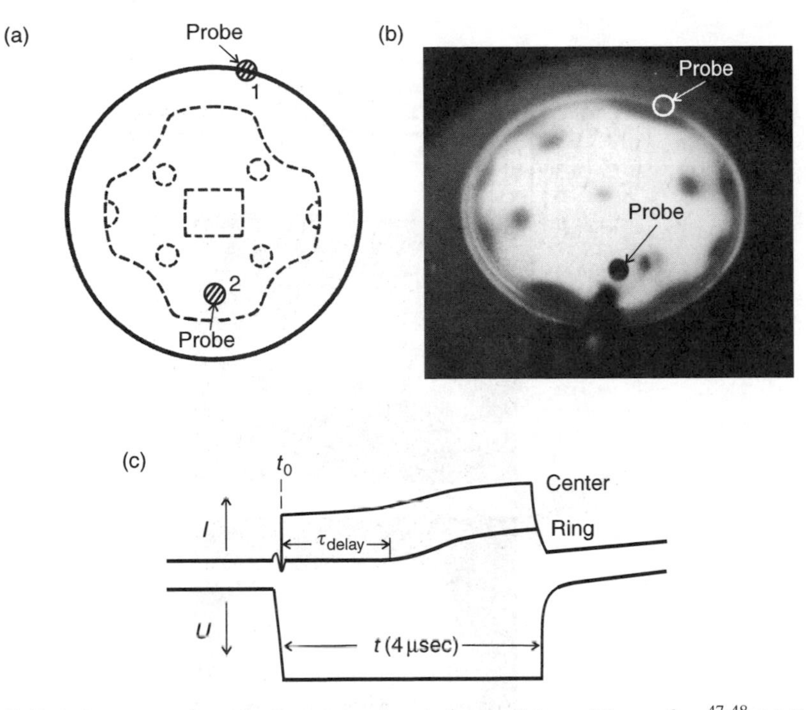

Figure 3.5. Emission current from the ring area and central area of the emitting surface.[47,48] (a) Schematic emission pattern for W and the probe-hole. (1, 2: different probe positions); (b) W emission pattern at maximum current density and the view of probe hole; (c) emission current versus time for the ring and central area with a fixed pulsed voltage.

3.3.1. Experimental Demonstration of Field Emitter Heating

Studies of the ring emission component of the total current have shown that it has a considerable time delay (Fig. 3.5). This delay may be considered as a direct proof of the emitter tip heating. If this is true, intentional heating of the tip by an external source will result in the shift of ring current to the beginning of the pulse.[48] Experimental results shown in Fig. 3.6a demonstrate emitter heating in the pre-breakdown phase.

Another means of ascertaining the role of emitter heating is by observing the behavior of adsorbed layers,[50] as surface migration of this coating can serve as a reliable indication of the emitter heating. In one experiment, a layer of Barium (Ba) is evaporated onto the side of the tip surface. As seen from Fig. 3.6b, up to the pre-breakdown phase, the emission image remains unchanged. Self-heating of the tip causes migration of Ba and the formation of a uniformly distributed layer (Fig. 3.6c).

The migration process can also be correlated with changes in the emission current and their occurrence during the voltage pulse see Fig. 3.6d. During the time interval $(t_0 - t_1)$, the emitter heats up to a temperature sufficient to initiate surface migration; during interval $(t_2 - t_3)$, the migrating layer passes the probe-hole and the collector current rises rapidly; during the time interval $(t_3 - t_4)$, the edge of the Ba deposit spreads over the entire portion of the surface covered by the probe-hole and the subsequent variation in collector current is

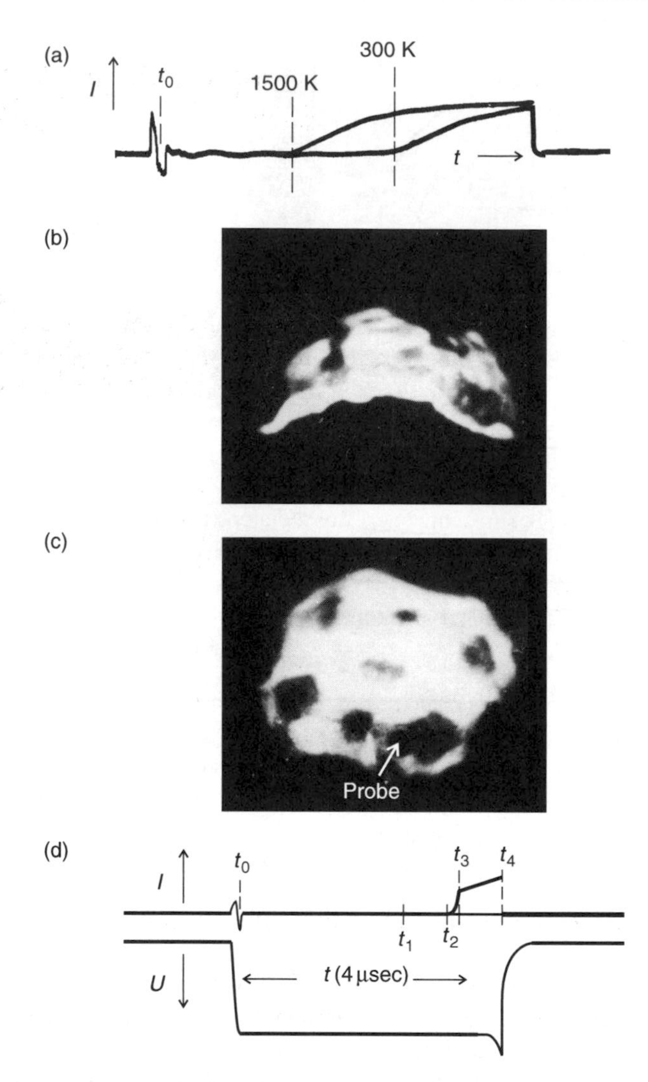

Figure 3.6. Experimental evidence of field emitter heating at maximum current densities. (a) Emission current from the ring at different temperatures $T_1 = 300$ K, $T_2 = 1500$ K; (b) and (c) migration of Ba across W surface in the pre-explosion phase (b: initial image, c: final image);[50, 223] (d) current through the probe-hole versus time at fixed pulsed voltage.

small. By measuring the time it takes the Ba layer to pass the probing orifice and estimating the magnification of the FE microscope, the migration rate and temperature (from the Arrhenius equation) can be determined. According to these estimates, Ba migration is initiated at the emitter temperature of ~2000–2500 K.

3.3.2. Analysis of Thermal Processes

The experiments[48, 50] conclusively demonstrate that with clean electrodes, pre-breakdown effects are due to Joule heating of the field emitter. Because of thermal instabilities in the field cathode tip, a transition from FE to explosive emission and vacuum breakdown occurs.[70] The emitter tip is heated as a result of two processes: Joule heating and heating due to the Nottingham effect.

3.3.2.1. Nottingham Effect

For many years it was believed that the major cause of emitter destruction was Joule heating. However, experiments by Swanson et al.[225] as well as calculations by Levine[226] indicate that the dominant contribution to the thermal balance during FE is due to a purely quantum-mechanical energy exchange process, referred to as the Nottingham effect.[227] According to this energy exchange mechanism, most of the electrons tunnel from energy levels below the Fermi level, that is, their energy is lower than that of the conduction band electrons that replace them (Fig. 3.7). As the incoming electrons are more energetic than the tunneling electrons, energy evolves near the surface (within roughly one electron mean free path). At high current densities, the energy associated with the Nottingham effect could exceed the energy due to Joule dissipation.

Together with Joule heating, Nottingham heating causes a rapid temperature rise in the emitter tip. It is noteworthy that after reaching a temperature referred to as the inversion temperature, the Nottingham effect first vanishes when the spectrum of emitted electrons becomes symmetric about the Fermi level (Fig. 3.7). Then, as the temperature increases beyond the inversion temperature, the Nottingham effect begins to cool the cathode because most of the emitted electrons come from energy levels above the Fermi level and, thus, extract energy from the cathode.

Joule heating and the Nottingham effect together can heat the cathode faster than each of them alone. As we will see, in this case, an interesting situation can arise because at some point the surface temperature drops considerably below the temperature inside the cathode.

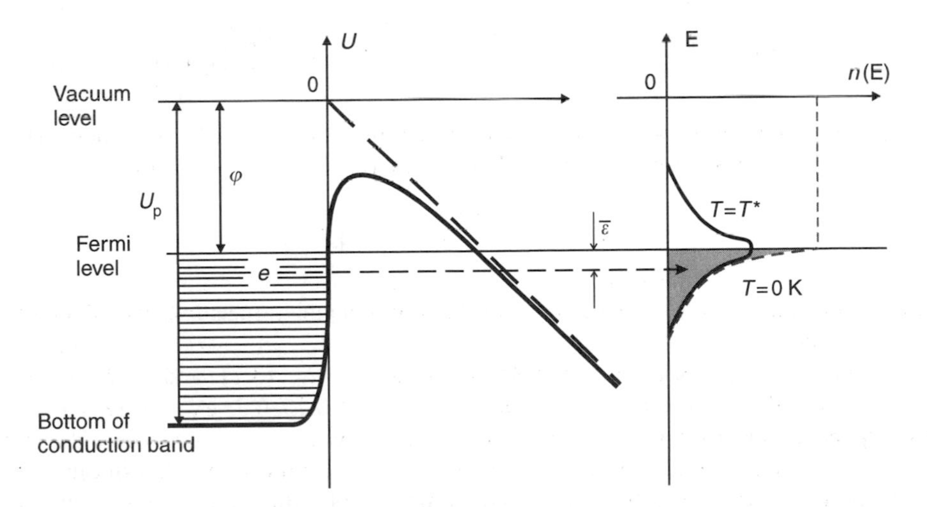

Figure 3.7. Illustration of Nottingham effect.

3.3.2.2. Theory of Nottingham Effect

The energy exchange in the Nottingham effect H_N is represented by the energy flow conveyed to the metal of taken from it by emitted electrons. This energy flow is equal to the number of electrons emitted in a unit time, that is, to the emission current i multiplied by the difference in average energies $\bar{\varepsilon}$ between the field emitted electrons $\bar{E}_{em}$ and the conduction band electrons $\bar{E}_{cd}$:

$$H_N = i \cdot \bar{\varepsilon}, \tag{3.8}$$

where $\bar{\varepsilon} = \bar{E}_{em} - \bar{E}_{cd}$.

In the Nottingham model,[234] it is assumed that the average energy of conduction band electrons ($\bar{E}_{cd}$) is equal to the Fermi energy ($\bar{E}_{cd} = \bar{E}_F$), and accordingly,

$$\bar{\varepsilon} = \bar{E}_{em} - \bar{E}_F. \tag{3.9}$$

Note that this assumption is true only for metals with spherical isoenergetic Fermi surfaces. For the complex band structure of transition metals, this condition is not an accurate assumption (for details see Seits[212] and Mott and Jones[229]). However, as it is very difficult to perform analytic calculations without this simplifying assumption, in the discussion below, it will be assumed that $\bar{E}_{cd} = \bar{E}_F$.

The average exchange energy per one emitted electron is

$$\bar{\varepsilon} = \frac{\int_{-\infty}^{\infty} \varepsilon \cdot P(\varepsilon)\, d\varepsilon}{\int_{-\infty}^{\infty} P(\varepsilon)\, d\varepsilon}, \tag{3.10}$$

where $\varepsilon \equiv E - E_F$ and $P(\varepsilon)\, d\varepsilon$ is the total energy distribution of emitted electrons relative to the Fermi level.

Here,[230]

$$P(\varepsilon) = \frac{j_0 \cdot f(\varepsilon) \cdot \exp(\varepsilon/d)}{d} \left[1 - \int_0^{2\pi} \exp\left(-\frac{E_m}{d} \right) d\Phi_p \right] \tag{3.11}$$

with j_0 being the emission current density as defined by the Fowler–Nordheim (FN) Eq. (1.8); d, a factor depending on φ and F and numerically (in eV) equal to

$$d = \frac{9.76 \cdot 10^{-9} \cdot F}{\sqrt{\varphi} \cdot t(y)} \ [\text{eV}]. \tag{3.12}$$

E_m is the maximum energy at a normal to the direction of emission and Φ_p the polar angle in the y, z plane.

The band structure is accounted for by the integral in Eq. (3.11) and for $E_m/d \gg I$ can be neglected. Under normal conditions of FE, $d \approx$ 0.15–0.25 eV, whereas E_m for most degenerate metals with high Fermi energies is a few eV, so in many cases the term involving the integral in Eq. (3.11) is negligibly small. An exception is transition metals with partially filled narrow d-bands. In some crystallographic directions, these metals have Fermi surfaces extending over small intervals thus making the integral in Eq. (3.11) an appreciable fraction of $P(\varepsilon)$.[225] Because exact values of E_{m} are not known, in approximate treatments, the contribution of the structural term is usually neglected. In this case, substituting the Fermi–Dirac distribution into Eq. (3.11) we have,

$$P(\varepsilon) = \frac{j_0}{d}\left[\frac{\exp(\varepsilon/d)}{1+\exp(\varepsilon/pd)}\right], \tag{3.13}$$

where

$$p = \frac{kT}{d} = 0.88 \cdot 10^4 t(y)\frac{\sqrt{\varphi}}{F}T \tag{3.14}$$

is a dimensionless parameter (φ, eV; T, K; F, V/cm). Substituting Eq. (3.13) into Eq. (3.10) and integrating gives an expression for $\bar{\varepsilon}$ first derived by Levine[226]

$$\bar{\varepsilon} = -\pi pd \cdot ctg(\pi p). \tag{3.15}$$

Analysis of Eq. (3.13) shows that the energy distribution becomes symmetrical for $p = \frac{1}{2}$. Distribution functions for this case as well as for the case $p(\varepsilon) = 0$ is given in Fig. 3.7.

The condition $p(\varepsilon) = \frac{1}{2}$ allows one to immediately obtain an expression for the inversion temperature:

$$T^* = \frac{d}{2K} = \frac{5.67 \cdot 10^{-5}}{\sqrt{\varphi} \cdot t(y)}F. \tag{3.16}$$

Therefore, assuming that E_{cd} is close to the Fermi energy E_{F}, at T^*, energy exchange due to the Nottingham effect is zero ($\bar{\varepsilon} = 0$). A plot of $T^*(F)$ is shown in Fig. 3.8.

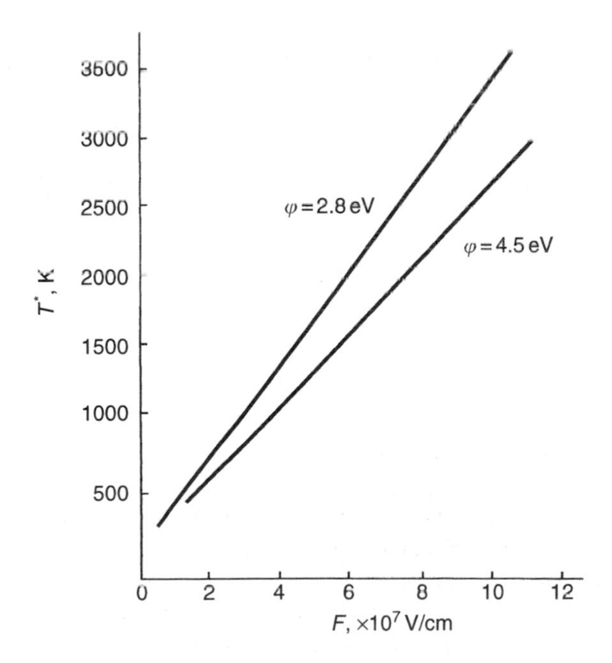

Figure 3.8. Variation of inversion temperature T^* with the electric field calculated using Eq. (3.16).

The steady-state temperature distribution along an emitter tip taking into account the Nottingham effect can be solved in the traditional way as it can be assumed that the energy exchange is confined to the near-surface region equivalent to the electron mean free path in a metal (10^{-6}–10^{-7} cm). Then for standard emitters where $r_e \geq 10^{-5}$ cm, the contribution of the Nottingham effect can be accounted for by boundary conditions:[223]

$$\chi_T \left. \frac{\partial T}{\partial r} \right|_{r=r_e} = -H_N(T_S) = -\pi pd \cdot ctg(\pi p) \cdot \frac{j}{e}, \tag{3.17}$$

where T_S is a temperature of the emitter apex where $r = r_e$.

For typical emitter tip cone angles ($\alpha < 30°$),

$$T(r)|_{r \to \infty} = T_0, \tag{3.18}$$

where T_0 is the temperature of the tip far from the apex. Neglecting radiation losses, the equation for thermal balance under steady-state conditions is

$$T'' + \frac{2}{r} T' + \frac{1}{r^4} \frac{j^2 r_0^4 \rho}{\chi_T} = 0, \tag{3.19}$$

where ρ is the resistivity constant, χ_T is the thermal conductivity coefficient, and r is the coordinate in the radially symmetric one-dimensional problem[33, 239] (Fig. 3.9). Integrating Eq. (3.19) with the boundary condition given in Eq. (3.17) gives

$$T = T_0 + \frac{H_N(T_S) \cdot j \cdot r_0^2}{\chi_T \cdot e \cdot r} + \frac{j^2 r_0^2 \rho}{\chi_T} \cdot \frac{r_0}{r} \left(1 - \frac{r_0}{2r}\right). \tag{3.20}$$

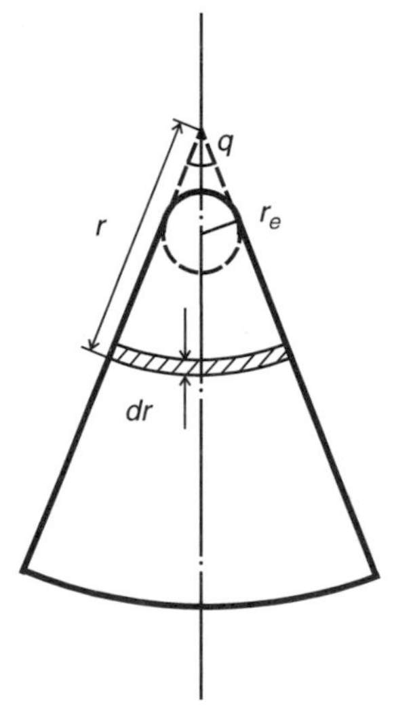

Figure 3.9. On the one-dimensional model of cathode tip heating by the FE current.

Then the temperature difference between the apex of the emitter tip T_e and its base T_0 can be represented by the formula

$$T_S - T_0 = \frac{H_N(T_S) \cdot j \cdot r_0}{\chi_T \cdot e} + \frac{j^2 \cdot r_0^2 \cdot \rho}{2\chi_T}, \tag{3.21}$$

where ρ is assumed independent of temperature and equal[231] to $\rho \approx \rho(0.8 \cdot T_S)$; $r_0 = r_e/\sin(\theta/2)$, r_e is the radius of the emitter tip's apex. In Eq. (3.21), the first term is due to the Nottingham effect and the second term is due to Joule heating. The relative contributions of these terms vary with temperature. At low temperatures, tip heating is mainly Nottingham heating (Fig. 3.10). At higher temperatures, approaching T^*, the Nottingham effect is close to the inversion point, and the dominant contribution will be Joule heating. This result is very important because it implies that for metals having melting points below the inversion temperature, emitter destruction will be the result of Nottingham heating. All metals with ordinary work function values (4.5–5 eV) that, therefore, require correspondingly high fields ($F > 5 \cdot 10^7$ V/cm) for high current density electron emission fall into this category, with the exception of the most refractory metals such as W, Mo, Ta, and Re.

Experimental verification of the Nottingham effect was made by Drechsler[232] and by Swanson et al. and Charbonnier et al.[225, 240] In general, experiment agrees with the theory. It is observed that the Nottingham effect plays a dominant role in emitter heating at intermediate temperatures. However, Drechsler[232] and Swanson et al.[225] observed considerable differences between the experimentally determined inversion temperature and the theoretical value. This was observed with clean tungsten ($\varphi \cong 4.5$ eV) and tungsten following the adsorption of electropositive atoms ($\varphi = 2.8$ eV). The experimental values were significantly lower than the calculated value—see Fig. 3.11. They[225] suggest that the discrepancy is due to a difference between the conduction electron energy E_{cd} and E_F. A difference of just a few hundredths of an eV between E_{cd} and E_F[233] can account for the observed lowering of T^*. Details of the mechanism of this process are not yet clear. Note, however, that with refractory metals, lower T^* values lead to increased emitter thermal stability due to the earlier inversion of the Nottingham effect and subsequent emitter cooling with increasing emission current. The effect is more pronounced at lower φ.

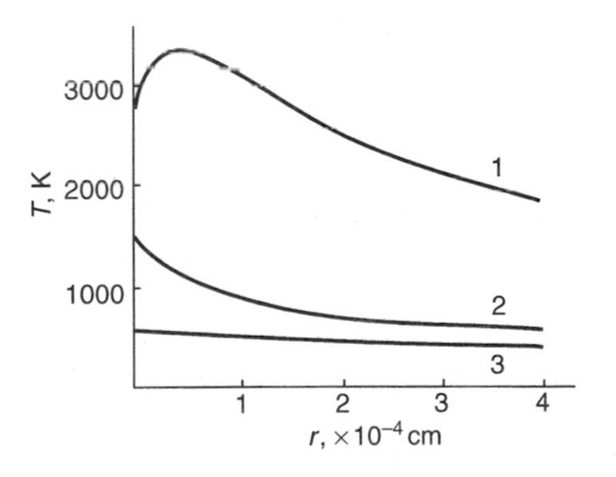

Figure 3.10. Temperature profile across emitter bulk in steady-state regime calculated for a one-dimensional model.[223] $F = 7.8 \cdot 10^7$ V/cm, $r_e = 10^{-4}$ cm. 1: Joule heating and Nottingham effect; 2: Nottingham effect; 3: Joule heating.

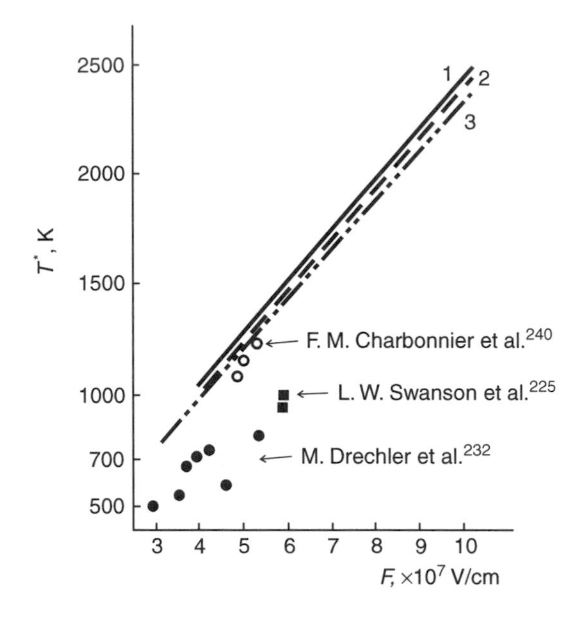

Figure 3.11. Comparison of inversion temperatures calculated using different potential barrier models.[233] 1: Image potential; 2: Cutler–Gibbons potential; 3: Zeitz–Vasilyev–Ostrum potential.

3.3.3. Three-Dimensional Analysis of Emitter Heating

The first theoretical analysis of the emitter tip self-heating was carried out by Dyke et al.[33] who calculated the value of the steady-state maximum current density j_m for a one-dimensional model of the emitter tip. Although this theoretical model was not completely accurate (the Nottingham effect and the dependence of the conductivity ρ on temperature were neglected), with arbitrary choice of the average value of r, satisfactory agreement with experiment was achieved.

Although the dependence of thermo-physical properties of the cathode material upon temperature were subsequently considered[239] while solving a one-dimensional time-dependent problem of the tip heating, the Nottingham effect, which is an important source of heating, was not accounted for. The time-dependent problem of emitter heating in the one-dimensional case (with the Nottingham effect included) was first studied by Litvinov et al.[235] and later by Nevrovsky and Rachovsky.[236]

Mitterauer[237] numerically solved the problem for a tip having the form of a prolate ellipsoid of rotation. In this case, thermal conductivity was accounted for only along the principle axis and the temperature and current density distribution were considered homogeneous for each cross-section perpendicular to the tip axis. This approach is, in fact, also a one-dimensional approach.

In one-dimensional models, the most important near-apex region of the cathode is not considered. Also, the actual distribution of the heat sources in the volume and over the surface, and the influence of the temperature on these sources are ignored in this model. Proper treatment requires the framework of a three-dimensional model.

The three-dimensional problem was solved by Glasanov, Baskin, and Fursey.[238] The calculations were conducted for a tip of the form suggested by Dyke,[39, 218] which describes well the typical geometry of real FE cathodes and enables one to calculate the field strengths. These calculations took into account[238, 245] Joule heating, Nottingham and Thomson[241] effects, and thermal radiation. The temperature dependence of the resistivity $\rho(T)$, heat capacity $c(T)$, and surface emissivity $b(T)$ were tabulated.[243]

The system of equations describing the kinetics of emitter heating by emission current includes the thermal conductivity equation

$$\delta \cdot c(T)\frac{\partial T}{\partial t} = \nabla\left[\lambda(T)\nabla T\right] + G(\vec{r}, t), \tag{3.22}$$

and the continuity equation for the current density in the bulk of the emitter

$$\operatorname{div} j(\vec{r}, t) = 0. \tag{3.23}$$

Here, $T = T(\vec{r}, t)$ is the emitter temperature; δ is the density, $\lambda(T)$ is thermal conductivity, $G(\vec{r}, t)$ is the volume density of heat evolution rate determined by the Joule and Tompson effects, where

$$G(r, t) = \rho(T) \cdot j^2(\vec{r}, t) + g(T) \cdot (\vec{j}(\vec{r}, t), \nabla T(\vec{r}, t)). \tag{3.24}$$

and $g(T)$ is a temperature-dependent Tompson coefficient.[242] $\vec{j}(\vec{r}, t)$ is the emission current density

$$\vec{j}(\vec{r}, t) = -\frac{1}{\rho(T)} \cdot \left[\nabla\Phi(\vec{r}, t) + \alpha(T)\nabla T(\vec{r}, t)\right], \tag{3.25}$$

where $\Phi(\vec{r}, t)$ is the electric potential, $\alpha(T)$ is a function related to $g(T)$.[242]

$$g(T) = -T\frac{d\alpha}{dT}. \tag{3.26}$$

Using Eqs. (3.23) and (3.25) one has an equation for the electric field potential $\Phi(\vec{r}, t)$ inside the emitter

$$\nabla\left[\frac{1}{\rho(T)}\nabla\Phi(\vec{r}, t)\right] = -\nabla\left[\frac{1}{\rho(T)}\alpha(T)\nabla T\right]. \tag{3.27}$$

Boundary conditions for these equations include Nottingham effect and thermal radiation from emitter surface

$$\lambda(T)\left.\frac{\partial T}{\partial n}\right|_{S} = \frac{j(F_S, T_S)}{e} \cdot \bar{\varepsilon}(F_S, T_S) - b(T_S) \cdot \sigma \cdot T_S^4, \tag{3.28}$$

where σ is the Stephan–Boltzmann constant; n is an external normal to the emitter boundary S; $j(F_S, T_S)$ is the emission current density depending on the emission surface temperature T_S and the electric field strength F_S at the surface. $\bar{\varepsilon}(F_S, T_S)$ is an average value of the energy evolved at the surface per emitted electron. This problem has been solved numerically using approximate formulae[69, 165] for $j(F_S, T_S)$ and $\bar{\varepsilon}(F_S, T_S)$.

In Fig. 3.12, sets of isotherms showing the dependence of temperature on time are plotted. The dependence of apex surface temperature T_S and the maximum tip temperature T_m on the time are shown in Figs. 3.13 and 3.14. Calculations showing the kinetics of heating a tip with axial symmetry with a form approximated by Dyke et al.[218] have also been performed.

The most important result of these calculations is the formation of the overheated core inside the emitter apex. The temperature at the surface may be well below the melting point while in the internal region (the kernel), it may be equal to several tens of thousands degrees kelvin (Figs. 3.12 and 3.13). This leads to enormous temperature gradients, $\sim 10^8$ K/cm and the generation of large thermoelastic stresses. The value of tangential stresses at the tip surface can exceed $2 \cdot 10^9$ Pa[246] that can destroy the tip before the melting point is reached.

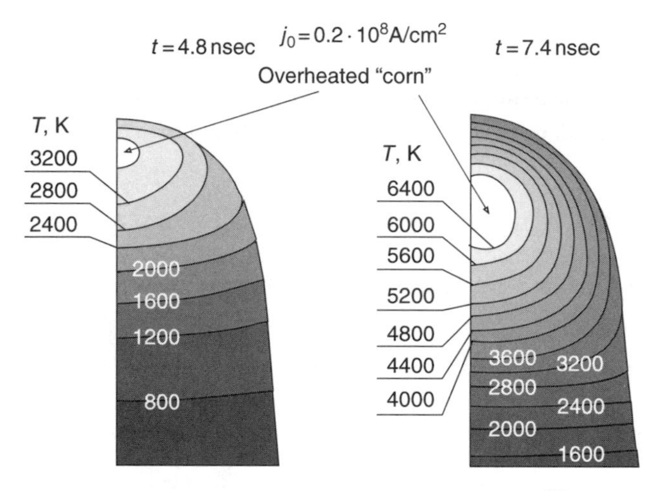

Figure 3.12. Temperature profiles in the emitter tip.[238]

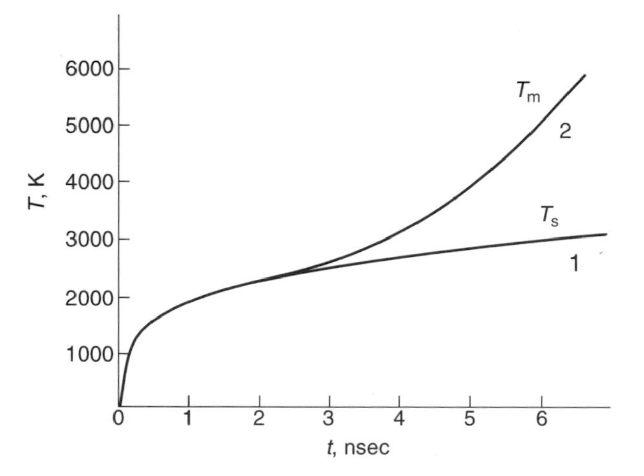

Figure 3.13. Temperature as a function of time.[238] 1: tip surface temperature; 2: maximum temperature in the emitter bulk.

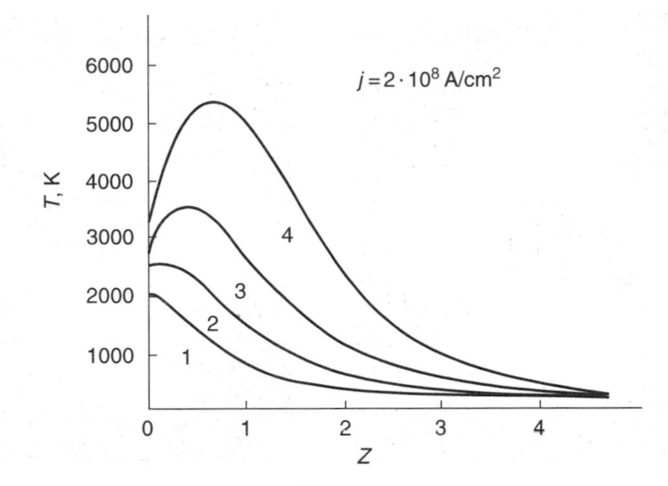

Figure 3.14. Temperature distribution along the emitter axis of rotational symmetry. 1: 1.6 nsec after application of electric field; 2: 3.5 nsec; 3: 4.0 nsec; 4: 7.4 nsec.

If the energy input rate is great enough, relative to the time required to destroy the tip by tangential stress, injection of a stream-liquid mixture into the vacuum gap can occur.

The kinetics of the heating process may be viewed in the following way. During the initial stage, the main contribution to heating of the emitter is due to the Nottingham effect, since the temperature of any point of the surface is lower than the inversion temperature. Eventually, the temperature of the emitting surface near the apex of the tip becomes greater than the inversion temperature, which induces Nottingham cooling and leads to a displacement of the temperature maximum from the surface into the bulk. Simultaneously, the rate of rise of the surface temperature decreases. However, complete stabilization of the temperature at the apex is not achieved.

3.4. BUILD-UP OF THE FIELD EMITTER SURFACE AT HIGH CURRENT DENSITIES: THERMAL FIELD SURFACE SELF-DIFFUSION

The efficiency of FE cathode heating increases with the appearance of micro-protrusions due to heat localization. This localization causes a further evolution of the micro-relief, a resulting local increase in field strength and thus FE current density and, finally, explosion of one or several of the resulting "micro-tips", destruction of the FE cathode, and the development of vacuum breakdown.

It has been shown[51, 52] that at high current densities and with sufficient pulse durations, rearrangement of the emitter surface takes place, whereby micro-protrusions and sharp ridges are formed.[247] The activation energy for surface diffusion is due to the heating of cathode by the electron current. If the electric field, and correspondingly the emission current, pulse duration is hundreds of microseconds to a few milliseconds or longer, then the rearrangement will always begin before the tip body heats to the critical temperature and the cathode explodes. The effects are cumulative as each subsequent high current density pulse creates protrusions, further enhancing the rearrangement effect.

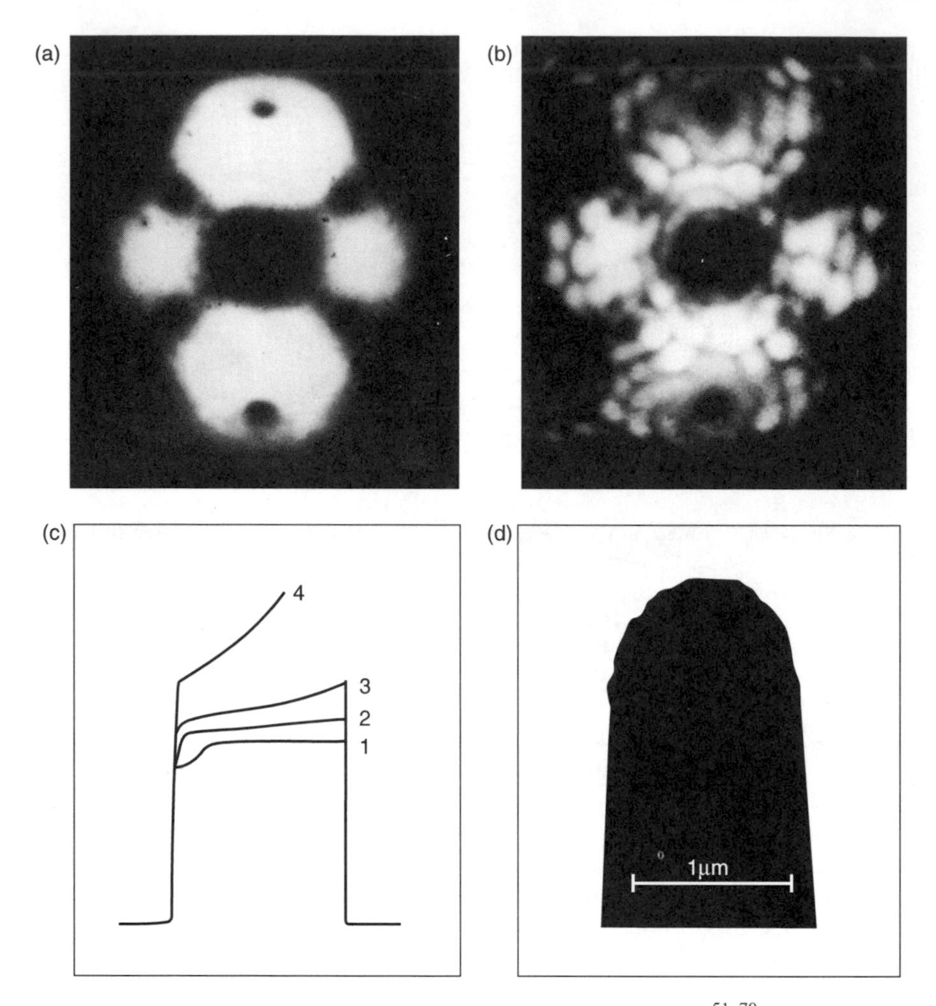

Figure 3.15. Surface migration with field emitter tips heated by the emission current.[51, 70] (a) and (b) FE patterns in the initial and final stages, respectively; (c) increase in the FE current during square wave voltage pulses of various amplitudes; (d) shadow-micrograph of the tip after current pulsing.

In earlier studies, this rearrangement process (diffusion) could not be detected directly because under such conditions, space charge is important and thus the image contrast is low. To study the build-up effects, after applying the maximum current density, "frozen" traces of the build-up were observed at lower current densities ($\sim 10^2$ A/cm)—Fig. 3.15. In the experiments[51, 52] FM Müller microscope was used to examine W, Mo, Nb, and Ta cathodes with tip radii of $r_0 \sim (1\text{–}5) \cdot 10^{-5}$ cm. It was found that the degree of emission-induced migration depends on the duration of the current pulse. Migration can be enhanced by repeated application of a voltage pulse for fixed durations, which causes "accumulation" of surface micro-roughness. As the surface micro-relief is changing, an appreciable increase in the emission current at fixed applied voltage occurs in the low current region. The rate of the current enhancement depends on the maximum current density and the initial emitter

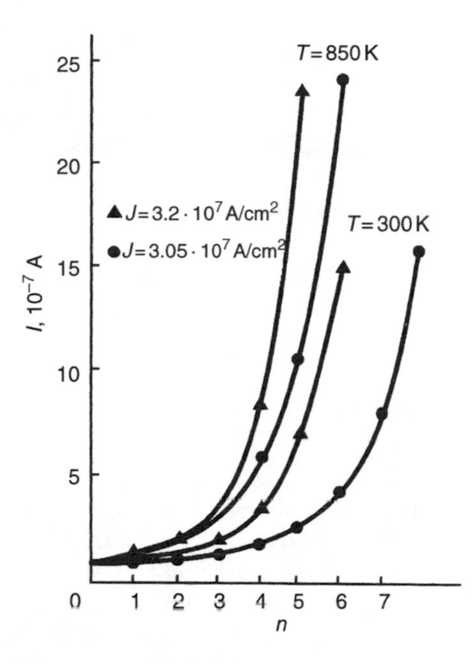

Figure 3.16. Variation of the FE current as a function of the number of applied voltage pulses for different values of j and T.[181] $\tau_{\text{pulse}} = 100\ \mu\text{sec}$.

temperature (Fig. 3.16). The increase in current as a result of repeated application of voltage pulses of the same amplitude continues until breakdown sets in.

Emission-induced migration is a key factor determining FE instability at maximum current densities with atomically clean cathode surfaces over a wide range of times ($t \approx 10^{-6}$–10^{-2} sec). The maximum current density is limited not only by thermo-physical parameters of the emitter bulk but to a greater extent by surface self-diffusion, a process controlled by the migration activation energy.[51, 52, 181]

3.5. THE HIGHEST FIELD EMISSION CURRENT DENSITIES ACHIEVED EXPERIMENTALLY

As shown above, the FE current density attainable in practice is limited by heating of the emitter tip. A number of experimental attempts to circumnavigate the limitation caused by the field emitter heating have been conducted. These attempts were based on the following ideas.

1. As heating is essentially an inertial process, the maximum current density can be increased by applying electric field pulses of very short duration. In this way, current densities up to $10^9\ \text{A/cm}^2$ were achieved using pulse durations of a few nanoseconds.[53, 54, 250]
2. Reduce Joule heating by using superconduction emitter tips. It was found that the superconductivity was destroyed by the high current densities. By cooling to temperatures of 2–4.2 K, j_{max} could be increased two- or threefold.[224]

3. Current densities of 10^9–10^{11} A/cm^2 were obtained in experiments with small emission spots[55, 244] and tips 10–30 Å in size.[56, 222] Because of small size of the field emitter, the electrons traverse it without scattering, so that no energy is dissipated by Joule heating. This will be discussed further in the next section.

3.5.1. Experimental Current Density Values

The limiting current densities obtained by various authors over the past years are presented in Table 3.1.

Table 3.1. Limiting current densities

Material	τ, sec	j_{max}, A/cm^2	Authors
W	dc mode	10^7	Martin et al.[35] and Dyke et al. (1960)[36]
W	10^1–10^{-3}	$2 \cdot 10^7$	Fursey and Kartsev (1970)[49]
Mo	10^{-3}	$8 \cdot 10^6$	Fursey et al. (1984),[51] Krotevich et al. (1985),[52] and Krotevich (1985)[181]
Nb	10^{-3}	$3 \cdot 10^6$	
LaB_6	$3 \cdot 10^{-6}$	10^7–10^8	Elinson and Vasiliev (1954)[251] and Elinson and Kudnitseva (1964)[252]
ZrC	$3 \cdot 10^{-6}$	10^7–10^8	
Ta	$4 \cdot 10^{-6}$	$5 \cdot 10^7$	Fursey and Tolkacheva (1963)[194]
Re	$4 \cdot 10^{-6}$	$(3–5) \cdot 10^7$	Fursey (1964)[195]
W	10^{-7}	$3 \cdot 10^8$	Mesyats and Fursey groups[53,54]
W	10^{-8}	$(5–6) \cdot 10^8$	
W	10^{-9}	10^9	

3.5.2. The Limiting Current Densities Attained Cooling

It is known that the maximum attainable FE current density is limited by Joule heating of the lattice and the Nottingham effect. Whereas the Joule heat is released throughout the bulk of the cathode, the heat produced by the Nottingham effect is localized to the near-surface region whose thickness is of the order of electron–phonon free path $\sim\lambda_{ep}$. Above the Debye temperature, this distance is of the order of 100 Å, that is, the Nottingham effect is a surface phenomenon. Lowering the emitter temperature causes a significant reduction in Joule heating and a considerable increase in λ_{ep}, which shifts Nottingham heating to a bulk effect. Therefore, the maximum FE current density can be increased by decreasing the emitter temperature.

By the Nottingham mechanism, after an electron is emitted, the remaining conduction band hole will scatter electrons into states above the Fermi level.[190] These electrons will thermalize in time τ_{ee}, which is a characteristic time of the electron–electron interaction. After time τ_{ep}, which is a characteristic time of the electron–phonon interaction, the energy will be transferred to the lattice. Note that the electron thermalization can only occur if $\tau_{ee} < \tau_{ep}$. At low temperatures, the thickness of the region in which energy transfer to the lattice takes place ($\lambda_{ep} \sim v_F \tau_{ep}$, v_F being the velocity of electrons on the Fermi surface), can be as large as a fraction of a millimeter in pure metals.

The complete theoretical analysis of such a complex situation is possible only in terms of kinetic theory. However, assuming $\tau_{ee} < \tau_{ep}$, a metal can be considered as a combination of two subsystems, one of electrons and the other of lattice atoms.[224]

The energy balance equations for these subsystems have the form:

$$c_e \cdot \frac{\partial T_e}{\partial t} = \nabla(\chi_e \cdot T_e) - A(T_e, T_i) + j_{fe} \cdot f(r), \tag{3.29}$$

$$c_i \cdot \frac{\partial T_i}{\partial t} = \nabla(\chi_i \cdot T_i) - A(T_e, T_i). \tag{3.30}$$

Here, T_e, c_e, and χ_e are the electron temperature, electron heat capacity, and thermal conductivity, respectively; T_i, c_i, and χ_i are respective quantities for the lattice subsystem; $A(T_e, T_i)$ is a function describing nonlocalized transfer of energy from the electron subsystem to the lattice, j_{fe} is FE current density. The last term in Eq. (3.29) takes account of the Nottingham energy released in the electron subsystem over the length λ_{ee}.

Despite the significant approximations in such an approach with an appropriate choice of $A(T_e, T_i)$ and $f(r)$, it can adequately reproduce the main features of the phenomenon: the spread of the electron–electron interaction over length λ_{ee} and the electron–lattice interaction over length λ_{ep}. Functions $A(T_e, T_i)$ and $f(r)$ were taken in the form:

$$A(T_e, T_i) = \alpha \cdot \int U\left[T_e(\vec{x}), T_i(\vec{x})\right] \cdot \exp\left\{-\frac{|\vec{r} - \vec{x}|}{\lambda_{ep}}\right\} \cdot \partial^3 x, \tag{3.31}$$

$$f(r) = \beta \cdot \exp\left\{-\frac{l}{\lambda_{ep}}\right\}, \tag{3.32}$$

where $U(T_e, T_i)$ is the usual expression for the rate of local energy transfer to the lattice;[248] α and β are constants derived from the condition that for $\lambda_{ep} \to 0$, Eqs. (3.31) and (3.32) transform to local relationships;[228, 248] l is a distance from point $\vec{r}$ to the emitting surface. Boundary conditions for the system of Eqs. (3.29–3.32) are the temperature stability inside the emitter and the absence of temperature gradients along the normal $\vec{n}$ to the emitting surface S:

$$T_i(r \to \infty) = T_e(r \to \infty) = T_0, \tag{3.33}$$

$$\left.\frac{\partial T_i}{\partial n}\right|_S = \left.\frac{\partial T_e}{\partial n}\right|_S = 0, \quad j|_S = j(T_e, F) \tag{3.34}$$

with initial conditions

$$T_i(t = 0) = T_e(t = 0) = T_0, \tag{3.35}$$

where T_0 is the initial temperature, $j(T_e, F)$ the known function[164] (see Eqs. (1.11) and (1.12), Chapter 1) of the electric field strength F and T_e on the emitting surface.

If spreading of only the electron–electron interaction is taken into account and A is taken in the form $A(T_e, T_i) = \nu(T_e - T_i)$, then neglecting the electronic heat capacity, we can obtain for a one-dimensional cathode:

$$T_e - T_i = \frac{B}{1-\gamma} \cdot \left[\frac{\exp\{-\sqrt{\gamma}\xi\}}{\sqrt{\gamma}} - \exp\{-\xi\} \right], \tag{3.36}$$

where $\gamma = \lambda_{ee}^2 \nu / \chi_e$; $\xi = x/\lambda_{ee}$; $B = \lambda_{ee} j \bar{\varepsilon} / \chi_e$; $\bar{\varepsilon} = \bar{E}_{em} - E_F$ is the average Nottingham energy per electron released in the electronic system during tunneling.

If $\lambda_{ee} \gg (\chi_e/\nu)^{1/2} \equiv \delta$, then $T_e - T_i = (\bar{\varepsilon} \cdot j/\nu\lambda_{ee}) \exp(-x/\lambda_{ee})$. Conversely, if $\lambda_{ee} \ll \delta$, $T_e - T_i = (\bar{\varepsilon} \cdot j/\nu\delta) \exp(-x/\delta)$.

As can be seen from these relationships, for $\lambda_{ee} \ll \delta$, the characteristic size of the heated region in the electron gas is $\sim \delta$.[228] For $\lambda_{ee} \gg \delta$, the heated region is much larger and the electron gas temperature much lower than in the former case. As a consequence, the lattice heating rate is much lower:

$$T_i - T_0 \approx \frac{\bar{\varepsilon} \cdot j}{c_i} \cdot \frac{t}{\lambda_{ee} + \delta}. \tag{3.37}$$

So, accounting only for heat spreading over the length λ_{ee} results in a much lower lattice heating rate.

The system of Eqs. (3.29–3.34) has been integrated numerically for a cone-shaped emitter. The calculation was carried out for $T_0 = 4.2$ and 300 K. The magnitude of the electric field F was chosen such that a steep current rise occurred during a time interval of 10 nsec.

Figure 3.17 shows the calculated time dependence of the emission current density. As seen in the figure, the maximum current density at $T_0 = 4.2$ K is a factor of 2.8 greater than at $T_0 = 300$ K. With increasing λ_{ep}, this ratio increases somewhat but λ_{ep} cannot be much in excess of r_0/α_0 (r_0 is the emitter radius and α_0 is the emitter cone angle).

An experimental study was carried out on tungsten single crystals cooled to liquid helium temperature (4.2 K).[224] The sample was placed in an evacuated tube and was

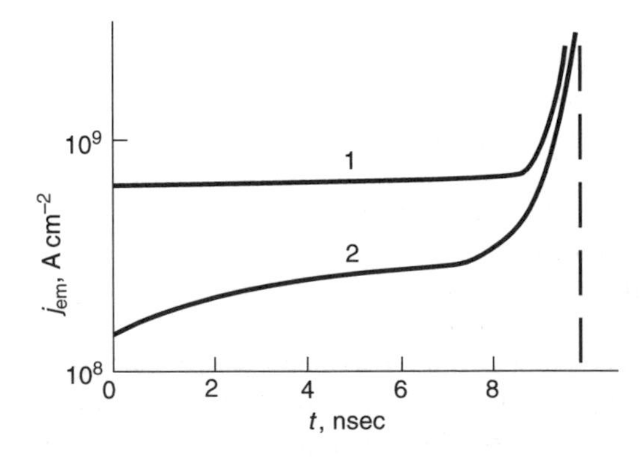

Figure 3.17. Calculated variation of the maximum current density $j_{em}(t)$: 1: $T_0 = 4.2$ K, 2: $T_0 = 300$ K, $r_0 = 0.2\ \mu$m, $\alpha_0 = 20°$.

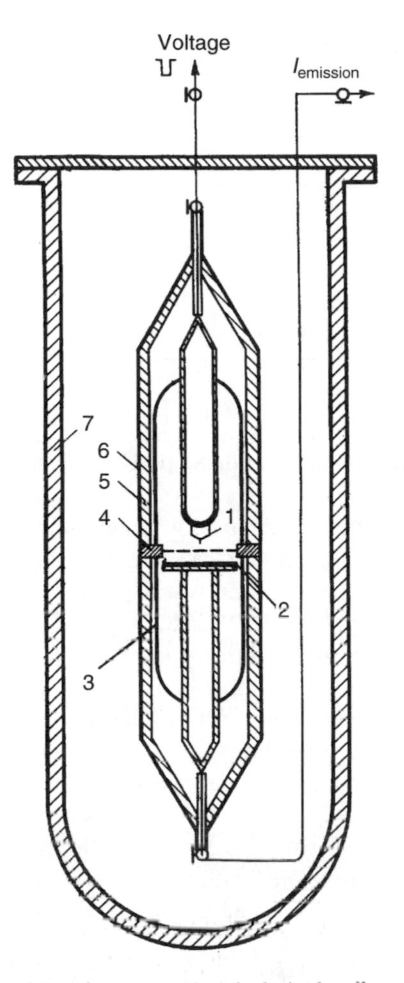

Figure 3.18. Schematic of the experimental apparatus. 1: cathode tip; 2: collector; 3: screen; 4: grid; 5: insulator; 6: matched coaxial screen; 7: Dewar vessel with liquid helium.

immersed in liquid helium (Fig. 3.18). The residual gas pressure in the working space of the device after immersion in liquid helium was 10^{-14}–10^{-13} Torr.

A modified Muller microscope permitted observation of the emission images and investigation of the FE characteristics in a nanosecond range. Most experiments were carried out with pulse durations of 10 nsec or in the dc mode. The tip surface was cleaned by heating to high temperature in the cryostat. The emitting area was determined using known methods[27] and the emitter radius was derived from current–voltage characteristics or from electron-optical images of the tip profile.

With samples cooled to 4.2 K, the maximum current density that could be extracted without damage to the tip was two to three times that could be extracted at room temperature. Measured current and current densities for various samples at 300 K and 4.2 K are shown in Table 3.2. The experimental results are in fair agreement with the theoretical model presented in the beginning of this section.

Table 3.2. Measured maximum current and current densities for various samples

$T_0 = 300$ K			$T_0 = 4.2$ K		
r_e, μm	I_m, A	j_m, A · cm^{-2}	r_e, μm	I_m, A	j_m, A · cm^{-2}
0.28	1.15	$3.7 \cdot 10^8$	0.14	0.79	$10.1 \cdot 10^8$
0.29	1.07	$3.2 \cdot 10^8$	0.15	0.73	$8.1 \cdot 10^8$
0.37	1.16	$2.1 \cdot 10^8$	0.18	1.17	$9.0 \cdot 10^8$
0.38	1.53	$2.6 \cdot 10^8$	0.25	2.0	$8.0 \cdot 10^8$
1.0	7.0	$1.75 \cdot 10^8$	0.29	3.39	$10.1 \cdot 10^8$

3.5.3. Current Densities from Nanometer-Scale Field Emitters

The highest current densities we obtained when emission was localized to very small (nanometer scale) areas of the emitter surface. Two types of experiments have been conducted. In the first type,[55, 244] zirconium (Zr) was deposited onto the surface of an ordinary tungsten emitter of size 10^{-5} cm (Fig. 3.19). In the second type,[56, 222, 244] nanometer-scale micro-protrusions having characteristic dimensions of 10–20 Å were formed on the emitter as a result of surface migration of W atoms in a strong electric field. In both cases, the local temperature rise, for a given emitter current density, was greatly reduced due to the cooling action of the remainder of the emitter tip. Note that with nanometer-scale emitting areas, there is practically no preferential energy deposition in the emitting region because its size, $\sim 10^{-7}$ cm, is substantially smaller than the electron–photon free pat. This means that the heat produced by the current flowing through the emitter tip is distributed over a much larger volume than that associated with the emission site alone.

The first experiments with selective adsorption of Zr on W were done by Fursey and Shakirova.[55] Emission patterns of Zr on W ⟨001⟩ are shown in Fig. 3.19c. The emission sites in Fig. 3.19c have radii of ~40–50 nm whereas the overall tip radius is 300–400 nm.

Ultra-small emitting spots, with radii of 10–15 nm (Fig. 3.19d) were formed by depositing a Zr–ZrO/W coating in an electric field ($\sim 5 \cdot 10^7$ V/cm).[244, 253] The smallest localized emission sites achieved had radii of ~3 nm. The dimension of the site was determined by comparing its image with that of the entire tip for which the radius had been determined independently.

The maximum current density extracted from the type of site shown in Fig. 3.19c in the pulsed mode (2–5 μsec duration pulses) was $\sim 5 \cdot 10^9$ A/cm.[255] For the sites shown in Fig. 3.19d, dc current densities up to 10^9 A/cm^2 were obtained.[244]

The first experiments with micro-protrusions were carried out by Shrednik's group.[56] In his study, dc mode current densities up to 10^9 A (in some experiments up to 10^{10} A/cm^2) could be obtained. Recent experiments[222] have shown that if very small isolated micro-protrusions are created on the surface, current densities greater than 10^{10} A/cm^2 and up to 10^{11} A/cm^2 could be achieved. Illustration of current measured at the nanotip by the probe is given in the Fig. 3.20.

Theoretical analysis[222] of emitter heating when the characteristic electron–phonon free path λ_{ep} is much greater than the size of the emitter has shown that for $2r_{em}/\lambda_{ep} < 0.6$,

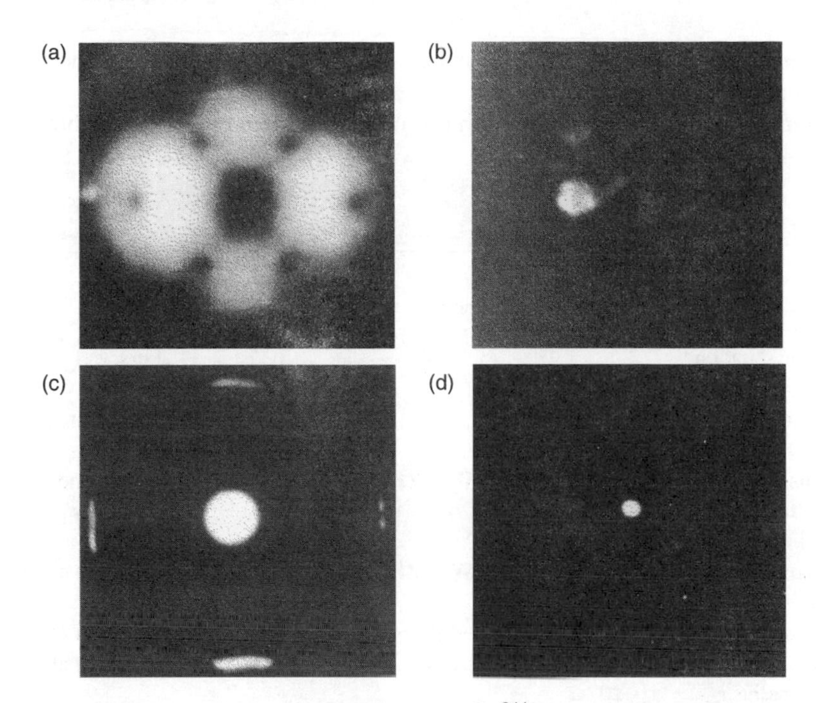

Figure 3.19. Nanoscale emission area on the field emitter surface.[244] (a) Initial FE image of W ⟨011⟩; (b) Image of a single microprotrusion on W ⟨011⟩ generated by the build-up process; (c) localization of FE due to selective Zr adsorption on the (001) face of tungsten; (d) image of an ultra-small emission site on the (001) face of tungsten.

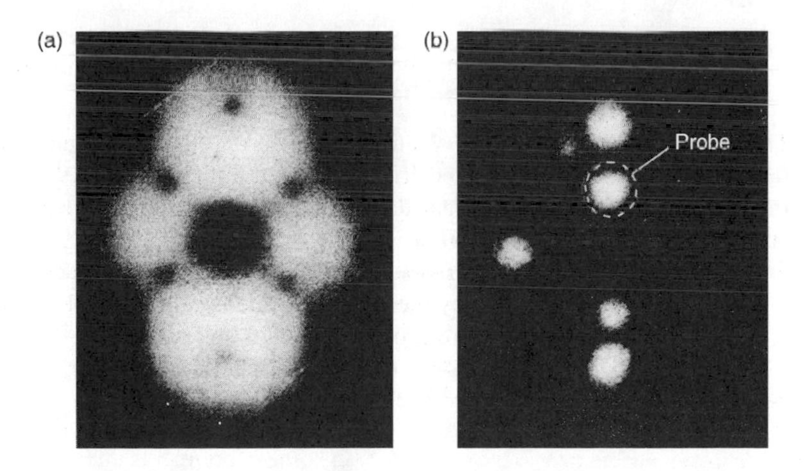

Figure 3.20. Probe experiments for W {011}: (a) initial FE image; (b) image of nanotips on the surface after build-up.

current densities $j > 2 \times 10^{10}$ A/cm^2 can be obtained in dc operation without destruction of field emitter.

Summary of the data for extremely high current densities is shown in Table 3.3.

Table 3.3. Summary of data for extremely high current densities

Conditions	Material	Record current densities, A/cm^2	Authors
Nanosecond duration ($\tau_{pulse} = 5$ nsec, $T = 300$ K)	**W**	10^9	Fursey et al. (1969),[53] Kartsev et al. (1970)[54]
Deep cooling ($\tau_{pulse} = 10$ nsec, $T = 4.2$ K)	**W**	10^9	Aksyonon et al (1979)[224]
Small emission spot (spot diameter $d \approx 100$ Å, $\tau_{pulse} = 4$ msec, $T = 300$ K)	**W + Zr**	$5 \cdot 10^9$	Fursey and Shakirova (1966)[55]
Super small emission spot ($d = 20$–30 Å, dc mode, $T = 300$ K)	**W + Zr**	10^9	Fursey et al. (1995)[244]
Nanotips (dc mode, $T = 300$ K)	**W**	10^9–10^{10}	Pavlov et al. (1975)[56]
Super small nanotips (emitter radius $r_{em} = 10$–20 Å, $\tau_{pulse} = 100$ msec, $T = 300$ K)	**W**	10^{10}–10^{11}	Fursey et al. (1998)[222]

3.6. RESUME

FE is a unique electron emission mechanism, in which no energy is required for emission:

1. Theoretically the entire flow of electrons incident on the solid/vacuum interface from inside the solid to be emitted. The theoretical limit is 2–3 $\times 10^{11}$ A/cm^2.
2. In practice, the maximum current density is limited by Joule and Nottingham processes in the near-surface region.
3. For current pulses having a duration of 10^{-6}–10^{-3} sec, thermal destruction of the emitter can be accelerated by thermally activated, field-assisted surface self-diffusion, and formation of micro-protrusions on its surface.
4. It has been experimentally shown that under special conditions, FE current densities can approach 10^{11} A/cm^2, close to the theoretical limit.

4

FIELD EMISSION IN MICROWAVE FIELD

4.1. INTRODUCTION

Field emission (FE) tips are promising in various microwave cathode applications. From a fundamental point of view, electron emission in microwave fields has a number of unusual characteristics.

The main features of FE in a microwave field are:

1. Electron tunneling occurs during very short time intervals ensuring adiabatic conditions up to frequencies of about 10^{14} Hz.
2. The dependence of electron emission on the electric field is nonlinear. Consequently, there is an opportunity to perform nonlinear operations (such as frequency transformation and multiplication, and microwave field power conversion), and directly produce electron bunches during the emission process.
3. Ion motion in microwave field reduces ion bombardment of emitter surface.
4. The electron energy distribution shows unique features.

An advantageous property of FE is that it can be switched on and off virtually instantaneously, due to the fact that no heating of the emitter is required for emission as it is with thermionic electron emission. Also, the cathode as an element of microelectronic system is radiation hard and its performance is temperature insensitive. Field cathodes can be used in high-power, high-current microwave field devices, and electron optical systems: magnetrons, energy conversion devices, high-voltage and high-resolution electron microscopy, and so on.

In this chapter, we discuss the physical mechanisms of the FE process and the resulting electron beam behavior in a microwave field. One of the few previous reviews of these topics is by Charbonnier et al.[89] There are also papers by Russian authors,[254, 255] which are not generally known to foreign researchers. Practical applications of FE in a microwave field have been briefly reviewed by Brodie and Spindt[87] and Busta,[88] and discussed at a number of conferences by Gray[256] and Smith and Gray[257] and, in more detail, by Parker[258] and Gulyaev et al.[259]

4.2. THE ADIABATIC CONDITION—TUNNELING TIME

In the presence of an ac field, the potential barrier between the solid surface and vacuum may change so rapidly that tunneling occurs in a time interval that is long relative to that during which the potential barrier at the surface changes. In this case, the FE process is no longer adiabatic and Fowler–Nordheim (FN) theory (which is based on the stationary Schrödinger equation) is no longer valid.

Adiabatic conditions as they relate to FE processes were first analyzed by Bunkin and Fedorov.[132] They showed that for frequencies up to roughly 10^{14} Hz, the current density is determined by the conventional FN equation with the intensity of electric field replaced by its instantaneous value.

A "physical" formulation of adiabatic conditions can be done in terms of tunneling time, τ_{tun}. The tunneling time is assumed to be the mean time of interaction between a tunneling electron and the potential barrier. This time parameter determines specific properties of the tunneling process.[132, 133, 260–262] For instance, when a tunneling electron passes through a time-dependent barrier, the characteristics of the tunneling process can vary dramatically if the conditions are not adiabatic.[132, 133] The role of the tunneling time τ_{tun} is important in tunneling with dissipation, wherein the electron "feels" the under-barrier "friction,"[260, 261] and also in image potential calculations.[262] It can be assumed that the electron motion becomes adiabatic, that is, the particles tunnel through a static potential $V(t)$, if the rate of change of the applied field is small relative to the tunneling time: $\omega|\tau_{\text{tun}}| \ll 1$, where ω is the frequency of the field oscillations.

The problem of tunneling time was first formulated by Wigner.[263] To determine the time, he described the particle as a wave packet. Scattering of such a packet by a spherical potential with a finite radius was considered, and the tunneling time was defined as the time interval between when the centers of mass of the incident and scattered wave packets crossed the surface of a sphere of radius R:

$$\tau_{\text{Wigner}} = \frac{2R}{v} + \frac{2}{v}\frac{\partial}{\partial k}[\delta(k)], \tag{4.1}$$

where v and k are the velocity of the incident particle and the wave vector respectively, and $\delta(k)$ is the phase shift between the incident and reflected waves.

This approach was used by Hartman[264] to describe tunneling through a rectangular barrier. Here the tunneling time, τ_{Hartman}, was calculated as the time interval between when the maximum in the wave packet passes through the left side of the barrier and the formation of the maximum in the transmitted wave packet at the right side of the boundary:

$$\tau_{\text{Hartman}} = \hbar\frac{\partial}{\partial E}[\delta_1(k)], \tag{4.2}$$

where $\delta_1(k)$ is the phase difference between the wave function $\Psi(t)$ near the barrier boundaries for a value of the particle energy E. It was shown later[134, 265, 266] that such an approach was not quite correct, because the wave packet was deformed in the barrier region. This imposes conditions on the shape of the packet and on the barrier thickness, which cannot be easily combined with one another. Hartman has shown[264] that in some cases, the calculations yield a negative tunneling time τ_{tun}, which has no physical sense.

An approach to calculating the tunneling time was suggested by Baz.[267, 268] It is based on a gedanken experiment related to the precession of tunneling electron spin in a weak magnetic field. Calculating the angle of spin rotation, ϕ, for the transmitted wave by solving the stationary Schrödinger equation on the assumption that the spin orientation of the incident wave is known, one gets the tunneling time, τ_{Baz}, as:

$$\tau_{\mathrm{Baz}} = \frac{\varphi}{\omega_{\mathrm{L}}}, \tag{4.3}$$

where ω_{L} is the Larmor precession frequency.

Baz's method was applied to the case of tunneling through a one-dimensional potential barrier.[269] The spin rotation angles for the transmitted wave in a plane perpendicular to the field were determined. Buttiker showed[270] that in the quantum case there are two angles of spin rotation in a magnetic field: the first angle $\varphi_{\perp}$ in the plane perpendicular to the magnetic field, and the second one $\varphi_{||}$ in the plane parallel to the field. It so happens that it is the time

$$\tau_{\mathrm{Buttikerz}} = \frac{\varphi_{||}}{\omega_{\mathrm{L}}}, \tag{4.4}$$

that is equal to the tunneling time defined for a rectangular barrier in the quasi-classical case.[134]

The most complete and accurate treatment of the tunneling time problem was by Sokolovski and Baskin.[135] They found that the time the electron interacts with the potential barrier can be calculated by Feinman's procedure of integrating along classical trajectories.[271] If the wave function $\Psi(t)$ corresponds to a state with a determined energy, the mean traversal time can be represented as

$$\tau_{\Omega}(E) = (t_2 - t_1)\int_{\Omega} |\Psi(r)|^2 d^3r, \tag{4.5}$$

where t_1 and t_2 are the times at which the particle is detected in its initial and final points, r_1 and r_2 respectively, in a finite region Ω.

It was shown that the tunneling time τ_{Ω} is a complex number, and, in this sense, is not observable. Nevertheless, it is intrinsically connected with the observable angles of spin rotation due to Larmor precession and with the adiabatic parameters of the time-dependent problem.[272] The relation between τ_{Ω} and the time derived by Baz's method[267, 268] can be shown[135, 273, 274] to be:

$$\varphi_{\perp} = \omega_{\mathrm{L}}\mathrm{Re}[\tau_{\Omega}(r_1, r_2, \tau)] \tag{4.6a}$$

and

$$\varphi_{||} = \omega_{\mathrm{L}}\mathrm{Im}[\tau_{\Omega}(r_1, r_2, \tau)], \tag{4.6b}$$

where for the whole angle of rotation,

$$\varphi = (\varphi_{\perp}^2 + \varphi_{||}^2)^{1/2} = \omega_{\mathrm{L}}|\tau_{\Omega}(r_1, r_2, \tau)|. \tag{4.6c}$$

A calculation of the tunneling time τ_{tun} through a one-dimensional barrier at a metal–vacuum boundary determined by $U(x) = U_0 - eFx - e^2/(16\pi æ_0 x)$ shows[254] that for a wide potential having a low transparency, $\mathrm{Im}[\tau_{\mathrm{tun}}(E)] \gg \mathrm{Re}[\tau_{\mathrm{tun}}(E)]$. Thus, it is sufficient to find the imaginary part of $\tau_{\mathrm{tun}}(E)$ to estimate the tunneling time. For $U_0 = 10\,\mathrm{eV}$, $E = 5\,\mathrm{eV}$,

β changing from 0 to 0.25 (F changes from 0 to 1.6 V/Å), the term $\mathrm{Im}[\tau_{\mathrm{tun}}(E)]$ can be approximated by the expression:

$$\mathrm{Im}[\tau_{\mathrm{tun}}(E)] \approx 1.5\lfloor m^{1/2}(U_0 - E)^{1/2}\rfloor/(eF)\,. \tag{4.7}$$

$\mathrm{Re}[\tau_{\mathrm{tun}}(E)]$ is obtained from the expression:

$$\mathrm{Re}[\tau_{\mathrm{tun}}(E)] = (3/2)\hbar/(U_0 - E). \tag{4.8}$$

For $F \approx 1$ V/Å: $\mathrm{Im}[\tau_{\mathrm{tun}}] \approx 10^{-15}$ sec and $\mathrm{Re}[\tau_{\mathrm{tun}}] \approx 2 \times 10^{-16}$ sec.

To conclude the discussion on tunneling time, we note that at present all the tunneling time estimates for FE yield values $\sim 10^{-15}$ sec. The tunneling time τ_{tun} is a fundamental parameter defining the applicability of FN theory to the case of microwave field excitation. Quantitatively, the applicability criterion is expressed through the adiabatic condition:

$$\gamma = \omega|\tau_{\mathrm{tun}}| \ll 1. \tag{4.9}$$

For practical devices, the parameter γ actually determines the upper frequency limit of $\sim 10^{14}$ Hz. This is the fundamental frequency limit for devices based on the direct interaction of an electromagnetic wave with a field emitter surface, for example, in resonators, etc. In practice, the actual frequency limit for devices is determined by such parameters as the interelectrode capacitance, inductance, and resistance.

4.3. EXPERIMENTAL VERIFICATION OF FOWLER–NORDHEIM THEORY

The first experiments demonstrating the validity of the main principles of static FE theory in the microwave fields were conducted by Dyke's group.[89] Current–voltage characteristics, retarding potential curves, and some conclusions from the FN theory associated with nonlinearity of the emission process were studied.[89] The incentive for this investigation was the application of FE cathodes to microwave power amplification and frequency conversion. Frequencies up to 10 GHz were studied.

It was shown[89, 275] that the field electron image of a tungsten (W) emitter is similar to that obtained in a static field. In more quantitative experiments, the field emitter was located directly in the gap of an X-band cavity. Assuming that the average emission current is then measured as a function of the microwave field input power P_{in}, the dependence of the logarithm of average current $\langle I\rangle$ on the input power P_{in} should be equivalent to the linear dependence $\ln(I/V^2) = f(1/U)$ predicted by the FN theory. This dependence has been demonstrated[89] over a range of the emission current varying by at least two orders of magnitude.

More accurate methods for the study of FE in microwave fields have since been.[60, 276] Such studies have combined FE microscopy in microwave fields, measurements of the dependence of the emission current density on field strength, and the investigation of the time dependence and of the retarding potential curves over time intervals of 10^{-6}–1 sec. Also, dc measurements on the specimens could be conducted. The range of emission currents investigated were increased by two to three orders of magnitude over those previously utilized[89, 275] and the vacuum in the experimental chamber could be varied between 10^{-5}

and 10^{-10} Torr. Experiments[60] were performed on W and W_2C emitters with $\sim$0.1 μm radii of curvature and on liquid metals either contained in narrow capillaries or as thin layers at a solid needle substrate.[277, 278]

The experiments were conducted using the arrangement shown schematically in Fig. 4.1. The FE cathode was located in the gap of a microwave prismatic resonator (H_{103} mode, natural frequency 4.7 GHz). The oscillation wavelength was 6.4 cm, the input power, P_{in}, was less than 1.5×10^3 W. The direct current components of the cathode current I_{cath} and the current of a beam collector I_{coll} placed in the beam drift space outside the resonator were measured.

The results of static and microwave field investigations were correlated with one another. The dependence $\ln(I) = f(P^{-1/2})$ and the FE image obtained in a microwave field are shown in Fig. 4.2a. One can see that the plot is linear over four orders of magnitude. An equivalent voltage in the resonator was calculated[60, 255] exactly for the well-defined resonator parameters. The transition from P_{in} to U permitted us to obtain a generalized FN dependence that is valid for both static and microwave fields (Fig. 4.2b). The small divergence from the linearity in the slopes is related to errors in the definition of equivalent voltage U.

The current in these experiments was varied between 10^{-10} and 10^{-5} A in the static measurements and between 10^{-5} and 10^{-2} A in the pulsed, quasistatic regime. In the microwave fields, an average current for a pulse was measured. This average current varied in the range 10^{-7}–10^{-3} A. The peak currents I_{peak} were approximately ten times greater than the average values. Current densities were determined from the expression $J = I/(\sigma r^2)$, where σ is a coefficient linearly dependent on the electric field.[279] For fields between 3×10^7 and 10^8 V/cm^2, the value of σ is 1–2.5 for a surface work function of 4.5 eV. We can see that

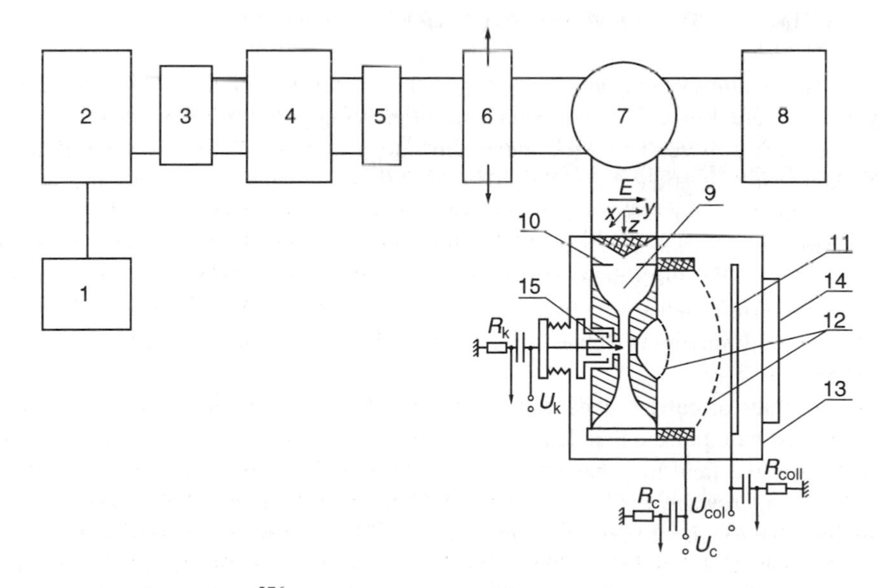

Figure 4.1. Microwave apparatus:[276] 1: Modulator, 2: excitation device, 3: attenuator, 4: power amplifier, 5: ferrite filter, 6: directional coupler, 7: commutator, 8: power measuring device, 9: cavity, 10: diaphragm, 11: screen, collector, 12: grids, 13: vacuum chamber, 14: optical window, 15: FE cathode.

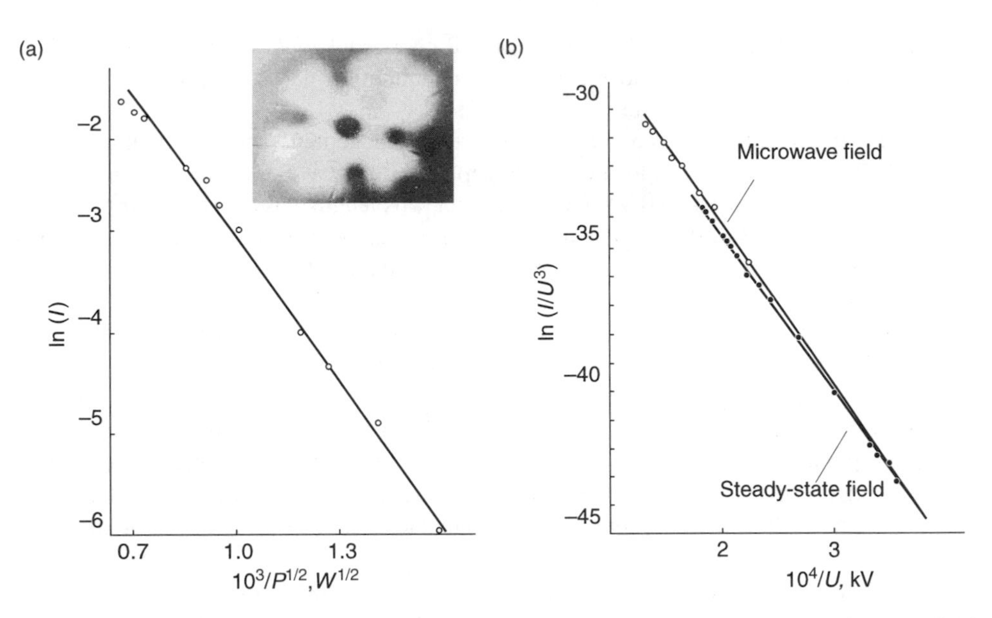

Figure 4.2. (a) Experimental emission characteristics for a W emitter in a microwave field. (b) The generalized FN dependence for static and microwave fields.[70, 255]

these experimental results indicate that the FN law is applicable to FE from dc to frequencies up to at least 10 GHz.

4.4. MAXIMUM FIELD EMISSION CURRENT DENSITIES

The maximum current densities, $j_{\max}$, which may be practically achieved with a field emitter, are limited by emission current heating induced thermal destruction of field cathode.[32, 70] Recall that the maximum current obtained with W field emitters is $\sim 10^7$ A/cm^2 for a static field, $\sim 5 \times 10^7$–10^8 A/cm^2 for microsecond pulses, and up to 10^9 A/cm^2 in the nanosecond duration pulse range[53, 54] (see Chapter 3).

Maximum allowable current densities in a microwave field have been measured.[255, 276, 280] With peak electric field values of 10^8 V/cm, maximum values of the average current density for pulse durations of about 10–100 μsec were 5×10^7 to 10^8 A/cm^2, and maximum values for a single pulse with a duration of $\sim 10^{-11}$ sec was 10^9 A/cm^2.

Such values of current densities are greater than those obtained under more lenient conditions in static pulse fields. It should be noted that the character of energy dissipation in the microwave field may have some peculiarities. The point is that the extraction of current in a microwave field takes up much shorter time (a time half a period of field oscillation) due to a nonlinear dependence of emission current on the field strength. For a frequency of 4.7 GHz,[255, 276] this time may be as small as 10^{-11} sec, which is shorter than the electron–phonon interaction time. Therefore, there may be a much smaller evolution of heat by the emitter than in static field. The difference in allowable current density does not appear to be much different: microsecond pulses $\sim 10^8$ A/cm^2 in the static pulsed case, and approximately the same current density in the microwave field with 10 μsec duration pulses. Also, at 4.7 GHz, the one time is 10^{-11} sec for $J \sim 10^9$ A/cm^2. This J is similar to that obtained for nanosecond duration pulses in the "static" case.

4.5. ION BOMBARDMENT OF THE CATHODE

The main cause of the instability of field cathodes is build-up processes on the surface in a strong electric field due to surface migration.[70] Virtually any mechanism activating the surface migration processes (ion bombardment, self-heating of the emitter by the emission current, etc.) may give rise to the appearance of microtips on cathode surface. The increased field at this microtip leads to increased emission current, cathode heating, and the eventual development of a vacuum arc.

Under practical vacuum conditions, surface migration is activated by ion bombardment of the cathode by ions formed from the residual gas atmosphere due to electron impact ionization. Early experiments with FE cathodes operated in microwave fields showed that emission stability was substantially higher than that in static fields.[281, 282] Evidently, the reason was a radical change in the character of ion bombardment in the microwave field.[60] The field distribution near the apex of the cathode tip is characterized by high gradients: about 10^{11}–10^{12} V/cm^2. While moving in an ac field with such a gradient, the ions "feel" a force directed toward the region of the lower field. This force causes the ions to drift into regions of decreasing field. As the field distribution near the apex of the emitter tip is nearly spherically symmetric, the equation of ion motion in an ac field of frequency ω is

$$\frac{d^2r}{dt^2} = \frac{e_\mathrm{i} F_0}{m_\mathrm{i}} \frac{r_\mathrm{e}^2}{r^2} \sin(\omega t), \tag{4.10}$$

where F_0 is the field at the cathode surface, r_e the field emitter tip radius, r the radial spherical coordinate, and e_i and m_i are the ion's charge and mass, respectively. Using Kapitsa's method,[283] it can be shown[60] that an ion of an energy E, less than E_max, never reaches the cathode surface where

$$E_\mathrm{max} = \begin{cases} 2^{-10/3}(e_\mathrm{i} F_0 r_\mathrm{e}^2)^{2/3}(m_\mathrm{i}\omega)^{1/3} & \omega < \omega_\mathrm{i} \\ 2^{-2}(e_\mathrm{i} F_0)^2/(m_\mathrm{i}\omega^2) & \omega > \omega_\mathrm{i} \end{cases} \tag{4.11}$$

and $\omega_\mathrm{i} = (2e_\mathrm{i} F_0/m_\mathrm{i} r_0)^{1/2}$. The dependence of E_max on ω is shown in Fig. 4.3a. The ion trajectories calculated by numerical solution of Eq. (4.10) for ions O^+ having an initial energy of 10 eV moving toward the emitter at different angles are shown in Fig. 4.3b.

The effect of eliminating ion bombardment in microwave fields has been observed experimentally[60] by comparing emission stability in microwave and static fields at a residual gas pressure of 10^{-5} Torr.

The experiment was aimed at comparing, under adequate conditions, the video pulse regime, having ion bombardment inevitably present, with the radio pulse regime, where the ejection of the ions from field nonuniformity occurs. To this end, the change of FE current with time was investigated in the both cases. The length of video and radio pulses was varied within a range of 10^{-5}–10^{-1} sec, and the pulse repetition, between single-shot

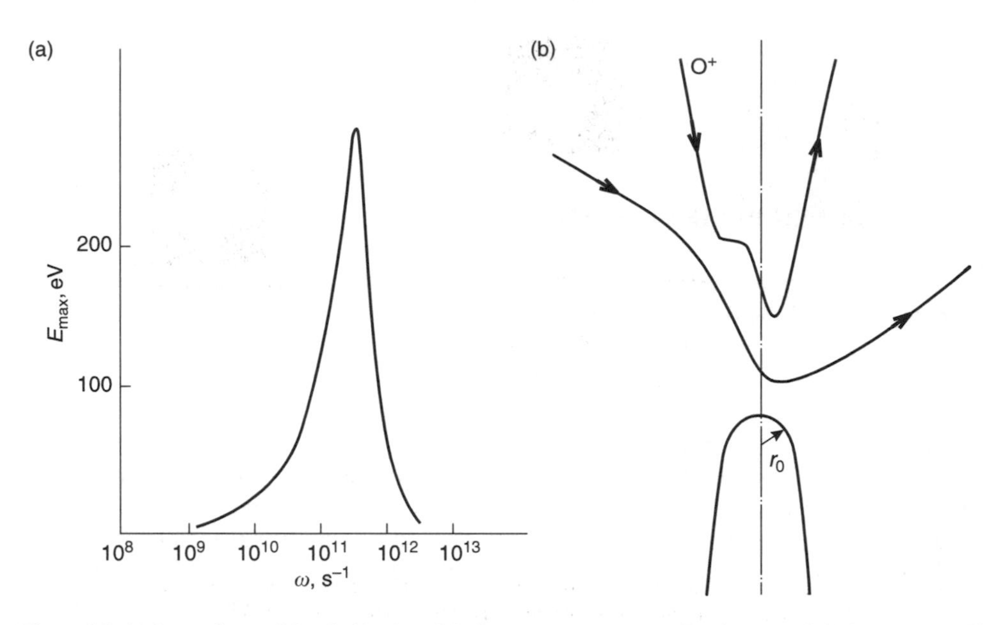

Figure 4.3. (a) Dependence of threshold value of the ion energy $E_{\max}$ upon the microwave field frequency ω. (If $E < E_{\max}$, ions do not bombard the cathode surface); (b) trajectories of O^+ ions; $r_0 = 0.5\,\mu\text{m}$, $\omega = 3 \times 10^{10}\ \text{s}^{-1}$, $F_0 = 6 \times 10^7$ V/cm.

pulses and a frequency of 10^4 Hz. The radio pulse stuffing frequency was 4.7 GHz. Vacuum was varied between 10^{-9} and 10^{-5} Torr. Test measurements to distinctly accentuate the "ejection effect" were carried out in a vacuum 10^{-5} Torr.

A significant difference in the emission current behavior for the microwave and static field was detected (see Fig. 4.4). In the microwave field, one can see that after an initial decrease in the emission current, due to adsorption processes, the current remains stable for more than 100 hr (Fig. 4.4, curve 2). In addition, the emission image did not change. For video pulses, an abrupt current rise was observed after a few minutes of operation at 10^{-5} Torr (Fig. 4.4, curve 1), and mobile emission centers were observed in the emission image (Fig. 4.4b). Irreversible changes in the emitter microgeometry and the emission current lead to explosion of the tip and vacuum breakdown. Besides facilitating devices applications, the absence of ion bombardment in microwave fields offers new opportunities for studying adsorption processes with a time independent emitter tip geometry.

4.6. ELECTRON ENERGY SPECTRA: TRANSIT TIME

Electron spectroscopy in an ac field reveals some unusual features in the energy distributions that are a result of the time varying field. It is very important to quantify the energy distribution of the electron beam in order to analyze the efficiency of a series of devices for which the FE cathode is located directly in the resonator. This problem was first studied by Dyke's group.[89] Retarding potential curves for the electron beam at the output of the resonator indicate that when a high static field is employed in combination with a microwave field transit time effects are negligible.[89]

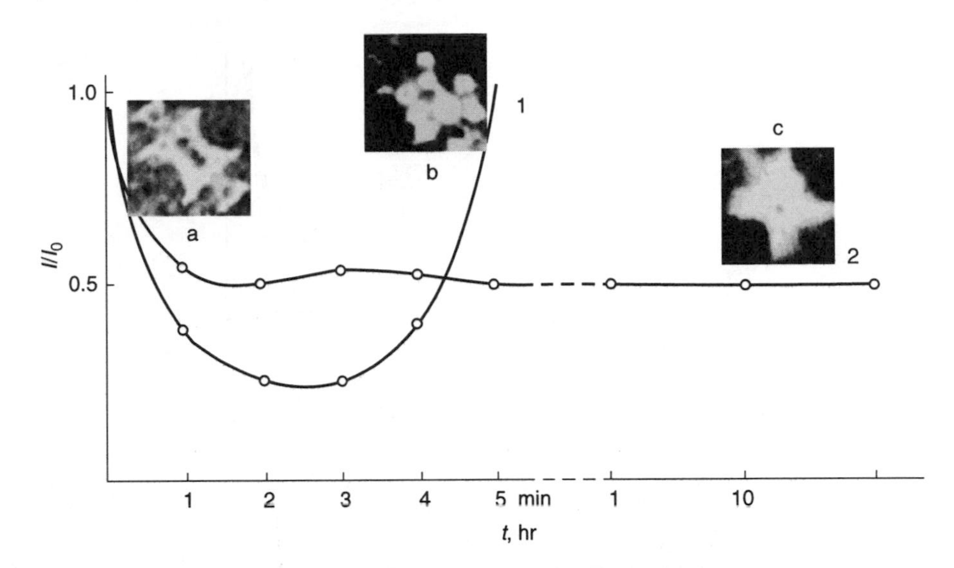

Figure 4.4. Field emission current stability for WC tip: 1: video pulse regime; 2: radio pulse regime. (The pulse current, I_0, is 200 μA for both regimes, pulse length, τ, is 100 μs, pulse repetition frequency is 100 Hz, radio pulse stuffing frequency is 4.7 GHz, and the residual vacuum pressure is 10^{-5} Torr.)

Transit effects have been investigated theoretically and experimentally.[254, 284] Complete calculations of the energy distributions for the ac fields are difficult, even using numerical methods. This difficulty is due to the large difference between the tip radius $r_0 = 0.1$–1 μm and the vacuum gap length $h = 1$–4 mm. Within a paraxial approximation, the problem can be solved. The calculated energy distributions of electrons at the resonator output are shown in Fig. 4.5a. Calculations show that the qualitative character of the curves is determined by the parameter $\alpha_1 = \omega H/v_1$. Here, v_1 is the maximum electron velocity in the near-emitter region (where the field gradient is high), and H is the length of the region with a homogeneous electric field distribution, F_0. The order of magnitude of parameter α_1 is equal to the transit phase of the emitted electron. The influence of the parameter $\alpha_2 = v_1/v_{osc}$ on energy distribution is not very significant. Here, v_{osc} is the velocity of oscillated electrons in the homogeneous resonator field.

Using special techniques,[276] we have investigated the energy distribution of electrons in microwave fields. In our studies, the gap h was 1–4 mm, the frequency was 4720 MHz, and the emitter radius was 0.1 μm. Retarding potential curves are shown in Fig. 4.5b. U_0 is the potential at which the collector current becomes equal to zero, and $I_{coll,\,max}$ is the collector current at a retarding potential $U_{ret} = 0$. The homogeneous field strength F_0 in the resonator determined by the input power was less than 2×10^4 V/cm. Curve 1 shows the measurement made in the static regime with an experimental error of less than 5 percent. The poor resolution is still sufficient for analysis of strong transit effects. The effects of a finite transit time on the energy distribution functions are shown in curves 2–4 in Fig. 4.5b. Here, α_1 varies from $\pi/2$ (curve 1) to π (curve 4). One can see from curve 2 that at $\alpha_1 = \pi/2$, only one group of electrons exists with the energy in the range of $0.5\varepsilon_{max}$, where ε_{max} is the maximum electron energy at the resonator output. The half width of energy distribution is equal to $0.2\varepsilon_{max}$ in this case (see curves 2 and 3 in Fig. 4.5a). When

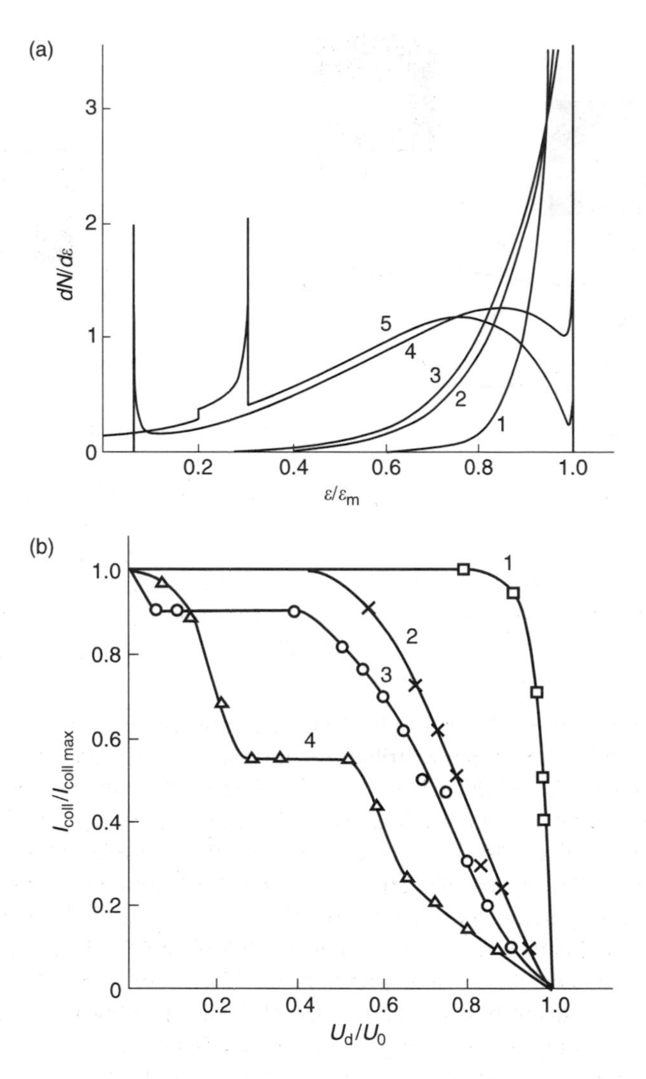

Figure 4.5. (a) Electron energy distribution function at the resonator output (tungsten cathode; $F_0 = 6 \times 10^7$ V/cm).[284] 1: $\alpha_1 = 0$, $\alpha_2 = 50$; 2: $\alpha_1 = \pi/2$, $\alpha_2 = 3$; 3: $\alpha_1 = \pi/2$, $\alpha_2 = 0.3$; 4: $\alpha_1 = \pi$, $\alpha_2 = 3$; 5: $\alpha_1 = \pi$, $\alpha_2 = 1$; (b) retarding potential curves. 1: static fields; 2: microwave fields: $h = 1.2$ mm, $U_0 = 3.6$ kV; 3: $h = 1.5$ mm, $U_0 = 4.0$ kV; 4: $h = 2.5$ mm, $U_0 = 6.8$ kV; h is the transit space of cavity; U_0 is the cut-off potential).

$\alpha_1 = \pi$ (curve 4 in Fig. 4.5b), at least two groups of electrons exist: One with high energies ($0.5\varepsilon_{max}$–ε_{max}), and another one with low energies (0–$0.3\varepsilon_{max}$). From curves 4 and 5 in Fig. 4.5, we can see that the primary cause of the appearance of low-energy electrons is transit effects. Thus, transit time effects have an important effect upon the electron energy distribution at the resonator output with FE cathodes even for rather short (~1 mm) gaps at several centimeter wavelengths. When $\alpha_1 = \pi$, a substantial fraction of the electrons in the beam have almost zero energy. Clearly, electron "bunching," that is, the formation of distinct groups of electrons is possible, as can be seen by the deep minimum in the retarding potential energy curves.

4.7. FIELD EMISSION FROM LIQUID SURFACES

FE from metals typically occurs in surface fields exceeding 10^7 V/cm. In practice, working at reasonable applied voltages requires forming the cathode into a small tip with a radius of curvature on the order of 10^{-4}–10^{-6} cm. These tips may be made by the traditional methods of electrochemical etching, etching in an oxygen atmosphere, and formation in a high electric field.[285] Interestingly, liquid metals allow one to generate very high electric fields at their surface with modest voltages due to an electrohydrodynamic instability that arises.[286, 287] For instance, typical dimensions of the submicron inhomogeneities produced with liquid-metal ion sources (LMIS) are on the scale of about 10^{-7} cm, which produces fields sufficient for field ion emission. It has been shown[288] that field electron emission from liquid metals may also be obtained by the same LMIS techniques if one reverses the emitter polarity.

It is also possible to form micro-protrusions of the surface of liquid conductors by using microwave fields to excite microcapillary waves on the surface.[277] Such an instability leads to micro-protrusions having characteristic dimensions of the order of 10^{-5}–10^{-6} cm, providing field enhancement sufficient to initiate electron emission at rather small average values of the microwave field in the resonator.

Such experiments have been conducted in a microwave field prismatic resonator (H_{103} mode, natural frequency of 4.7 GHz, see Section 4.3). The oscillation wavelength was 6.4 cm and the input power P_{in} was less than 1.5×10^3 W. Capillary emitters filled with Ka–Na alloy or with fusible indium alloys were used. Also, a thin (roughly 1 μm) layer of Bi–Sn–Pb alloy deposited at a solid copper needle was studied. Materials having low melting points (less than 150°C) were chosen to exclude effects due to thermionic emission.

The emission characteristics for various cathodes are shown in Fig. 4.6. The field of resonator F is proportional to $P_{\text{in}}^{1/2}$ and that measurements were taken using single pulses. The pulse durations were chosen to be less than 10^{-5} sec in order to prevent a transition of FE to the explosive emission.[70] However, it should be noted that, as distinct from emitters having a high melting point (e.g., with W), the operational characteristics of liquid-metal cathodes are restored after a breakdown event. Lastly, it was found that the emission characteristics of liquid-metal cathodes in microwave fields are reproducible over a wide range of pressures, up to 10^{-5} Torr.

Note that the spread in experimental points is less for the microwave fields[278] (Fig. 4.6) than for the static fields.[288] This is due to the reproducible response of the liquid surface to pulses of microwave field power.

It was shown[278] that there is a finite time lag, τ_{d}, for the development of the FE process at the surface of liquid metal cathodes. τ_{d} is related to characteristics of the system that govern the formation of the electrohydrodynamic instability; field strength, surface tension forces, etc.

The excitation of FE from liquid-metal cathodes by static and microwave fields has been compared.[278] It was found that in static fields, FE is not initiated even at fields more than

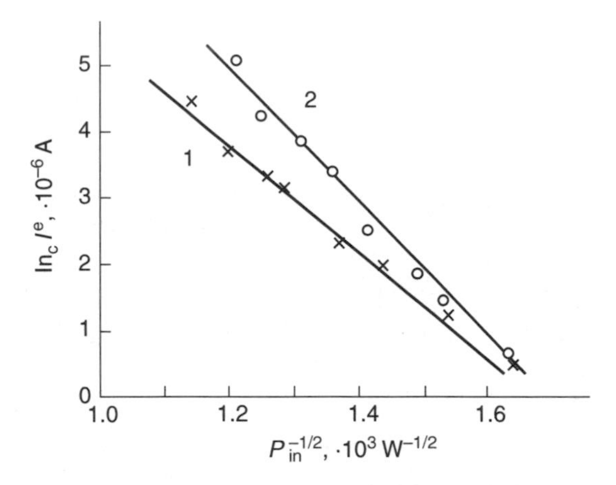

Figure 4.6. FE characteristics for liquid-metal surface in microwave field. 1: K–Na alloy: $\tau_p = 5\,\mu s$, $I_{max} = 90\,\mu A$ ($d_{capill} = 90\,\mu m$); 2: Bi–Sn–Pb alloy: $\tau_p = 12\,\mu s$, $I_{max} = 180\,\mu A$; τ_p is the pulse duration.

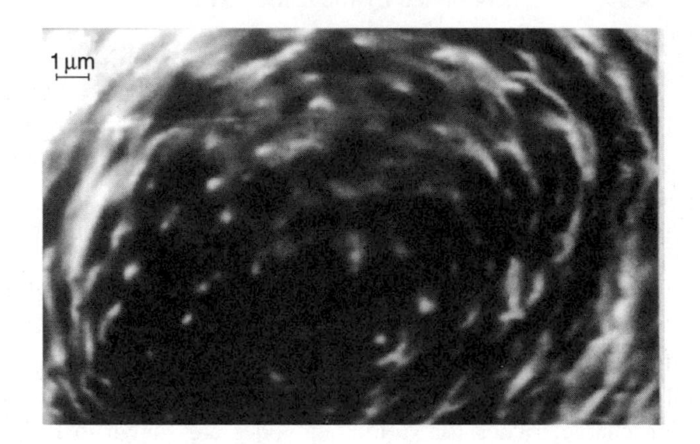

Figure 4.7. Frozen signs of the microprotrusions created on the surface of a liquid-metal cathode in a microwave field.

an order of magnitude larger than those at which the FE is observed with microwave excitation. Therefore, the threshold value of the microwave field required for the development of instabilities of a conducting liquid surface are smaller than the critical Tonks–Frenkel field.[286, 287] Detailed calculations of the threshold values for microwave field excitation corroborate this conclusion.[277]

Direct experimental evidence of the occurrence of microcapillary waves and of the formation of fine micro-protrusions at a liquid-metal surface in microwave field has also been obtained.[289] Visualization was achieved by freezing the micro-protrusions on the surface of a thin (1–2 μm) alloy layer deposited on a Cu needle (Fig. 4.7). A detailed discussion of initiating the FE processes from liquid-metal surface is given in Fursey et al.[290]

4.8. RESUME

We summarize the main points concerning the fundamental physical aspects of operating the FE cathodes at microwave frequencies.

1. FN theory accurately describes the FE process up to frequencies of $\sim 10^{14}$ Hz. The condition for validly of the adiabatic approximation to FE is $\omega|\tau_{\text{tun}}| \ll 1$. Theoretical estimates of the tunneling time give $\tau_{\text{tun}} \sim 10^{-15}$ sec.
2. FN theory has been experimentally validated at microwave frequencies over 4–5 orders of magnitude in FE current. Other consequences of the theory, including the shape of the electron energy distributions have also been verified.
3. Maximum measured FE current densities are on the order of 10^9 A/cm^2 at microwave frequencies.
4. Ion bombardment of the cathode surface is virtually eliminated at microwave frequencies by the formation of an effectively repulsive potential for the ions near the cathode surface. The result is a significant increase in the stability and operational lifetime of the cathode when compared to dc operation at pressures of the order of 10^{-5} Torr.
5. The non-homogeneity of the electric field has significant effects on the electron beam energy distribution in macroscopic microwave devices. Transit time effects begin to modify the electron energy distribution at frequencies greater than 10^9 Hz with millimeter-range gaps. In vacuum microelectronics devices, such effects may be neglected at up to frequencies of 100 GHz as the gaps are typically of the order of 1–10 μm.
6. Stable FE from the surface of liquid metals has been demonstrated. Microcapillary waves are formed at the liquid-metal surface, their crests forming an array of field emitters. The dimensions of the surface instabilities can be of the order of nanometers.
7. The unique properties of FE offer many intriguing opportunities for applications in: (a) "macroscopic," high-current devices: beam-bunching injectors, energy converters, high-power traveling wave tubes, klystrons, magnetrons, etc.; (b) high-energy electron optical devices; in particular, very-high resolution electron microscopes having beam energies of the order of several MeV; (c) vacuum microelectronics devices (frequency multipliers, amplifiers, traveling wave tubes, miniature electron sources for electron optical systems used for microscopy, holography, lithography, etc.).

5

FIELD EMISSION FROM SEMICONDUCTORS

5.1. INTRODUCTION

The interest in field emission (FE) from semiconductors is stimulated by the following unique features of this process:

1. FE provides a source of valuable information on the physics of solid state surfaces. FE and ion emission microscopy, which enable visualization of solid state surfaces with high resolution and magnification yield data on a number of important characteristics of the surface such as the work function, ionization potential of surface atoms, electron affinity, surface tension, evaporation energy of the surface atoms, and activation energies for surface transport processes. Knowledge gained by these techniques in studies of surface phenomena on metals are widely known.[13, 14, 19, 23, 29, 285] Initially the application of these techniques to semiconductors presented many problems due to the difficulty of cleaning the surface of a semiconductor emitter and obtaining a symmetric emission image. In this chapter the issue of semiconductor tip surface cleaning is given special consideration.

2. Distinct from metals, a semiconductor offers numerous ways of varying the characteristics of the emission process by control of carrier concentration in the emitter bulk, thus making unique electron devices possible. In recent years a number of ideas have been generated which might lead to new systems in vacuum electronics. It appears possible to have a limited carrier concentration in the near-surface region of a semiconductor and control the FE process by generating or injecting carriers into this region.

3. Application of semiconductors in vacuum microelectronics is aided by the fact that for some semiconductor materials, such as Si, the basic technology of fabricating complex structures has been developed.

4. Local deviations from electrical neutrality in the emitter bulk could create an electric field near the surface thereby reducing the threshold externally applied field for electron emission.[342–344]

5. Finally, a semiconductor microcrystal/field-emitter can produce an extremely high electric field (10^8 V/cm) near its surface, which cannot be produced by any other means. Thus it may be possible to study surface processes and transport: adsorption, desorption, catalysis, surface diffusion in strong electric fields.

FE from semiconductors, relative to metals, is a much more complicated process due to the low carrier concentration in the emitter bulk. The low carrier concentrations allow for field penetration into the semiconductor, causing band bending and nonlinearity of the current–voltage characteristics in Fowler–Nordheim (FN) coordinates. Under certain conditions these same features can make the FE current very thermal- and photosensitive.

The capture of free carriers by traps can cause a deviation from the electrical quasi-neutrality in the emitter bulk and lead to modifications of the field distribution at the surface. A semiconductor emitter is very difficult to prepare because surface cleaning, sharpening, and heating can lead to irreversible changes in its initial properties. These irreversible changes can be difficult to identify and lead to experimental results that are difficult to reproduce. For many years this problem prevented systematic studies of the field electron emission in semiconductors and thereby the development of an adequate model and quantitative theory of the phenomenon.

Reproducible results can be obtained only if one begins with a clean surface using condition and treatment procedures that do not change the sample's initial properties. We consider this matter very important and therefore assigned to it two subsections, 5.2 and 5.4. In this chapter we will concentrate on the most reliable results obtained on emitters with atomically clean surfaces and reproducible initial properties.

In earlier reviews some experimental observations were presented. As regards theory, a number of rather abstract mechanisms were suggested, which, as was shown afterwards, either had no effect on the process at all or represented factors of second or third order.

Data on FE from semiconductors was first reviewed in Elinson and Vasil'ev[25] and the most substantial review was given in Fischer and Neumann.[149] Comprehensive efforts to elicit general properties of the phenomenon and give a consistent account of the suggested mechanisms were undertaken in Fursey and L'vov[150] and Baskin et al.,[151] as well as in a fairly recent review by Modinos,[29] a paper by Fursey and Baskin,[342] and in recent works by Fursey.[152]

5.2. EMITTER SURFACE CLEANING

In FE experiments, especially with semiconductors, it is most important in sample preparation not to change its initial properties during its cleaning and obtaining FE patterns.

The most effective technique of cleaning a semiconductor tip of surface impurities is desorption in a strong electric field, either in vacuum or in a hydrogen atmosphere. The latter has been used in cleaning Ge,[291] Si,[292] GaAs,[293, 345] and CdS.[307] Field-stimulated reactions between the impurity atoms and hydrogen on the sample surface greatly facilitate the desorption process. However, as hydrogen diffusion into the sample bulk can change the characteristics of the emitter, vacuum desorption cleaning is preferable. Until recently, cleaning of the emitter surface by field desorption was an extremely laborious process. The main problem was the "explosion" of the emitter tip often in the final stage of cleaning. Emitter destruction was initially associated with fatigue phenomena.[291, 294] Later, it has been shown[295–297] that emitter instability is the result of the fact that the electric field needed to remove well bound surface impurity atoms is typically much higher than the field required to evaporate the tip material. Thus impurity removal can often result in an abrupt, explosive-like evaporation of the semiconductor material and destruction of the tip.

Stability of the semiconductor tip during field-desorption cleaning is sensitive to the temperature and the duration of sample heating during sample pretreatment and vacuum chamber outgassing.[291] For example, it has been found that the field-desorption cleaning of Ge in vacuum is feasible only if the sample temperature does not exceed 673 K.[297] Lowering of the pretreatment temperature to 473 K in most cases allowed for contamination removal without destruction of the semiconductor tip.

The reasons for the change in the properties of Ge samples during heating have been analyzed.[294–297] The formation of an adherent coating occurs as a result of conversion of various germanium oxides to GeO_2. Heating of the tip to temperatures above 673 K, even for a short time stabilized the oxide layer to the point that it is impossible to remove by field desorption without destroying the tip. An FE pattern of Ge with an adherent oxide layer at the periphery of the emitting region is shown in Fig. 5.1.

Lowering the temperature to 473 K during pre-treatment facilitated surface cleaning in the case of Si as well. At this temperature the surface of the tip can be cleaned by field-desorption alone.

Figure 5.2a shows a FE pattern of a Ge tip oriented in the $\langle 011 \rangle$ direction and cleaned by field desorption in ultra-high vacuum.[297] During preliminary treatment the temperature did not exceed 473 K. The FE pattern is low-contrast and details of the crystal structure are poorly resolved.[291, 294, 296] After a brief heating to 423–673 K for 5–10 sec a distinct symmetry emerged as shown in Fig. 5.2b. An interesting feature of the freshly cleaned Ge surface is the bright spots in areas of the cube faces (Fig. 5.2a). This local emission enhancement was observed in the past[294] and attributed to the migration of atoms towards the apex of the tip where the field strength is highest. However, it has been shown[295, 297–300] that (100) faces in other locations as well (even those at the edge of the emitting region) have enhanced emission, Fig. 5.2a. Investigations of FE from different areas of the emitting surface of Ge [297] have been used to determine the difference in work function values between cube faces and the central part of the emitting surface which was found to be 0.3 eV. Such a large difference in work function contrasts with previous results obtained using other methods. This discrepancy is likely due to the fact that Ge surfaces cleaned using field differ significantly from those cleaned via ion bombardment.[300]

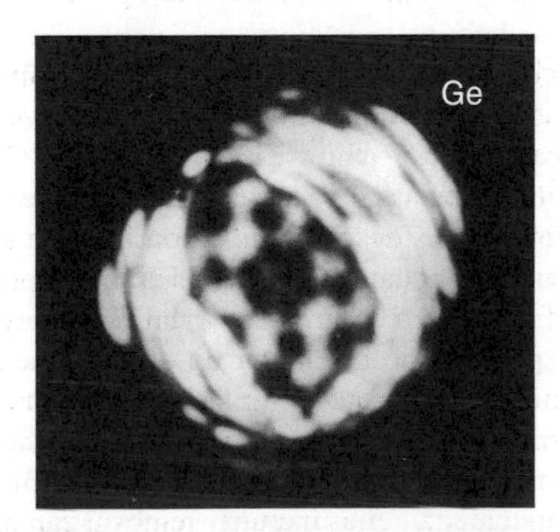

Figure 5.1. FE micrograph of a Ge tip with an adherent oxide at the periphery of the emitting region.[295]

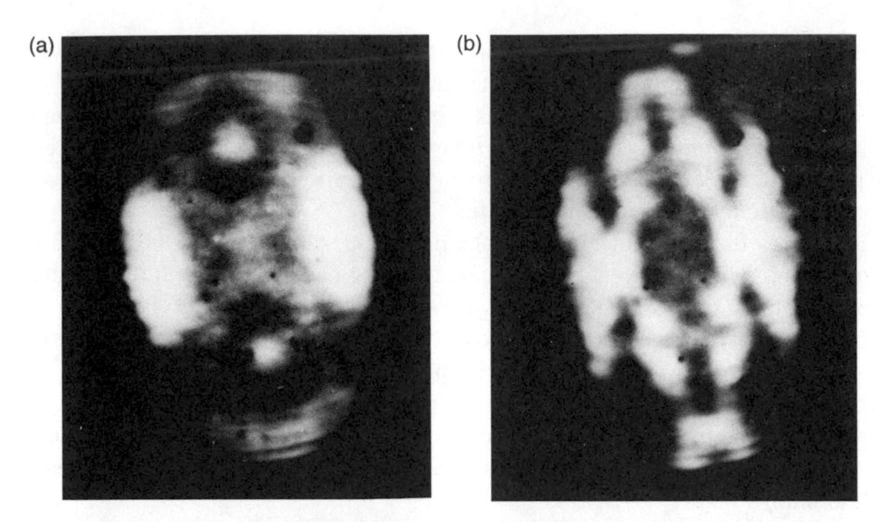

Figure 5.2. FE pattern of Ge[297] after (a) field desorption; (b) thermal anneal.

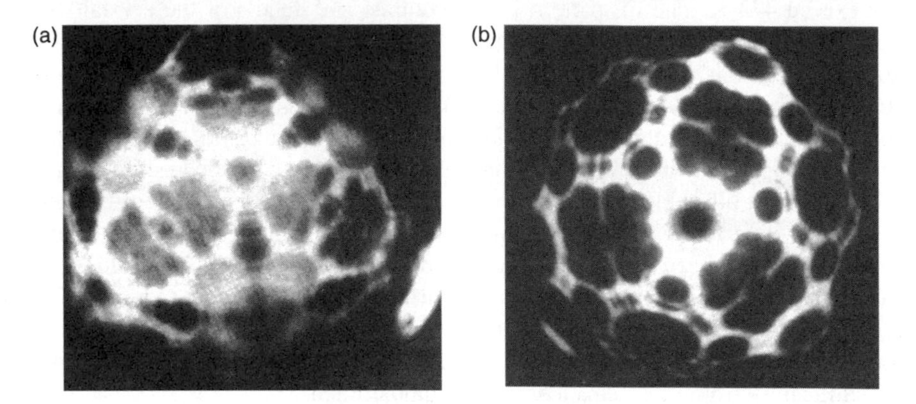

Figure 5.3. FE patterns of Si:[292] (a) annealed at 1610 K; (b) annealed at 1650 K.

The first FE patterns of silicon were obtained by D'Asaro[301] after heating the tip to high temperatures. Later, better quality FE images were obtained by Allen[294] and Perry.[302] They both used extensive heating and field desorption of the tip sample. Field-emission patterns of Si cleaned by field desorption in vacuum are shown in Fursey and Ivanov[303] and Fursey and Egarov.[304] Excellent FE patterns of Si cleaned by field desoption in hydrogen were published by Busch and Fischer.[292] These authors have also obtained superb FE patterns of Si after tip annealing (Fig. 5.3). FE patterns have also been obtained for some compound semiconductors: GaAs,[293, 345, 346] InAs,[305] InSb,[306] and CdS.[307]

Field ion imaging of semiconductors is very difficult because image fields are often higher than the evaporation field of the emitter material. Nevertheless, such patterns have been obtained. Melmed and Stern[308] and Melmed and Givargizov[309] have taken high quality field ion patterns for Ge, Si, GaAs, and InAs (Fig. 5.4) using neon and hydrogen as the image gases and employing a micro-channel plate intensifier. These field ion

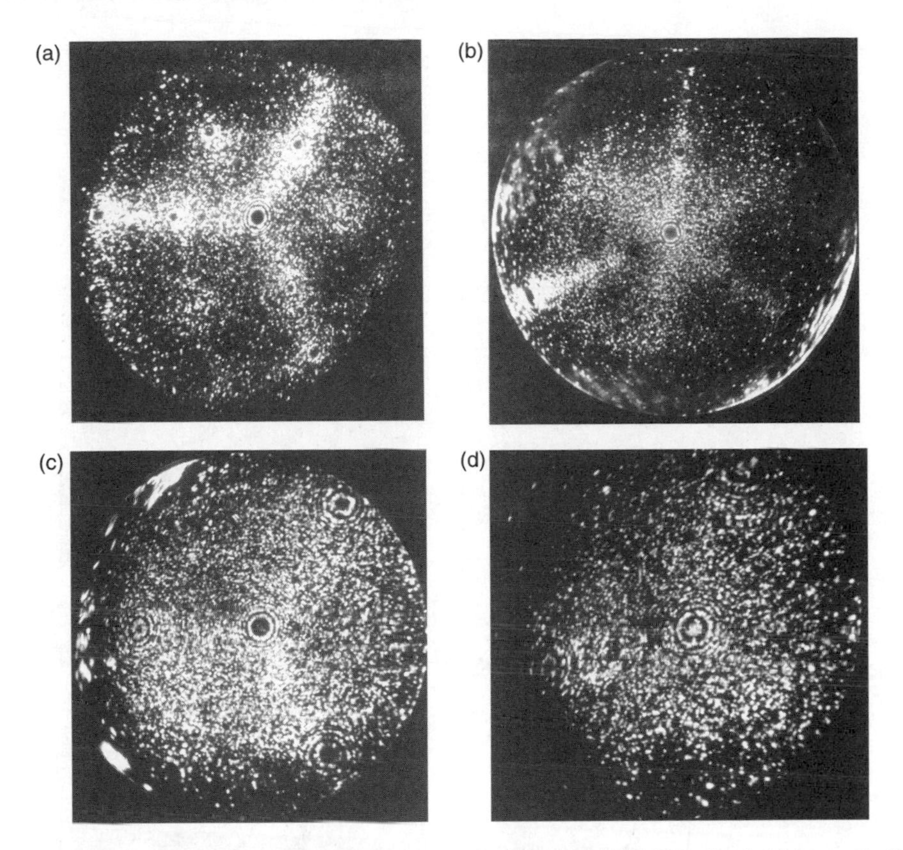

Figure 5.4. Field ion micrographs at ~30 K (courtesy A. J. Melmed and E. I. Givargizov): (a) *n* type Ge ⟨111⟩ imaged with neon (~62 nm diameter image); (b) Si, 111 imaged with neon (~56 nm diameter image); (c) B-GaAs ⟨111⟩ imaged with hydrogen (~56 nm diameter image); (d) InAs ⟨111⟩ imaged with hydrogen (~44 nm diameter image).

micrographs were obtained for semiconductor whiskers grown by the vapor–liquid–solid (VLS) method.[310] The individual whiskers were removed from substrates and epoxyed to tungsten wire supports. Electropolishing was used to sharpen the whiskers for field ion microscopy.[309]

5.3. CURRENT–VOLTAGE CHARACTERISTICS

The relationship between the FE current and electric field (or voltage) has been the subject of a number of investigations.[29, 149] It was found, in contrast to metals, that in semiconductors with low carrier concentrations in the conduction band (high-resistivity *p*- and *n*-types) the current–voltage dependence in FN coordinates is nonlinear. Some evidence of nonlinear current–voltage characteristics appear in early studies by Apker and Taft who investigated CdS.[311]

The first comprehensive study of the nonlinear current–voltage characteristics was carried out by Sokol'skaya and Shcherbakov.[312, 313] These experiments showed that the

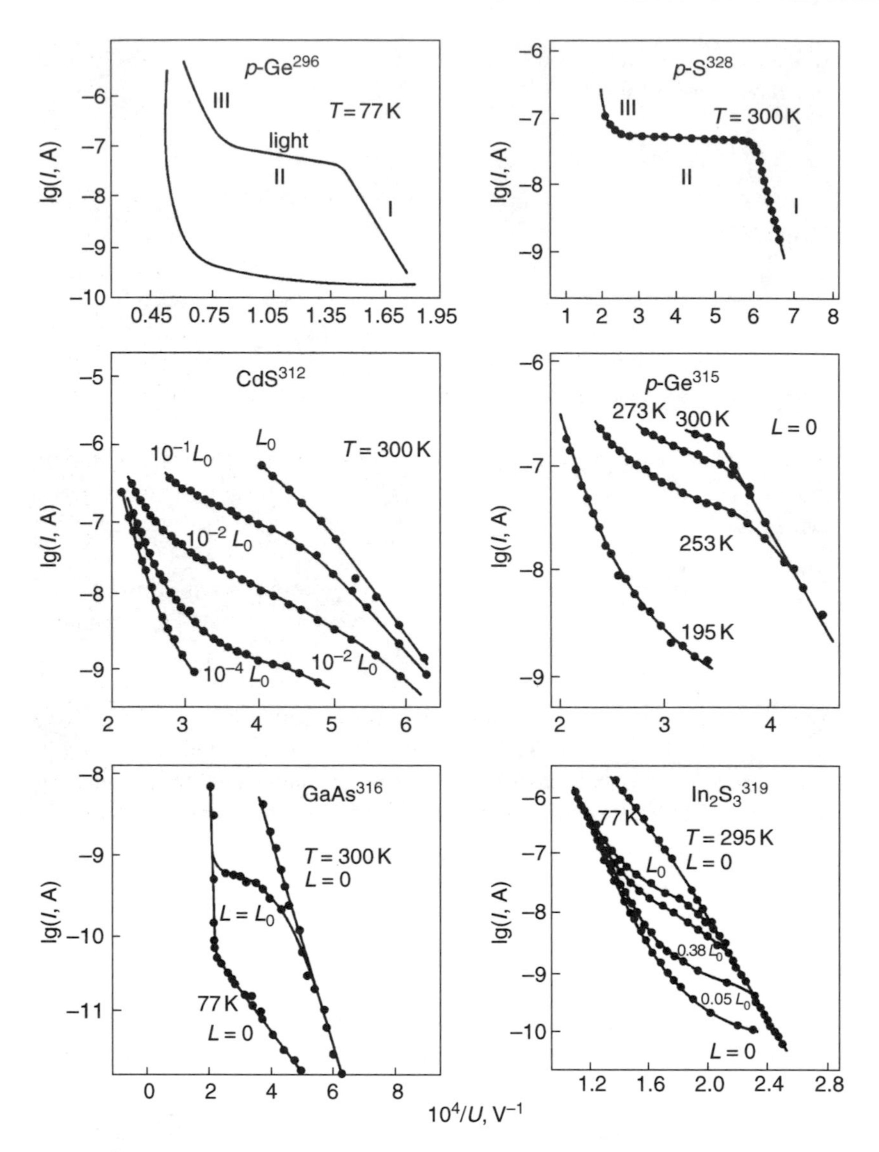

Figure 5.5. Nonlinear current–voltage characteristics of various semiconductors; L_0 = maximum light intensity.[151, 152]

current–voltage curve can be divided into three regions (Fig. 5.5): region I—linear current–voltage dependence; region II—slow current variation ("saturation" region)—it is in this region of the current–voltage characteristic that a strong photo- and thermal sensitivity of the FE current has been observed; region III—rapid rise of the current with voltage (in some cases more rapid than in region I).

Later studies showed that the current–voltage characteristics of this type are typical of a wide variety of semiconductors, including Sb_2S_3,[314] high-resistivity Si,[315] *p*-type Si,[292] low-resistivity Ge,[296] *p*-type GaAs,[316, 317] In_2S_3,[318, 319] CdS,[312, 313] and CdP_2.[319] These observations have been summarized[151] (see Fig. 5.5).

5.4. ON PRESERVING THE INITIAL SURFACE PROPERTIES OF A FIELD EMITTER

It was noted that the shape of current–voltage characteristics of a semiconductor emitter can be significantly changed during sample preparation as well as during the course of the experiment itself. It was found that noticeable changes to the initial, intrinsic properties of emitter material take place during its heating. This is associated with the contaminants remaining on the side of the tip surface after electrolytic etching. At elevated temperatures in vacuum these contaminants diffuse into the tip bulk and modify the doping. In some cases highly conducting channels form on the sides, electrically shunting the sample's bulk.

Changes of the initial properties of field emitters were noted in early studies of Si.[294, 302] It was assumed in these studies that during cleaning of the silicon tip by annealing at 1500–1600 K, rapid diffusion of acceptor impurities into the sample occured, which led to the formation of a highly degenerate surface *p*-layer, irrespective of whether the material was *n* or *p* type.

It has also been shown that the FE properties of Si, Ge, and some other semiconductors change markedly as a result of treatment at much lower temperatures.[295–297, 303] Thermal treatment of Ge and Si samples at 700 K linearized the current–voltage characteristics and eliminated any photo- or thermal sensitivity. Perceptible changes in the I–V characteristics are observed even after heating to only 570 K. Heating to this temperature weakens the current saturation in region II and shifts the appearence of the region to higher current levels (see Fig. 5.6).

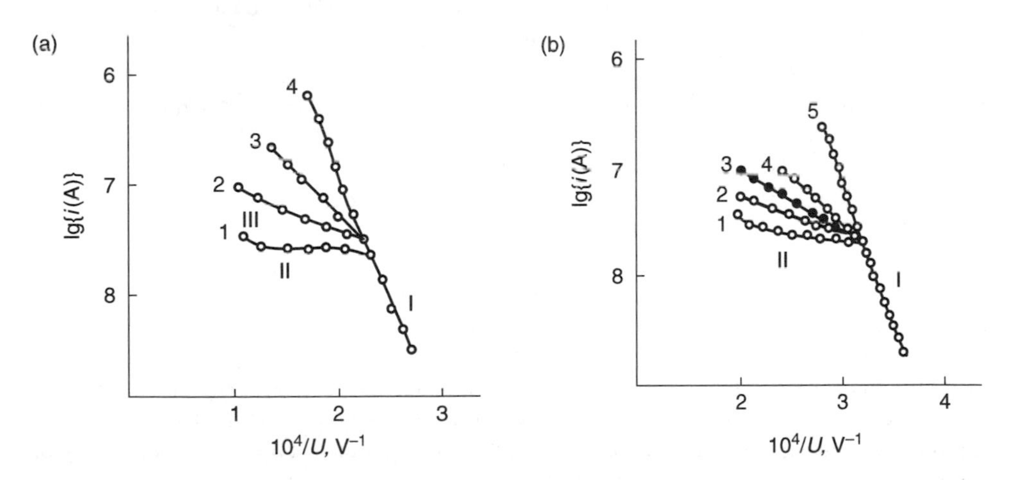

Figure 5.6. Influence of thermal treatment on the current–voltage characteristics:[321] (a) change in initial emission properties of a *p*-type Si cathode (14 Ω · cm) after heating. *T*°C: (1) 150; (2) 250; (3) 300; (4) 350; (b) CVC for a *p*-type Si cathode (14 Ω · cm) after different thermal treatments. (Previously, the emitter was electrochemically etched creating stable complexes with metallic ions.) *T*°C: (1) 250; (2) 300; (3) 350; (4) 400; (5) 500.

The initial properties of a semiconductor emitter are not affected unless it is heated to above 420–500 K. *In this case the current–voltage characteristics do not change, the photo- and thermal sensitivity remains the same, and the experimental results are completely reproducible.*

It has been found experimentally[321] that irreversible changes of the properties of semiconductor field cathodes heated to temperatures in excess of 400 K are principally caused by metal ions depositing at the emitter surface during electrolytic tip sharpening. These are the metals which modify the sample surface conductivity upon heating. To reduce the concentration of metal impurities,[321] tip treatments in solutions of organic agents that form complexes with copper and similar metal ions in base solution[322] were effective. These complexes can subsequently be easily removed by rinsing with water. The experimental results are shown in Fig. 5.6b. These experiments demonstrate that the initial properties of semiconductor emitters subjected to thermal treatment are not changed so strongly if preventive measures are taken.

5.5. VOLTAGE DROP ACROSS THE SAMPLE AND THE FIELD DISTRIBUTION IN THE EMITTING AREA

As yet no theories satisfactorily describe field electron emission from semiconductors. Shapes of the current–voltage characteristics predicted by the Morgulis–Stratton theory[323–325] do not agree with experimental curves.

In a number of mechanisms that reproduce the essential features of the observed current–voltage characteristics, that the prevailing four points were suggested are as follows:

1. The nonlinearity of the current–voltage characteristics is due to the voltage drop ΔU_c across the emitter. This idea, first suggested by Apker and Taft,[311] was later used to explain the saturation observed in region II of the current–voltage characteristics.[319, 326] However, direct measurements by the retarding potential method, demonstrated that this interpretation is not correct. Actually the shape of the current–voltage curves in the majority of p-type semiconductors[328, 329] as well as in high-resistivity n-type semiconductors[327] is only slightly affected by ΔU_c. The extent to which ΔU_c affects the shape of the current–voltage curves for Ge and Si[328, 329] can be seen in Fig. 5.7.

2. In a number of experiments the emission images change size when going from region I to region II of the current–voltage characteristic.[292, 296, 334] Throughout region II the emission pattern shrinks in size as the voltage is increased. In region III the size of emission pattern typically stays constant and sometimes increases and becomes more than in region II. The effect of image shrinkage is dependent upon the electron concentration in the conduction band and the emitter shape. A similar experimental study of the changes of emission image size with the applied voltage have been carried out on Ge and Si.[328, 329]

Changes in the size of the emission image reflects changes in the structure of the emitter space-charge region as it spreads. The field around the tip apex becomes more uniform and the equipotential surfaces less curved leading to lower magnification values in the FE microscope. Emission image shrinkage is illustrated schematically in Fig. 5.8. Variations of the size of emission image with applied voltage U and the voltage drop across the emitter are shown in Fig. 5.7.

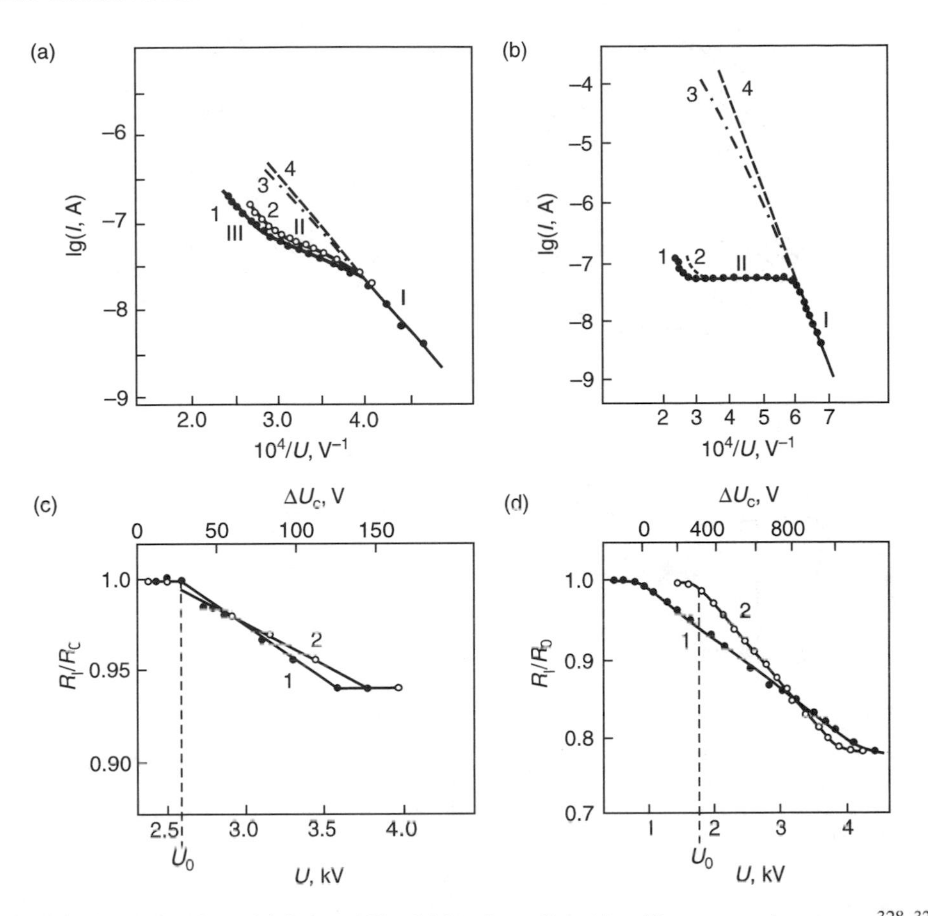

Figure 5.7. Illustration of a weak influence ΔU and $\Delta\beta$ on the nonlinearities of the current–voltage curves:[328, 329] (a) Current–voltage characteristics of Ge at 253 K (no illumination): (1) typical experimental curve; (2) with correction to the voltage drop; (3) with correction to the variation of $\beta(V)$; (4) FN plot. Current regions I, II, and III visible in curve 1; (b) analogous to (a) for p-type Si (3 $\Omega \cdot$ cm); (c) (1) relative change in the size of the emission pattern with applied voltage; (2) relative change in size of the emission pattern with tip voltage (Ge at 253 K not illuminated); (d) analogous to (c) for p-type Si (3 $\Omega \cdot$ cm).

Emission image shrinkage means that in region II of the current–voltage characteristic the field geometric factor $\beta = F/U$ is decreasing. If this fact is not taken into account in calculations of the current–voltage characteristics, the values of the field strength F near the emitter surface will be overestimated. From the measured relative sizes of emission image, R/R_0, the distortion of the field profile near the surface can be estimated.

Assuming

$$\frac{R}{R_0} \approx \frac{\beta}{\beta_0} \tag{5.1}$$

and taking account of the fact that R/R_0, as follows from experiments (see Fig. 5.8), is decreasing linearly with U in region II of the current–voltage characteristic, we have

$$\frac{\beta}{\beta_0} = 1 - k(U - U_0). \tag{5.2}$$

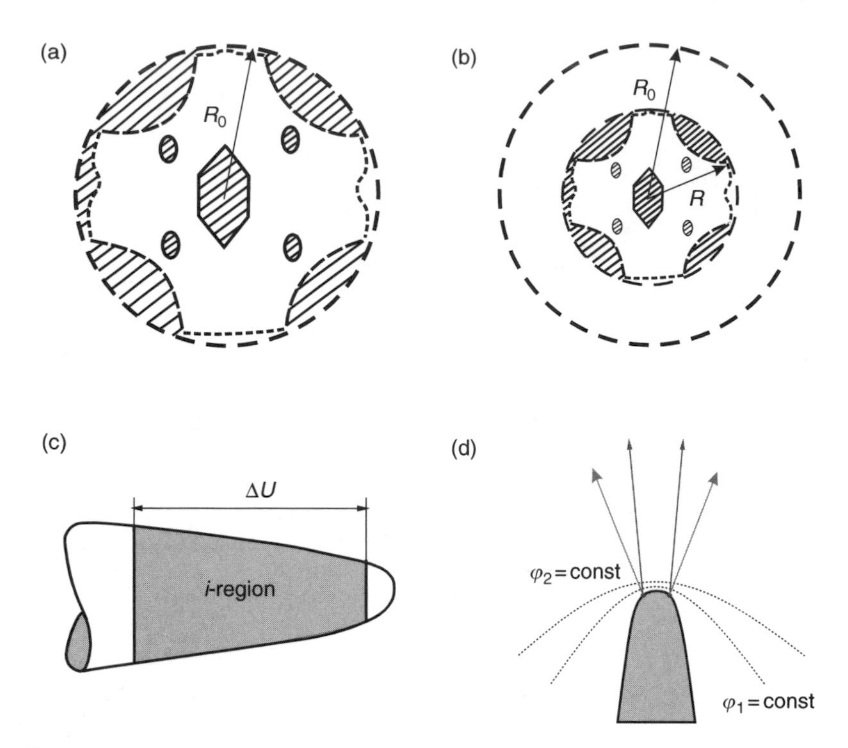

Figure 5.8. Schematic of the effect of a shrinking emission image: (a) initial emission image of the semiconductor, R_0 = initial image radius; (b) shrunken emission image, R—image radius; (c) ΔU is the voltage drop across the tip over the space-charge region depleted of carriers; (d) the potential and field distribution near a semiconductor tip with field penetration.

Here β_0 is a U-independent geometrical factor in region I of the current–voltage characteristic; U_0 is the voltage at which the emission picture starts to shrink; k is the slope of the function $R/R_0 = f(U)$; $k \approx 10^{-4}$–$10^{-6}\ \text{V}^{-1}$.

Assuming that in the range of anode voltages exceeding U_0 the dependence of the field emission current density j on field strength F follows FN relationship, the expression for the FE current I as a function of U that accounts for variations in β can be written as

$$I = A(U) \exp\left\{\frac{B\varphi^{3/2}\vartheta(y)}{\beta_0 U[1 - k(U - U_0)]}\right\}, \tag{5.3}$$

where φ is the effective work function, B is a constant, $\vartheta(y)$ is the Nordheim function of argument $y = [(æ - 1) \times e^3\beta(U)]^{1/2}/(æ + 1)^{1/2}\varphi$ and æ is a permittivity.

The FE current curves for varying field-voltage proportionality factor are plotted in Fig. 5.7a. As seen in the figure (curve 3), $\beta(U)$ causes only an insignificant deviation from the FN linear relationship. Thus, the fact that $\beta = \beta(U)$ is evident in an emission image as the "shrinkage" effect but does not explain the observed kink and the saturation region in the experimental current–voltage curves.

3. Elinson et al.[330] explained the slow rise of current at voltages above a certain value in an assumption that the average electron energy in this voltage range exceeds the electron

affinity due to heating of electrons by the internal field. In this case the potential barrier transparency is at its highest and the emission is no longer impeded. However, emission current saturation in samples with large electron affinity, $\approx$3–4 eV, and a relatively small energy gap (Si, Ge, and others) cannot be explained by heating of electron gas or emission over the barrier because the energy of electrons is limited by the energy gap ε_g.

4. An analysis of the experimental results for p-type semiconductors with cleaned emitting surfaces indicates that the most probable cause for saturation of the current–voltage curves is limited by the carrier generation rate in the bulk of the sample. The current saturation is the result of a number of interdependent processes, which reduce the electron concentration in the near-surface region of the semiconductor, increase field penetration into the sample, change the field geometry near the emitter, and increase the voltage drop across the crystal bulk. Under such conditions the emission current is limited by the supply of carriers to the surface and is somewhat dependent on the barrier transparency and on the applied potential up to the abrupt rise of the carrier concentration in region III. The same processes may be responsible for nonlinearities of the current–voltage characteristics in high-resistivity n-type samples. An attempt at a theoretical consideration of such a mechanism for p-type semiconductors has been undertaken.[335] This approach has been elaborated and extended to include the case of n-type semiconductors.[151]

5.6. THEORY OF THE FIELD ELECTRON EMISSION FROM SEMICONDUCTORS

Despite the considerable accumulation and interpretation of experimental data on field emission from semiconductors, and establishment of some general relationships, a qualitative theory of the process has not yet been developed. Different interpretations have been put forward about individual facts or effects, such as the appearance of various slopes in current–voltage characteristics, the effect on emission of the heating of the electron gas, the role of possible quantization of electron states in the emitter near surface region, and so on. Yet data on FE kinetics have not been treated theoretically. However, a simple review of past and present theoretical work would warrant the conclusion that FE from semiconductors is qualitatively understood.

As shown above, the nonlinear current–voltage characteristics of FE are inherent in an extensive class of semiconductors. As a rule, these are p- or n-type semiconductors with a low concentration of electrons in the conduction band.[149, 150] In the saturation region, the emission current shows photo- and thermal sensitivity, the size of emission image changes, and the voltage drop across the crystal increases.

The principal aim of the theoretical research on this problem is to reveal the mechanism leading to the saturation region and the region of subsequent rapid current growth, to ascertain the role of bulk defects and surface states, and to identify the energy states from which the electrons tunnel into vacuum. Electron emission from the conduction and valence bands of n-type semiconductors has been described by Stratton[325] without accounting for current saturation. Arthur[296] directed his attention to the formation of a degenerate inverse layer in the near-surface region of p-type semiconductor emitters and suggested explaining saturation of the FE current by analogy with the saturation of the reverse current in a p–n junction.

A very detailed model was analyzed theoretically by Yatsenko.[338] Here the system of anode-vacuum gap/p-type field emitter was considered as a kind of metal–dielectric–semiconductor (MDS) structure, whose leakage current is essentially the emission current. Numerically calculated saturation currents of these structures are a few orders of magnitude lower than those observed experimentally. Later,[339] an attempt was made to explain this difference by the high generation rate on the lateral surface of the emitter. The current values obtained agree with the experimental values to within an order of magnitude.

A theory which takes into account the mutual influence of the space charge region and the current flowing through it has been developed.[151] A small deviation of the electron gas from thermodynamic equilibrium was assumed. This is equivalent to assuming a short effective lifetime of electron-hole pairs. The calculated values of the emission saturation current agree with the experimental values to within an order of magnitude. The possible mechanisms which reduce the effective lifetime are a high surface generation–recombination velocity and an increased trapping cross-section in the strong electric field.

The influence of screening of the applied field by the charge of surface states was studied by Stratton.[325] Fischer[340] has calculated probabilities of ionization of the surface states by strong electric field and the values of emission current.

The influence of surface states on FE from p-type semiconductors has been analyzed,[338] and the considerable effect of screening by the charge of surface states in the presence of a stationary emission current was explained assuming that the surface states exchange electrons with the energy bands.

5.6.1. Stratton's Theory

In the range of relatively low currents the Morgulis–Stratton theory[323, 324] gives quite an adequate description of the experimental results: a linear increase in the logarithm of the current with inverse value of applied voltage, constant size of the emission image, and lack of photosensitivity. This theory assumes that due to partial penetration of the electric field into the near-surface region of emitter the free electron concentration in this region rises considerably and the electron gas is degenerate. Degeneration is characterized with a parameter ϑ which is the separation of the Fermi level from the bottom of the conduction band. One of the basic assumptions of the theory is that only an insignificant fraction of the electrons arriving at the potential barrier tunnel through it into vacuum. Therefore, the emission process can be treated in a "zero current" approximation and the Fermi level considered to be the same throughout the sample.

Calculation of the emission current density as a function of field strength is carried out using the FN method.[6] The current–voltage characteristics in lg j versus F^{-1} coordinates, in all cases, considered of emission from the valence or conduction bands for an arbitrary degree of the electron gas degeneration in the near-surface region yield straight lines. Their slopes depend on the electron affinity χ, the forbidden band width ε_g (for emission from the valence band), ε_a the acceptor level, and the degeneration parameter ϑ, which equals ϑ_s near the surface (Fig. 5.9).

Stratton has shown that in the presence of considerable concentrations of positively charged surface centers the current–voltage characteristics may display characteristic kinks. Their origin is explained as follows: at low applied field strengths the surface charge field bends the bands of allowed energies upwards so that initially emission comes from the

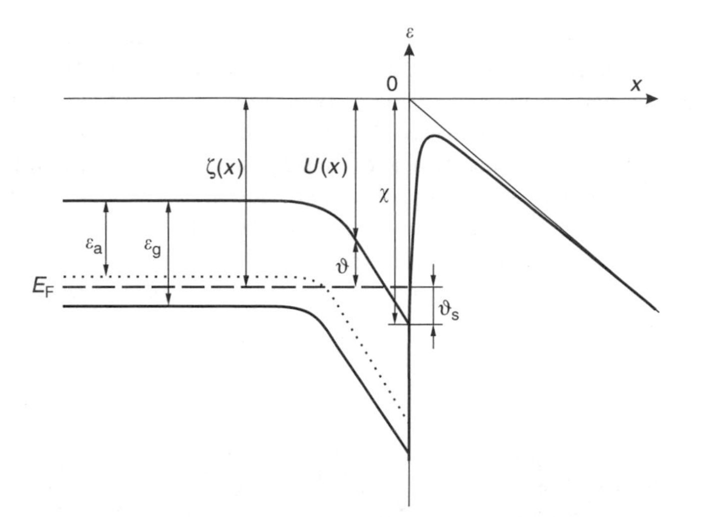

Figure 5.9. Energy diagram of the near-surface region of a semiconductor in the "zero-field" approximation.[151] (ϑ—degeneration parameter; ϑ_s—degeneration parameter near the surface; χ—electron affinity; $U(x)$—energy at the bottom of the conduction band; ζ level of electrochemical potential.)

valence band. When the applied field becomes high enough to compensate for the surface charge field the band bending changes sign, the concentration of the conduction band electrons rises, and emission from the conduction band becomes dominant. In the current–voltage characteristics the transition from valence band emission to conduction band emission will be seen as a dramatic increase in current. *However, curves of this shape have not been observed in experiments.*

A key feature of the measured nonlinear current–voltage characteristics with highly resistive n-type semiconductors as well as p-type semiconductors (Fig. 5.5) is the presence of a saturation region at high fields which is at odds with Stratton's predictions. Thus it appears that the nonlinearities in the experimental current–voltage characteristics are not related to the influence of the positive surface charge.

A more rigorous analysis,[325, 332] accounting for the effect of the semiconductor's energy band structure on the electron emission process and corrections for the effective mass as well as the surface states, yielded no new predictions as to the general shape of current–voltage characteristics.

Although Stratton's theory cannot be applied to field electron emission from semiconductors with low equilibrium concentrations of conduction band electrons in high fields, qualitatively it is in fair agreement with experiment for most semiconductors in low electric fields (region I of the current–voltage characteristic [Fig. 5.5]).

5.6.2. Modeling a Semiconductor Emitter: Statement of the Problem

The latest concepts ion modeling the semiconducting field emitter, that are supported by experimental observations have been developed.[151, 335, 341]

Consider an extrinsic semiconductor with an arbitrary concentration of acceptor and donor impurities. The contact of the semiconductor to the substrate is assumed to be located at such a distance that its influence on the processes in the near-surface regions of the tip may be neglected. It is further assumed that the carrier concentration at any point in the semiconductor is determined by the quasi-equilibrium level of the electrochemical potential. The penetration of the field into the semiconductor and the resulting change of the carrier concentration in the near-surface region is allowed for in the band bending terms and the problem is solved in a one-dimensional approximation.

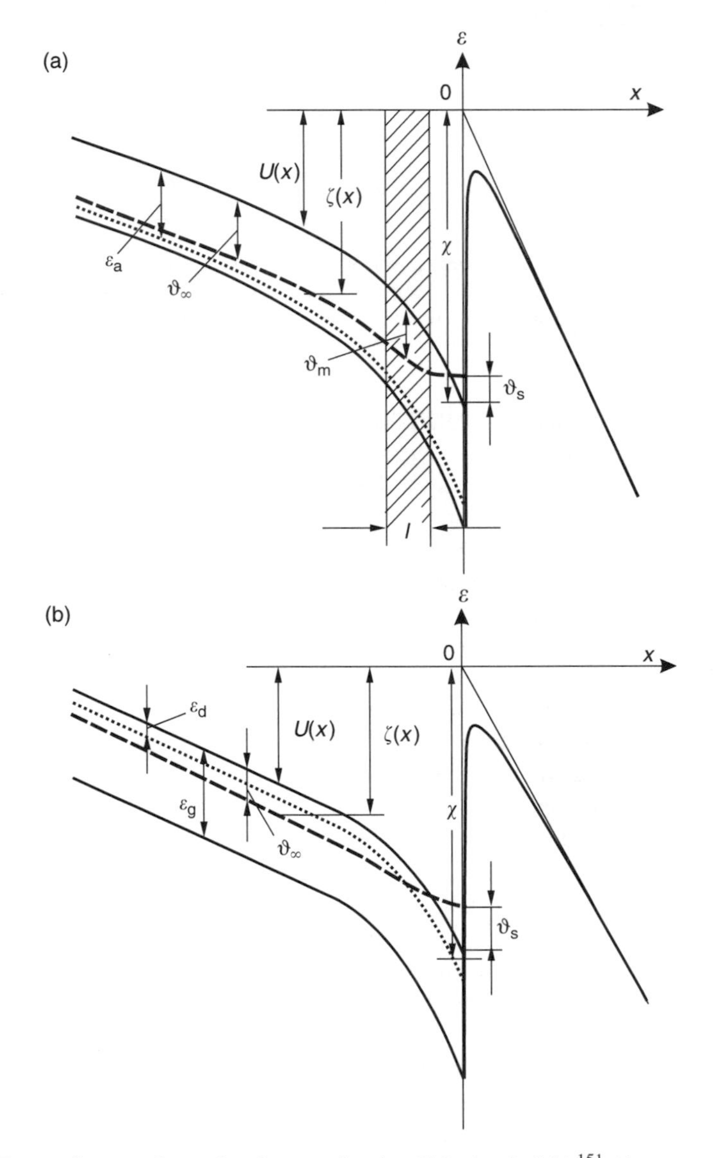

Figure 5.10. Energy diagram of a semiconductor emitter in a high electric field:[151] (a) p-type semiconductors; (b) n-type semiconductors.

The energy diagram of the near-surface region of a semiconductor in an electric field is shown in Fig. 5.10. Here $\vartheta(x)$ is a function characterizing the position of the bottom of the conduction band relative to the Fermi level at any point in the crystal, ϑ_s the value of this function at the surface, ϑ_∞ its value in the sample bulk, $U(x)$ and $\zeta(x)$ are the energies of the bottom of the conduction band and of the electrochemical potential level, respectively, which are referred to an electron at rest at infinity, χ is the electron affinity, ε_a and ε_d are the energies of the acceptor and donor levels, referred to the bottom of the conduction band.

For the emission current density from the conduction band the following expression, obtained by Stratton,[324] may be used (for a more general case see Christov[332]):

$$j_{em} = \frac{e^3 F_s^2}{8\pi h\varphi} \exp\left[-\frac{4}{3}\frac{\gamma}{F_s}\varphi^{3/2}\Theta(Y)\right]\left[1 - \exp\left(-\frac{2\gamma}{F_s}\right)\varphi^{1/2}\vartheta_s - 2\frac{\gamma}{F_s}\varphi^{1/2}\vartheta_s \exp\left(-\frac{2\gamma}{F_s}\right)\varphi^{1/2}\vartheta_s\right]; \quad \vartheta_s > 0 \tag{5.4}$$

$$j_{em} = \frac{4\pi me(kT)^2}{h^3} \exp\left[-\frac{4}{3}\frac{\gamma}{F_s}\chi^{3/2}\Theta(Y)\right]\exp\left(\frac{\vartheta_s}{kT}\right); \quad \vartheta_s < 0$$

where

$$\gamma = 2\pi(2m)^{1/2}/eh; \quad \varphi = \chi - \vartheta_s;$$

$\Theta(Y)$ is the Nordheim function of the argument

$$Y = \left(\frac{æ - 1}{æ + 1}\right)^{1/2} \frac{(e^3 F_s)^{1/2}}{\varphi}$$

in which, though, the degeneration parameter ϑ_s is not merely a function of the field penetration into semiconductor under condition of zero (negligible) current, as in Stratton's theory, but calculated by simultaneously solving Poisson's equation

$$\frac{d^2U}{dx^2} = -\frac{4\pi e^2}{æ}(n - p + N^- - N^+) \tag{5.5}$$

and the equation for the density of current j flowing through the sample and equal to the emission current density j_{em}:

$$j = -(\mu_n n + \mu_p p)\frac{d\zeta}{dx} \tag{5.6}$$

Here n, p, N^+, and N^- are the concentrations of electrons, holes, and charged impurity centers, respectively, μ_n, μ_p are the electron and hole mobilities, F_s the applied field strength, and æ the high-frequency dielectric constant.

Assuming a quadratic dependence of the electrons and hole energies on the quasi-momentum, we can write

$$n = A\mathcal{F}_{1/2}\left(\frac{\vartheta}{kT}\right); \quad p = \alpha A\mathcal{F}_{1/2}\left(-\frac{\vartheta + \varepsilon_g}{kT}\right) \tag{5.7}$$

where ε_{g} is the energy gap, $\alpha = (m_p/m_n)^{3/2}$, $A = 4\pi(2m_nkT/h^2)^{3/2}$, m_n and m_p are the effective masses of electrons and holes, $\mathcal{F}_{1/2}(\eta)$ the Fermi integral of the order $\frac{1}{2}$ defined in the usual way.

We assume that in formulae (5.7) T is the lattice temperature. The concentrations of the ionized donor and acceptor centers will be, respectively,

$$N^+ = N_{\text{d}}\left[1 + 2\exp\left(\frac{\vartheta - \varepsilon_{\text{d}}}{kT}\right)\right]^{-1} \tag{5.8}$$

$$N^- = N_{\text{a}}\left[1 + 2\exp\left(\frac{\varepsilon_{\text{a}} - \vartheta}{kT}\right)\right]^{-1} \tag{5.9}$$

The simultaneous Eqs. (5.5) and (5.6) are solved taking into account the boundary conditions

$$\left.\frac{dU}{dx}\right|_{x\to -0} = eE(x)\big|_{x\to -0} = -\frac{eF_s}{æ}, \tag{5.10}$$

$$\left.\frac{dU}{dx}\right|_{x\to -\infty} = \left.\frac{d\zeta}{dx}\right|_{x\to -\infty} = -\frac{j}{\mu_n n_\infty + \mu_p p_\infty}, \tag{5.11}$$

while the band structure of the semiconductor and the doping level (i.e., n_∞, n_∞, N_{d}, and N_{a}) are assumed known. Expression (5.11) may serve as a boundary condition because in the simultaneous solution of Eqs. (5.5) and (5.6) j is assumed to be an independent parameter equal to the emission current $j = j_{\text{em}}(F_{\text{s}}, \vartheta_{\text{s}})$. The calculations are much simplified if a dimensionless energy parameter $y(x) \equiv \sigma(x)/kT$ is introduced, whereupon the space charge density electric ρ is expressed as

$$\rho(y) = e(n - p - N_{\text{d}}^+ + N_{\text{a}}^-) \tag{5.12}$$

and the "effective" total density of mobile charges

$$b(y) = n + \frac{\mu_p}{\mu_n}p. \tag{5.13}$$

It is possible to transform the simultaneous Eqs. (5.4)–(5.6), using y as an independent variable instead of x. By boundary condition (5.11) and denoting $E(y)$ the field strength inside the sample, we obtain

$$e^2E^2(y) - \frac{j^2}{\mu_n^2 b^2(y_\infty)} = \frac{8\pi ekT}{æ}[H(y) + G(y)], \tag{5.14}$$

where

$$H(y) = \int_{y_\infty}^{y} \rho(y)\,dy, \tag{5.15}$$

$$G(y) = j\int_{y_\infty}^{y} \rho(y)\frac{dy}{e\mu_n b(y)E(y) - j}. \tag{5.16}$$

Here $y_\infty \equiv y(x)|_{x\to-\infty}$ and $y_s \equiv y(x)|_{x\to-0} = \vartheta_s/kT$. If $x = 0$ is chosen as the upper integration limit, then using condition (5.10) expression (5.14) may be rewritten in the form

$$\frac{e^2F_s^2}{æ^2} - \frac{j^2}{\mu_n^2 b^2(y_\infty)} = \frac{8\pi ekT}{æ}[H(y_s) + G(y_s)]. \tag{5.17}$$

Equations (5.14) and (5.17) are very convenient for a qualitative discussion of the dependence of emission current density on the applied field strength, crystal temperature, and doping level for both p- and n-type semiconductors.

Together with Eq. (5.4) they allow for a numerical evaluation of the current–voltage characteristics. To determine the spatial distribution of the charge density, field strength, and the position of the quasi-Fermi level in a semiconductor, it is necessary to know the dependence of the space coordinate x on y. This dependence follows from Eq. (5.6) written in integral form

$$x = kT\int_y^{y_s} \frac{e\mu_n b(y)}{e\mu_n b(y)E(y) - j}\,dy \tag{5.18}$$

similar to integral (13) in Stratton.[324]

5.6.3. Zero Current Approximation

For low electric fields, where the electron emission currents are still small, the influence of the current flowing through the sample may be neglected. Assuming in (5.14) and (5.17) that the density of the current flowing through the sample is zero (the "zero current approximation"), we obtain

$$e^2E^2(y) = \frac{8\pi e}{æ}kT\,H(y) \tag{5.19}$$

and

$$\frac{e^2F_s}{æ^2} = \frac{8\pi e}{æ}kT\,H(y_s) \tag{5.20}$$

It should be remembered, of course, that the emission current density j_{em} is, in fact not zero.

The integral $H(y)$ defined by (5.15), is taken in the explicit form

$$\begin{aligned} H(y) = eA\Bigg\{&\frac{2}{3}\mathcal{F}_{3/2}(y) + \alpha\frac{2}{3}\mathcal{F}_{3/2}(-y-\omega_g) \\ &+ \frac{N_a}{A}\ln\left[1+\frac{1}{2}\exp(y-\omega_a)\right] + \frac{N_d}{A}\ln\left[1+\frac{1}{2}\exp(\omega_d - y)\right]\Bigg\} \end{aligned} \tag{5.21}$$

where $\omega_q \equiv \varepsilon_q/kT$. This result is similar to that obtained earlier by Seiwatz and Green.[336]

It can be shown that in not very heavily doped semiconductors ($N_a\omega_g/A \ll 1$ and $N_d/A \ll 1$ the first term in (5.21) becomes dominant once $y = \vartheta/kT > -2$.

Using this fact, we obtain from (5.20) and (5.21)

$$F_s^2 = 8\pi æ AkT\,\tfrac{2}{3}\mathcal{F}_{3/2}(y_s) \tag{5.22}$$

The condition $\vartheta_s = 0$ means that the bottom of the conduction band is at the Fermi level and the electron gas near the semiconductor surface is practically degenerate (Strictly speaking, the degeneration sets in already at $\vartheta_s \approx -2\,kT$).

Substituting into (5.22) the value of $\frac{2}{3}\mathcal{F}_{3/2}(0) = 0.7685$, and with T expressed in K, we obtain a degeneration criterion for the electron gas near the semiconductor surface

$$F \geq F^{(d)} \equiv 1.44 \times 10^6 æ^{1/2} \left(\frac{m_n}{m}\right)^{3/2} \left(\frac{T}{300}\right)^{5/4} \text{ V/cm} \tag{5.23}$$

For the case of Ge and Si ($æ_{Ge} = 16$, $æ_{Si} = 12$) at $T = 300$ K and $(m_n/m)_{Ge} = 0.55$, $(m_n/m)_{Si} = 1.08$, we have $F^{(d)}_{Ge} = 3.7 \times 10^6$ V/cm; $F^{(d)}_{Si} = 5.7 \times 10^6$ V/cm.

The calculations above indicate that the fields at which degeneration of the electron gas at the emitter surface sets in are considerably lower than the fields necessary for an appreciable FE, that is, in the stationary case the emission always comes from the degenerate conduction band.

From Eq. (5.23) it is seen that the degeneration criterion does not depend on the type of conduction and the doping level because with approaching degeneration the concentration of electrons in the near-surface region becomes, in fact, independent of the electron concentration in the bulk.

At fields corresponding to $y \geq 6$, it is possible to make use of the asymptotic behavior of the Fermi integral $\mathcal{F}_{3/2}(y) \approx \frac{2}{5}y^{5/2}$; this approximation is accurate to within 15 percent. In this case expression (5.22) is simplified and we get a relationship equivalent to the one derived by Stratton[324]

$$\vartheta_s = \left[\frac{15h^3}{128\pi^2 æ(2m_n)^{3/2}}\right]^{2/5} F_s^{4/5} \tag{5.24}$$

The current–voltage characteristics plotted using the relationships (5.4) and (5.22) are straight lines. Distributions of the field strength and the electron concentration in the sample bulk under zero current may be obtained from expressions (5.18) and (5.7).

Numerical calculations for p-Ge at $N_a = 10^{15}$ cm^{-3} and $T = 300$ K show that the field strength and the electron concentration change drastically at a depth of $\approx 10^{-6}$ cm, which is approximately an order of magnitude smaller than the emitter radius. This means that in the region of small FE currents of the emitter surface may be considered nearly equipotential and the geometrical factor $\beta = F_s/U$ constant and independent of the applied field, what is in agreement with experiment (cf. Section 5.5).

5.6.4. The "Nonzero Current" Approximation with p-Type Semiconductors

Both n- and p-type semiconductors in a high electric field have different structures in the space charge region. The carrier concentration in n-type semiconductors smoothly decreases away from the crystal surface (Fig. 5.10b), while in p-type semiconductors a depletion region is formed directly after the near-surface inversion layer (Fig. 5.10a), in which the free carrier concentration and, accordingly, the conductivity are much lower than in the bulk. In p-type semiconductors the current through the sample and thus the emission current are essentially space-charge limited.

In the region of minimum conductivity the level of the electrochemical potential is approximately in the middle of the forbidden gap and $y = y_m = -\omega_g/2 + \ln\sqrt{a}$, where

$a = (\mu_p/\mu_n)(m_p/m_n)^{3/2}$. According to (5.13) the effective free carrier concentration is b_m where

$$b_m \equiv b(y_m) = A\sqrt{\pi a}\exp\left(-\frac{\omega_g}{2}\right) \tag{5.25}$$

The position of the Fermi level and $b(y)$ are related by formula $j \sim b(y)d\zeta/dx$. Obviously, for a given current value $d\zeta/dx$ is maximum at $y = y_m$ and, generally, remains so throughout the depletion region. At the surface ($y = y_s$) and in the bulk of the sample $(y = y_\infty)b(y)$ is large and the electrochemical potential changes slowly. Schematically, the position of the quasi-Fermi level in the energy diagram for a p-type sample through which current is flowing is presented in Fig. 5.10a. As $d\zeta/dx$ approaches dU/dx, the integral of $G(y)$ in Eq. (5.14), depending on j rises sharply and the zero current approximation ceases to be valid. By means of estimates it can be shown that with increasing current in Eqs. (5.14) and (5.17), first of all the term containing $G(y_s)$ should be considered, because the contribution of the term proportional to j^2, is approximately two orders less for the fields which satisfy the condition $e^2F_s^2/æ \approx 8\pi eKTG(y_s)$. It follows from expression (5.16) that $G(y_s)$ is large if the denominator of the integral expression is small, that is, in the region where $b(y) = b_m$. The denominator of the integral expression for $G(y)$ is always positive as it represents the diffusion current. At any arbitrary point of the semiconductor emitter the limiting value of the current density is given by

$$j \leq e\mu_n b(y)E(y) \tag{5.26}$$

with a minimum value at $y = y_m$.

Noting also that $E(y) \leq F_s/æ$, we can write

$$j \leq e\mu_n b(y_m)\frac{F_s}{æ} = e\mu_n A\sqrt{\pi a}e^{-\omega_g/2}\frac{F_s}{æ} = j^{\text{lim}} \tag{5.27}$$

The value of the emission current is limited by the supply of carriers from the depletion region. In the current–voltage curves this limitation is seen as current "saturation", that is, slower current increase with applied field (Fig. 5.5). The values of the limiting current densities calculated from (5.27) are for Ge ($N_a = 10^{15}$ cm^{-3}, $T = 300$ K) $j^{\text{lim}} \approx 2\times 10^4$ A/cm^2, and for Si ($N_a = 10^{15}$ cm^{-3}, $T = 300$ K) $j^{\text{lim}} \approx 1$ A/cm^2 which is considerably less than the critical current densities leading to the emitter destruction. The limiting current density depends slightly on the semiconductor doping level.

From formula (5.27) one can derive an important relationship between the limiting current density and temperature. With $\omega_g = \varepsilon_g/kT$ and noting that the strongest dependence on T is contained in the exponent, and neglecting the weak dependence of the coefficient A on temperature, we write

$$\frac{d(\ln j^{\text{lim}})}{d(1/kT)} \approx -\frac{\varepsilon_g}{2}. \tag{5.28}$$

The result shows the important role of the carrier generation in the depletion region and of the high temperature sensitivity of the semiconductor field emitter at $j_{\text{em}} = j^{\text{lim}}$. It is quite clear that the conductivity of the depletion region may be increased by the generation of carriers by light (see Fig. 5.5).

5.6.5. The Effect of High Internal Fields

In our calculations the process of carrier generation by strong electric field, which results in the avalanche current rise prior to breakdown is not considered (region III of the current–voltage characteristic, Fig. 5.5). The internal field affects the emission current in the saturation region mainly by changing the carrier mobility. This dependence may be approximated by the formula $\mu(E) = \mu_0(E_0/E)^\delta$, where the exponent δ is typically in the range $0 \leq \delta \leq 1$ (depending on the particular mechanism that limits the carrier mobility), μ_0 is a constant equal to the carrier mobility in field E_0, and E_0 is the field strength value above which this mechanism becomes dominant. With most semiconductors in the fields of 10^3–10^4 V/cm, the major limiting mechanism is generation of optical phonons by the electrons, in which case $\delta = 1$.[337] For weaker fields the dominant mechanism may be scattering of carriers by acoustic phonons, with $\delta = \frac{1}{2}$.

As shown by the calculations, in the region of transition to saturation the internal fields in p-type crystals may exceed $E_0 \approx 10^3$–10^4 V/cm, so that in estimations of the limiting emission current density by Eq. (5.27) it may be assumed that $\delta = 1$. Then

$$j^{\text{lim}} = eA\sqrt{\pi a}[\mu_0 E_0] \exp\left(-\frac{\varepsilon_g}{2kT}\right) \tag{5.29}$$

In this case j^{lim} does not depend on the field strength.

It can be demonstrated that j^{lim} is very close to the saturation current density (the latter can be determined in a straightforward manner only under complete saturation), and the transition region from the linear current–voltage relationship to saturation is very narrow. If the carrier mobility in this case is proportional to E^{-1}, then complete saturation of the current should be expected. On samples of p-Ge an almost complete saturation has been observed.[296] Full saturation has been observed on samples of p-Si[328] (Fig. 5.5). In the case of slowly varying $\mu_n(E)$, the transition to saturation occurs in higher fields and the current density in this region increases slightly with the applied field.

The results of numerical calculations of the current–voltage characteristics accounting for the function $\mu_n(E)$ are discussed in Section 5.6.7.

5.6.6. Field Emission from *n*-Type Semiconductors

In n-type samples the diffusion current density in any arbitrary cross section is large in comparison with the total current density j. Therefore the function $H(y)$ is small everywhere, and the term $\mathrm{G}(y)$ dominates in Eq. (5.17). The condition of the applicability of the zero current approximation for an n-type semiconductor can be written in the form

$$e^2 F_s^2/æ^2 \gg j^2/\mu_n^2 n_\infty^2.$$

This condition means that the density of current flowing through the sample should be much smaller than the limiting current density

$$j^{\text{lim}} = \frac{eF_s}{æ}\mu_n n_\infty \tag{5.30}$$

It is evident that at $j \approx j^{\text{lim}}$ the FN-type $j(F)$ function reduces to Ohm's law if $\mu(E)$ is ignored.

It follows from (5.30) that the saturation current density is proportional to the majority carrier concentration n_∞ in the sample bulk which in turn depends on the donor concentration N_d through the relationships

$$n_\infty = \begin{cases} \left(\frac{2\pi m\, kT}{h^2}\right)^{3/4} N_d^{1/2} e^{-|\varepsilon_d|/(2kT)} & \text{at } T < T_0 \\ N_d & \text{at } T > T_0, \end{cases} \tag{5.31}$$

where $T_0 = 750(m/m_n)(N_d/10^{18})^{2/3}$.

According to (5.31), at $T > T_0$ the saturation current density $j^{\lim}$ does not depend on the temperature and at $T < T_0$

$$\frac{d(\ln j^{\lim})}{d(1/kT)} \approx -\frac{|\varepsilon_d|}{2}. \tag{5.32}$$

As $|\varepsilon_d| \ll \varepsilon_g$, the temperature dependence of the saturation current density in n-type semiconductors is much weaker than in p-type semiconductors (cf. formula (5.28)). The results of numerical calculations of current–voltage characteristics for n-type emitters follow in Section 5.6.7.

5.6.7. Results of Numerical Calculations

The simultaneous Eqs. (5.5) and (5.6) may be reduced to a first-order nonlinear differential equation of the form

$$\frac{dE}{dy} = \frac{4\pi e}{æ} \frac{kT\rho(y)}{E(y) - j/\mu_n b(y)} \tag{5.33}$$

with the boundary condition

$$\mu_n E|_{y\to y_\infty} = \frac{j}{b(y_\infty)} \tag{5.34}$$

where y_∞ is determined from the relation $\rho(y_\infty) = 0$.

Equation (5.33) was solved numerically. In the calculation, the current density $j_{em} = j$ was assumed to be a parameter. With the prescribed emission current density for fixed values of the parameter j, $E(y_\infty)$ was calculated and, by Eq. (5.33), $E(y)$ derived (see Fig. 5.11). From the relationships for the emission current density in Eq. (5.4), F_s may be determined as a function of j_{em} and y_s: $F_s = F_s(j_{em}, y_s)$. The values of y_s and F_s corresponding to the specified emission current density are determined graphically from the equation $F_s(j_{em}, y_s)/æ = E(y_s)$. The current–voltage characteristics calculated by this method are given in Fig. 5.12, curves 1 and 2.

As seen in Fig. 5.12a, the transition to saturation in the second case begins at higher voltages and the corresponding saturation current density is increased by roughly one order of magnitude. The importance of the dependence of carrier mobility μ_n on the field strength is clear. In the second case, the saturation region is much flatter, which is in agreement with the results of the quantitative analysis.

Note that the calculated values of the saturation current density $j^{\lim}$ are considerably smaller than the measured values. This discrepancy is likely due to the fact that the emitter

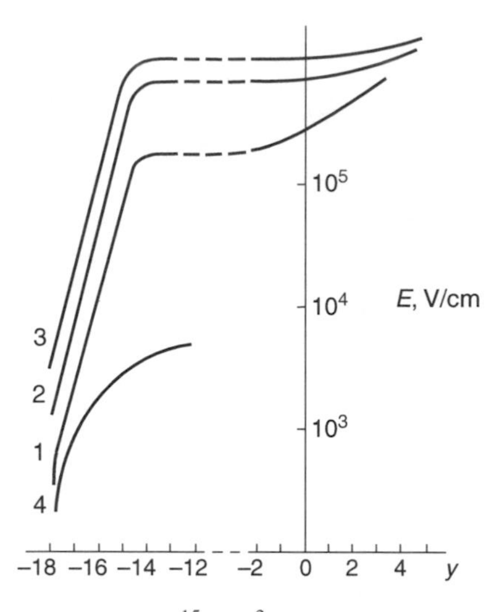

Figure 5.11. Relationship $E(y)$ (p-Ge, $N_a = 10^{15}$ cm^{-3}, $T = 300$ K) for several values of the parameter j:[151] (1) $j = 0$; (2) $j = 100$ A/cm^2; (3) $j = 150$ A/cm^2; (4) $j = 200$ A/cm^2.

model is one-dimensional. Preliminary calculations show that $j^{\lim}$ for conical emitters is one or two orders larger.[341]

Calculation results of current–voltage characteristics for n-type samples are shown in Fig. 5.12b. Here a strong dependence of the saturation current on the impurity concentration is observed.

The structure of the space charge region at nonzero current through the sample may be characterized by the distribution of free carriers in the space charge $\rho(x)$ and the parameter $y(x)$ across the sample. Calculations show that the initial decrease and subsequent growth of the concentration of free charge carriers $b(x)$ are rather sharp. Therefore, the characteristic dimension of the depletion region l in a p-type sample may be defined as a distance between the points in which the free carrier concentration is double the minimum concentration. Figure 5.13 clearly shows broadening of the depletion region with increase of the field F_s (p-Ge, $N_a = 10^{15}$ cm^{-3}, $T = 300$ K). When the current density saturates, the broadening of the depletion region continues solely because of more field penetration into the semiconductor. The result is in good qualitative agreement with the measurements of the characteristic dimension of the depletion region by light probe method.[296] The electron concentration n_s near the semiconductor surface first slowly increases as the applied field is increased, then over rather wide range of F_s it remains essentially constant and eventually drops sharply as F_s approaches the value corresponding to the transition to saturation when the near-surface region is depleted (Fig. 5.14).

Analysis shows that the main features of the FE current–voltage characteristics are related to the low concentration of free electrons and the existence of capture centers, including deep traps, inside the semiconductor.[151, 335, 338, 341]

FE from p-type semiconductors is difficult to analyze. The first difficulty involves accounting for bipolar conductivity and, hence, generation and recombination processes in the emitter. The simultaneous solution of the Poisson equation, the generation–recombination equations, and the equation for the current density in the sample is necessary.

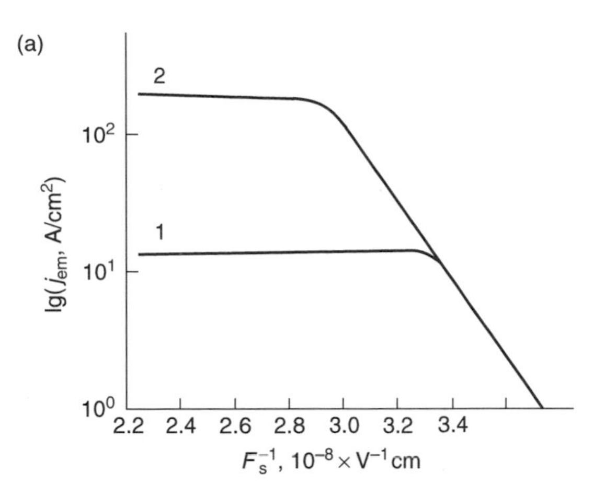

Figure 5.12. (a) Theoretical current–voltage characteristics calculated for p Ge with the assumption of various relationships between mobility and strength of the internal field in a crystal of p-Ge, $N_a = 10^{15}$ cm^{-3}, $T = 300$ K;[151]

$$(1)\quad \mu_n(F) = \begin{cases} \mu_n(0); & F < 10^2 \text{ V/cm} \\ \mu_n(0)\left(\dfrac{10^2}{F}\right)^{1/2}; & 10^2 \text{ V/cm} < F < 10^4 \text{ V/cm} \\ \dfrac{1}{10}\mu_n(0)\left(\dfrac{10^4}{F}\right)^{1/2}; & F \geq 10^4 \text{ V/cm} \end{cases}$$

$$(2)\quad \mu_n(F) = \begin{cases} \mu_n(0); & F < 10^2 \text{ V/cm} \\ \mu_n(0)\left(\dfrac{10^2}{F}\right)^{1/2}; & F > 10^2 \text{ V/cm} \end{cases}$$

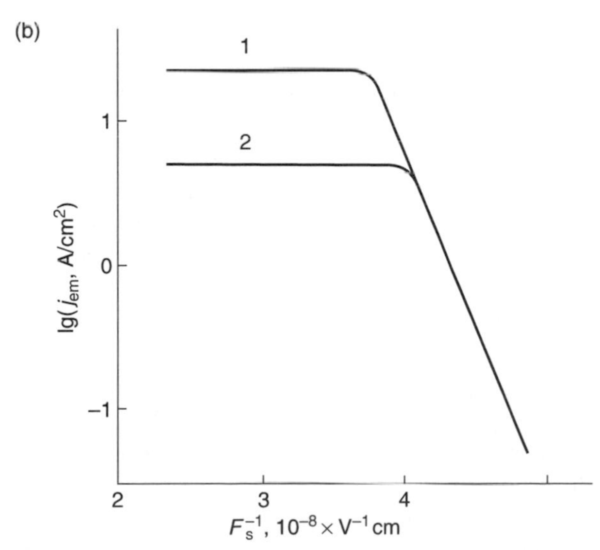

Figure 5.12. (b) Theoretical current–voltage characteristics calculated for sample of n-Si with (1) $N_d = 2 \cdot 10^{12}$ cm^{-2}, (2) $8.5 \cdot 10^{12}$ cm^{-3}

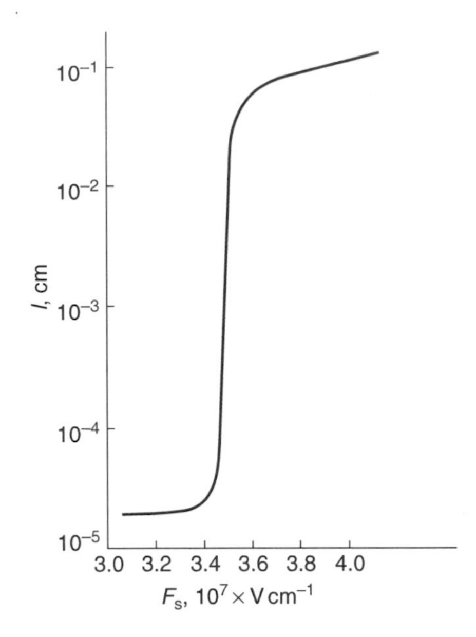

Figure 5.13. Change of the extent of the depleted region, l, versus external field F_s (p-Ge, $N_a = 10^{15}$ cm^{-3}, $T = 300$ K).[151]

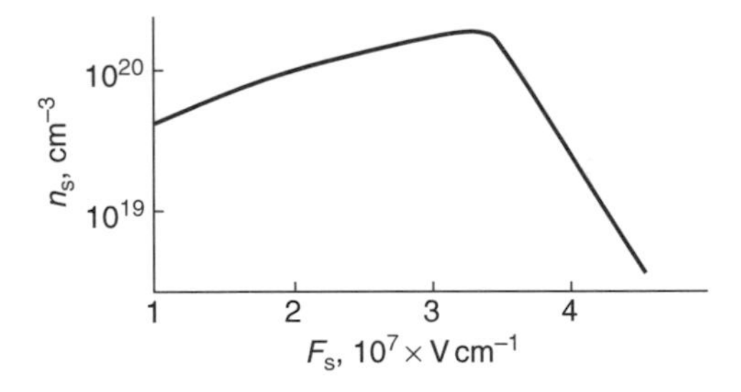

Figure 5.14. Concentration of conduction electrons in the near-surface region, n_s, versus field strength F_s (p-Ge, $N_a = 10^{15}$ cm^{-3}, $T = 300$ K).[151]

It provides an opportunity to take into account the mutual influence of processes of field penetration in a semiconductor and a finite conductivity current flow through a specimen. If

$$j < e(\mu_n \mu_p)^{1/2} n_i F_s / æ \tag{5.35}$$

an approximation of a zero conductivity current is reasonable. In this equation μ_n and μ_p is the mobility of electrons, holes, æ the dielectric penetrance of the semiconductor, F_s the field strength on the field emission border, n_i the electron concentration in a pure semiconductor, L the length of the tip. This applies to region I current–voltage characteristic.

If F_s increases, the current density can reach values, for which the near-surface layer is exhausted, and the process of emission begins to depend upon inflow of electrons from the bulk of the emitter. Our analysis[151, 341] shows that if the FE current is comparable with the maximum attainable electron inflow rate from the depth of a specimen, it is possible to separate three areas in a crystal: near-surface n-region, following it i-region and deep p-region, where the concentration of holes is approximately equal to the concentration of ionized acceptors. Of course, such a division is reasonable only if the thickness of these areas is much greater than the thickness of the transition region between them. It is possible to show, that this requirement is equivalent to satisfying the inequalities

$$N_a^{1/2} < F_s/(4\pi æ kT)^{1/2} < N_a^{3/2}/n_i. \tag{5.36}$$

So, a similar division is impossible for semiconductors with conductivity, close to the intrinsic semiconductors, as well as for semiconductors with high concentration of impurities.

Calculations show, that the generation–recombination processes in the i-area give the main contribution to the value of the saturation current. The decisions obtained made it possible to describe two limit cases of system behavior:

1. negligible recombination in the i-region;
2. high speed recombination, when a small deviation from the state of thermodynamical equilibrium of the electron–hole plasma under balance takes place.

When recombination rates in the i-region can be ignored, recombination does not influence upon the value of the saturation current. The saturation current is determined by the generation rate of carriers in the i-area. In the case of large recombination rates, the current is limited not by generation of carriers in the i-area, but by the final conductivity of this area. Calculations[151, 341] have shown that, at room temperature, the saturation current density for a p-Si emitter with a cone angle of $\approx 10°$ cannot exceed 10^3 A/cm^2. This value is close to experimentally observed values.

For FE from p-semiconductors, the saturation current is determined not only by the generation and recombination rates in the i-area, but also by the volume of this area.[151] The presence of traps, on which significant charge can reside, leads to a change in this volume and, thus, to a change in the saturation current. It has been shown that accounting for electron charge on traps leads to thermal sensitivity of the saturation current. For a cylindrical emitter,

$$\frac{d(\ln j_s)}{d(1/kT)} = -\frac{\varepsilon_g}{4} - \frac{\varepsilon_t}{2}, \tag{5.37}$$

and for a conical emitter,

$$\frac{d(\ln j_s)}{d(1/kT)} = -\frac{3}{4}\varepsilon_t - \frac{1}{8}\varepsilon_g, \tag{5.38}$$

Here ε_g is the width of a forbidden band, ε_t the activation energy of traps.

FE of n-type semiconductors has a specific character. For low fields, the emission current is determined by the barrier transparency D. However, when the field strength F_s and, accordingly, D increases, the FE current is found to be restricted by the inflow of electrons from the emitter bulk. So as for p-type semiconductors, the external field penetrates into the bulk of an emitter up to contact, causing injection of electrons into a conductivity band. Therefore, "screening" of an external field takes place on the injected charge. If

$$F_s/4\pi æ e \gg N_d L, \tag{5.39}$$

where L is the length of emitter tip, the role of injection of electrons from contact is a main factor and the emission current is restricted by the space charge of the carriers in the semiconductor. In this case

$$j \sim F_s^2/L. \tag{5.40}$$

If

$$F_s/4\pi æe \sim N_d L, \tag{5.41}$$

the injection is small and the emission current is limited by the finite conductivity of the emitter's material and the FE current density is

$$j \sim e\mu_n F_s N_d/æ \tag{5.42}$$

The investigation of the influence of deep trap states with concentration N_t and activation energy ε_t upon FE has shown[342, 351, 352] that traps influence on FE only when

$$N_t > N_d, \quad N_t > N_c \exp(-\varepsilon_t/kT), \tag{5.43}$$

where N_c is the density of states at the bottom of a conduction band. Therefore only deep traps with high concentrations influence the FE processes. This would be expected in broad-band n-type semiconductors (e.g., CdS).

Calculations show that, generally, in semiconductors of both n- and p-type there are five regions of current–voltage characteristics:

1. FN region;
2. First saturation region, $j \sim \mu_n F_s \exp(-\varepsilon_t/kT)$;
3. Second saturation, $j \sim \mu_n (F_s^2/L) \exp(-\varepsilon_t/kT)$;
4. A region in which the current increases rapidly;
5. Third saturation, $j \sim \mu_n F_s^2/L$.

The region of rapid current increase (4) is not associated with avalanche multiplication of the carrier, but with the limited filling of traps.

The existence of these regions is shown by experimental data for single crystal CdS. Figure 5.15 shows the current–voltage characteristics plotted in $\lg I_a = f(\lg U_a)$ coordinates. These coordinates are chosen to help identify the power law regions of the curves $I_a \sim U_a^n$. The FN region corresponds to extremely low emission current values and was not observed on this sample because emission current was very small and the sensitivity of the apparatus was not enough for current registration. What is seen is that at low voltages the current–voltage characteristic have regions of linear ($n = 1$) and quadratic ($n = 2$) dependence of emission current on the applied voltage. Saturation of the emission current following the region in which the current rises rapidly with voltage shows that no avalanche breakdown takes place in the emitter. When the anode voltage is decreased, the near-surface electric field supporting the emission is maintained by charged traps. Therefore, in semiconductors with a high ($>10^{16}$ cm^{-3}) concentration of deep trap states, the current–voltage characteristics show appreciable hysteresis, that is, the curves $I_a = f(U_a)$ taken with increasing and decreasing anode voltage do not coincide. After complete removal of the anode voltage the deep traps remain charged for long time periods. When the voltage is reapplied, the emission current returns to the value it had before saturation (a memory

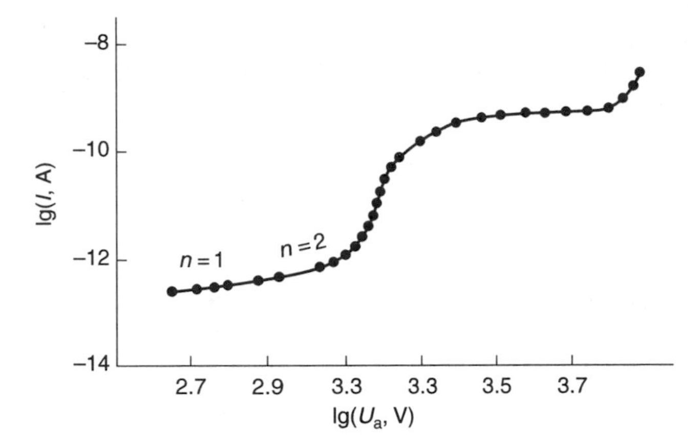

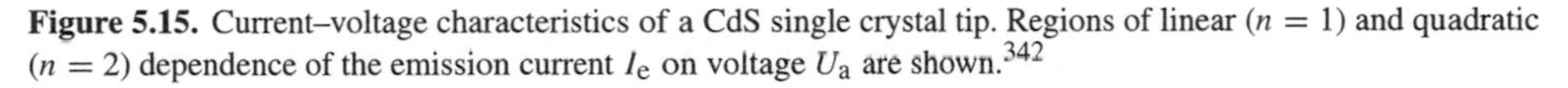

Figure 5.15. Current–voltage characteristics of a CdS single crystal tip. Regions of linear ($n = 1$) and quadratic ($n = 2$) dependence of the emission current I_e on voltage U_a are shown.[342]

effect). Possibly, this phenomenon can be utilized in designing power-dependent storage devices.

Recently it was reported[461–468] that nonlinear current–voltage characteristics similar to those in Fig. 5.5 have been observed for diamond-like films. The hysteresis or divergence of the characteristics measured with increasing and decreasing voltage was observed as well. This indicates that the processes taking place during field electron emission from diamond-like films may have much in common with space-charge region effects in wide band gap semiconductors.

5.7. TRANSITION PROCESSES IN FIELD EMISSION FROM SEMICONDUCTORS

The experiments discussed above were conducted in the slow registration mode. That is, recording of the emission characteristics usually took at least a few minutes. Therefore the observations were of steady-state FE processes. Although electron tunneling into vacuum is essentially instantaneous, the establishment of a stationary distribution of charge carriers and field in the sample is characterized by finite relaxation times. Because the electron concentration in the conduction band is not large, when the electron supply is depleted by extracting large current densities, a bulk depletion region is formed and it is the relaxation processes in this region that determine the character of the field electron emission. Some data on the inertial characteristics of FE process in semiconductors are given by Arthur[296] and Sokolskaya and Klimin.[364] Arthur observed the occurrence of hysteresis during the measurement of transient current–voltage characteristics in *p*-type Ge. Sokolskaya and Klimin noted a lag in the variation of field electron emission current from CdS during pulsed illumination.

Stetsenko[365] studied photo-field emission from *p*-Si (7×10^3 $\Omega \cdot$cm) in a pulsed mode. It was shown that the current pulses lagged the voltage pulse front by a time interval τ_d which varied from 10 to 100 μsec depending on temperature, intensity of illumination, and the voltage pulse amplitude; decreasing as these parameters were increased. τ_d was also found to be very sensitive to the value of the fixed bias voltage. In addition, the emission current monotonically decreased with time after reaching some peak value. Stetsenko suggested that this behavior was related to the presence of positively charged surface states having a concentration of $\sim 10^{13}$ cm^{-2}. The kinetics of field electron emission in high-ohmic Ge and Si under emitter illumination with rectangular light pulses of a few millisecond duration has also been studied[363] and a considerable delay of the photoresponse to illumination was detected. Characteristic current rise times varied from 10^{-5} to 10^{-3} sec depended on the density of traps in a sample. In this study, a light probe with a spot size of less than 0.2 mm was utilized. Local illumination of isolated tip areas permitted the size of the depleted region to be determined. It was shown that the extent of photosensitive area is many orders' magnitude larger than the emitter radius (about 1 mm in the silicon samples studied).

5.7.1. Time Variation of the Field Electron Emission

Establishment of a stationary concentration in the emitter bulk following application of a pulsed anode voltage is evident in the current kinetics.[304, 362, 366, 367] On atomically clean Ge and Si surfaces the current in region I of current–voltage characteristics lag by a time interval τ_d that decreases with increasing anode voltage U_a (Fig. 5.16). The kinetics in region II are of the most interest: The current I_a first reaches a maximum and then drops to some steady state value. As U_a is increased, the maximum in I_a is reached sooner and the ratio, G, of the maximum current to its steady-state value increases. By the end of region II the value of $G \approx 10$. The time variation of I_a reflects the process of forming the steady-state structure of the space charge in the semiconductor emitter.

In region I the current shows no variation with time and for a square voltage pulse a square current pulse is obtained. This behavior is to be expected because emission in this region only depends on the transparency of the potential barrier, and not on any bulk properties. The carrier generation rate for a given value of emission current is sufficient to maintain the degenerate state of the electron gas near the emitter surface. As seen in oscillograms 2–5 of Fig. 5.16a and oscillograms 5–7 of Fig. 5.16b the current drops with time. This drop appears to be associated with the depletion of electrons in the near-surface layer of the emitter due to a number of interrelated factors: (1) reduction in the number of electrons incident on the potential barrier; (2) reduction of the effective field strength F near the tip apex by redistribution of voltage between the tip and the vacuum gap; and (3) a decrease in F due to a change of the space charge configuration. The characteristic time is determined by the time taken to establish a new diffusion–drift equilibrium under conditions of current flow through the sample.

In region III of the current–voltage characteristic a current rise is observed at constant U. This current rise is seen in the beginning of region III (oscillograms 6, 7 of Fig. 5.16a and oscillogram 8 of Fig 5.16b), and is probably evidence of carrier multiplication in the high internal field of the crystal. Rather sluggish response of the emission current to the change in carrier concentration due to field-generated carriers can also be explained by the slow rearrangement of the space charge in the sample, in which the presence of traps can play an important role. We note that the current increase at the end of region III causes heating and, in most cases, destruction of the emitter.

The results of studies of the kinetics and residual relaxation effects indicate the important role that transient processes in the sample bulk play in establishing steady-state

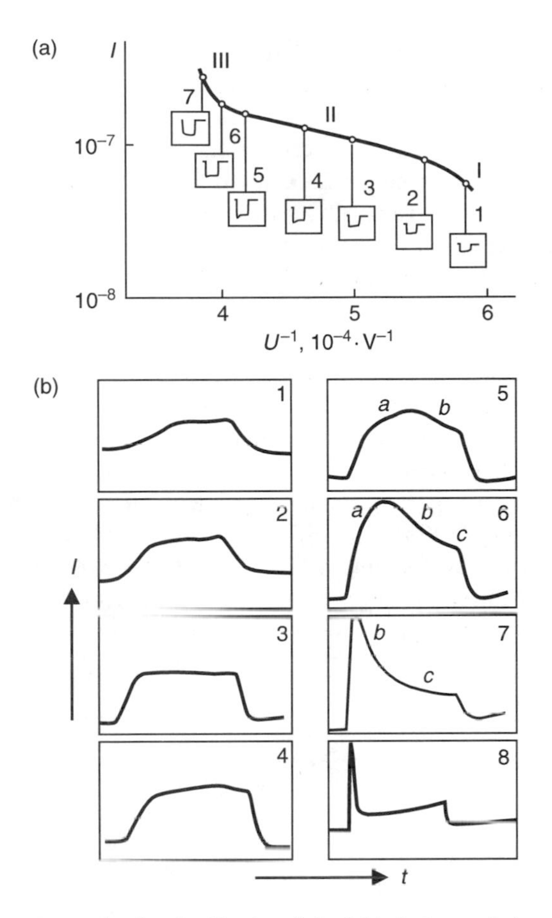

Figure 5.16. Oscilloscope traces showing the kinetics of the field electron emission: (a) *p*-type Ge (voltage pulse duration $\tau_{pulse} = 5$ msec);[366] (b) *p*-type Si; traces 1 and 2 correspond to region I of the I–V characteristics; traces 3–5 to the transition between regions I and II; trace 6 to region II; traces 7 and 8 to the transition between regions II and III ($\tau_{pulse} = 100$ msec).[362, 366]

emission. The absence of residual effects in region I and their presence in region II indicates that the processes are related to the high electric field in the sample.

The kinetics of FE from clean semiconductor surfaces has been investigated with *n*-type silicon.[304, 362] Si samples doped to a wide range of concentrations and having resistivity values of 10, 50, 300, 800, and 2000 $\Omega \cdot cm^{-1}$ were studied. It was shown that the variation with time of the FE current in *n*-type semiconductors (Si) is qualitatively similar to that observed in *p*-type samples.

5.8. THE STABLE SEMICONDUCTOR FIELD EMISSION CATHODE

As we have seen, experiments and theoretical studies show that the appearance of the region of slow current variation is related to the limited supply of carriers from the semiconductor bulk. It follows that beyond a certain emission current, the current is unaffected by a further increase in barrier transparency. Direct evidence of this can be inferred from the insensitivity of the electron emission to surface contamination, in particular, the residual gas pressure.[350]

Studies of this insensitivity were[350] carried out on single crystal p-type silicon with resistivities of 14 and 200 $\Omega \cdot$cm and on 2000 $\Omega \cdot$cm n-type silicon. The vacuum was varied between $5 \cdot 10^{-9}$ and $5 \cdot 10^{-5}$ Torr with air by a leak value. The experimental procedure involved: (1) evacuation to $\approx 5 \cdot 10^{-9}$ Torr; (2) cleaning of the sample surface by field desorption; (3) collection of I–V data in UHV; and (4) collection of I–V data in various residual pressures of air.

The key results are shown in Fig. 5.17. It can be seen that the emission current is nearly independent of the residual gas pressure in the current range corresponding to region II of current–voltage characteristic (Fig. 5.17b, curve 1) and independent of time (Fig. 5.17b, curve 2). Under these same vacuum conditions large current fluctuations occurred in region I of current–voltage characteristic. Note that the extent of the saturation region of current–voltage characteristic is highly dependent on the doping level of material (cf. curves 1 and 2 in Fig. 5.17a). For p-Si with $\rho = 2000\ \Omega \cdot$ cm, the extent of this region was >23 kV which is more than 4 times the range of voltage variation corresponding to the linear part of current–voltage characteristic. This result agrees with the theoretical premise that the extent of the depleted region[151, 341] and, correspondingly, the internal field

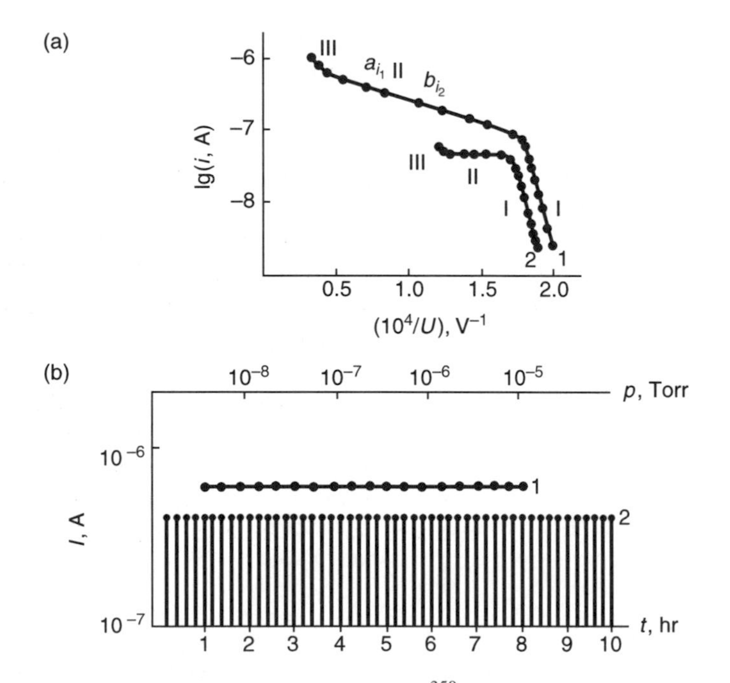

Figure 5.17. FE current as a function of residual gas pressure;[350] (a) experimental I–V characteristics of p-Si: 1—small cone angle (2000 Ω cm); 2—large cone angle (3 Ω cm). (b) FE current as a function of residual gas pressure (1) and time (2) (at $p = 10^{-5}$ Torr). Current values correspond to points a (curve 1) and b (curve 2) of the current–voltage characteristic in part (a).

gradient and "breakdown strength" of a sample, is inversely proportional to the doping level.

These results support the proposition that under certain conditions (region II of the current–voltage characteristic) the electron emission is independent of the barrier transparency and that bulk properties play the dominant role in the emission process. These results also indicate that under such conditions the FE current is not sensitive to the surface states introduced by adsorption of residual gases.

5.9. ADSORPTION ON SEMICONDUCTOR SURFACES

The ability to visualize adsorption and diffusion processes on different crystal surfaces with a 20–30 Å resolution makes the FE microscopy a unique instrument studying such phenomena. The delay in using field electron microscopy to investigate adsorption on semiconductors was due to considerable difficulties involved in producing a clean semiconductor emitter surface. Advances in the semiconductor surface cleaning process[291, 293, 297, 300, 310, 353, 354] have allowed for some studies of adsorption.

The adsorption of Ba,[356] Au and Ti[357, 358] and SiO[355] on Ge has been investigated. The adsorption of Ga and As on GaAs has also been studied.[293]

Bakhtizin and Stepanov[355] carried out first experiments on SiO adsorption on Ge. Arthur[293] studied adsorption of Ga and As on GaAs. A number of studies on oxygen adsorption on Ge was performed.

5.9.1. Oxygen Adsorption on Ge

The first studies of oxygen adsorption on atomically clean Ge were carried out by Ernst,[300, 353] Ivanov and Zabotin,[354] Mileshkina and Bakhtızın.[359] The most significant observation was the highly anisotropic adsorption of oxygen on different Ge crystal planes.[300] preferential adsorption was observed on the (100) plane and in the vicinity of (111) plane on the (123), (144), (112), and (113) planes. These experiments were later confirmed by Ivanov, Rozova, and Fursey.[347] The average sticking coefficient $\bar{s}$ was found to be 0.1 from measurement of the time intervals required for the formation of a manolayer at a given oxygen pressure.[300, 353]

Using a more sophisticated oxygen source[354] we reproduced these results of O_2 adsorption on p-type Ge.[347] Emission images on Ge surface are shown in Fig. 5.18. It was found that O_2 adsorption affects the nonlinear portion of current–voltage characteristics by lowering the saturation current and appreciably reducing the extent of saturation region. The field desorption of oxygen resulted in complete recovery of the initial shape of the current–voltage characteristics.

5.9.2. Electroadsorption

Electroadsorption suddenly accelerate O_2 adsorption on Ge at a particular electric field value F_s.[348, 349] It was discovered that electroadsorption effects appear at field values near those required for Ge evaporation. The threshold field F_s can be calculated from the

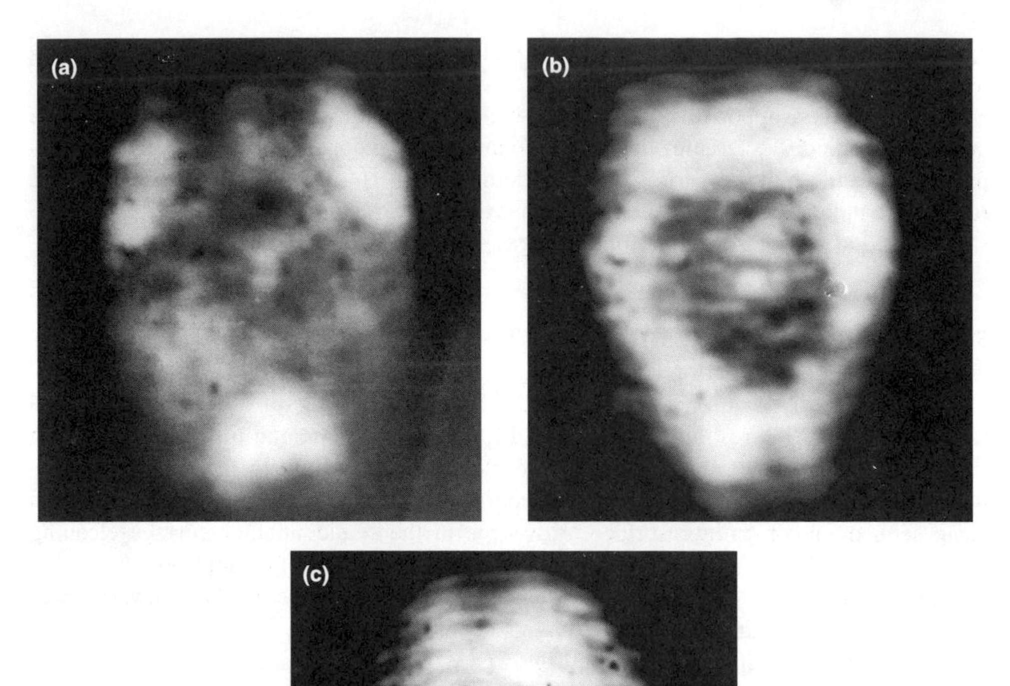

(c)

Figure 5.18. Adsorption of O_2 on Ge:[347] (a) FE pattern of Ge ⟨111⟩ with a clean surface after the field evaporation of contaminations. (V_D = 24.7 kV, V_{FE} = 5.8 kV); (b) O_2 adsorption on the Ge ($P_{O_2} = 10^{-7}$ Torr, t_{exp} = 20 min, U_{FE} = 6.7 kV); (c) thick layer O_2 on the Ge. ($P_{O_2} = 5 \times 10^{-6}$ Torr, t_{exp} = 100 min, U_{FE} = 7.4 kV).

equation $K = U_s/U_e = F_s/F_e$, where U_e is the voltage required for Ge evaporation, U_s is the voltage at which the acceleration of desorption takes place, and F_e is the known value of Ge evaporation field. The effects of electroadsorption are illustrated in Fig. 5.19.

The character of the O_2 coating on Ge varies under high field conditions (compare Figs 5.18 and 5.20).[349] It is possible to produce dense, well bound layers having thicknesses of a few monatomic layers. The highest O_2 adsorption rates are observed in the vicinity of (001), (111), and (113) faces as evidenced by the dark spots in the emission pattern (Fig. 5.20).

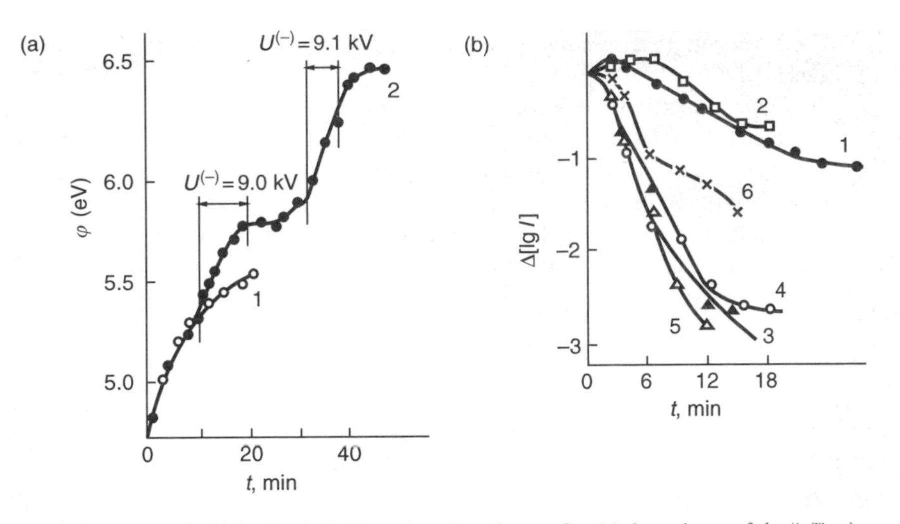

Figure 5.19. Influence of a high electric field on O_2 adsorption on Ge: (a) dependence of the "effective work function" on the time of oxygen exposure: (1) without field; (2) with pulsed electric field. $U_{ads} = 9$ kV, $U_{ads} = 0.46 \cdot U_{desorp\ of\ Ge}$.[348] (b) Current versus time during O_2 adsorption on Ge ⟨111⟩: (1) $K = 0$; (2) $K = 0.4$; (3) $K = 0.45$; (4) $K = 0.6$; (5) $K = 0.7$; (6) $K = 0.8$. ($K = U_{ads}/U_{desorp\ of\ Ge}$).[349]

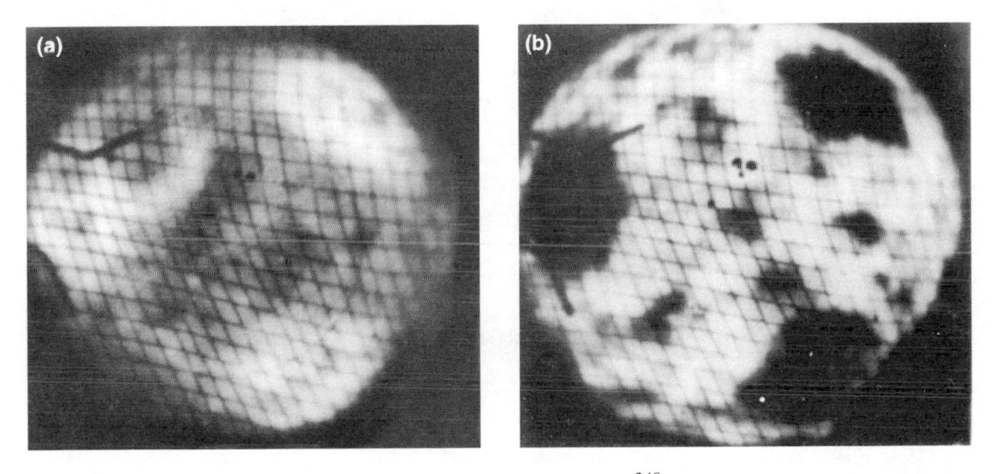

Figure 5.20. Adsorption of the O_2 in a high electric field with Ge ⟨111⟩:[348] (a) clean Ge surface obtained by field desorption ($U_D = 13.25$ kV, $U_{FE} = 3$ kV, $I_{FE} = 2.5 \times 10^{-7}$ A); (b) surface after exposure to O_2 for 20 min at $U = 6$ kV ($P_{O_2} = 10^{-6}$ Torr, $U_{FE} = 3.63$ kV, $I_{FE} = 2.5 \times 10^{-7}$ A).

5.10. RESUME

This chapter outlines the most important studies of FE from semiconductors. The main conclusions are:

1. In recent years, the technology of preparing atomically clean semiconductor surface has been developed and FE and field ion microscopy of semiconductors

has advanced. Perfect emission images of a number of semiconductors have been obtained.

2. The conditions have been established under which FE experiments must be conducted in order to preserve the initial properties of the semiconductor sample.
3. Key features of FE from semiconductors have been observed experimentally and described theoretically.
4. Research methods to study adsorption on semiconductor surfaces have been developed and very interesting results have been obtained.

6

STATISTICS OF FIELD ELECTRON EMISSION

6.1. FORMULATION OF THE PROBLEM

Several fundamental physical processes involved in field electron emission can be elucidated through measurements of the statistical distribution of "elementary emission events," that is, the statistical distribution of the emission of either a single electron or group of correlated electrons. In the general case, with some probability P_i, one may observe some number N_i of electrons for a given emission event. The function $P(N)$, is the distribution of N over different events.

Investigations of secondary and photoelectron emission have shown that groups containing different numbers of correlated electrons have been observed in the emission current. The information obtained in these experiments enables one to obtain not only the fundamental parameters of the process (the secondary emission coefficient, quantum exit, fluctuations) but also aids in elucidating the physical mechanisms underlying the emission phenomenon.

The unique feature of field emission (FE) is that electrons tunnel into vacuum without external excitation. Investigation of the emission statistics associated with a non-excited electron gas not only allows us to quantitatively evaluate the applicability of the single electron approximation to the FE process, but also aids in elucidating fundamental aspects of solid-state electronics; in particular, those involving electron–phonon interactions and correlation effects that yield information about electron–electron interactions (see Chapter 1, Section 1.3.3).

In a number of papers, the possibility of groups of electrons tunneling into vacuum, correlated in space and time, have been discussed. For instance, Young[191] suggested that the anomalously large value of the work function of the tungsten (W) {110} plane may be due to tunnelling electron pairs. The possibility of pair tunnelling was explained either by (1) fluctuations in the transparency of the potential barrier, (2) lattice thermal oscillations,[191] or (3) the adsorption of residual gas molecules.[370] Lee and Gomer[121] have shown that pair tunneling would explain the "high energy tail" observed in field electron energy distributions (see Chapter 1, Fig. 1.7). Lastly, the statistics of FE from semiconductors contains information about the process of the creation of current carriers when field penetration is involved and the search for electron pairs in FE from superconducting materials is of special interest.

6.2. METHOD OF INVESTIGATION

The methodology of measuring emission spectra statistics is shown in Fig. 6.1. The FE tip serves as the cathode. The emitted electrons, after passing through a hole in the first anode, are accelerated by an auxiliary electrode to increase their energy in order to facilitate detection. Upon arriving at the detector a signal is produced with an amplitude proportional to the number of electrons. The pulse amplitude analyzer counts the number of events associated with the arrival of groups of electrons containing one, two, three, or more electrons.

The first direct measurement of statistical FE events was reported by Herman.[371] The emitted electrons were accelerated to 15 keV whereupon they entered a gas-filled proportional counter. The absence of electron groups was observed with an accuracy of 1 percent. Later, observing field electron emission from spontaneous microprotrusions on metal surface, groups of two, three, and four electrons, were detected[370] with the help of a proportional semiconductor detector. The pairs were several percent of the total number of electrons. Note that in these previous investigations,[370, 371] the emitter surface was not cleaned, and the measurements were carried out only under "technical" vacuum conditions ($\sim 10^{-6}$ Torr).

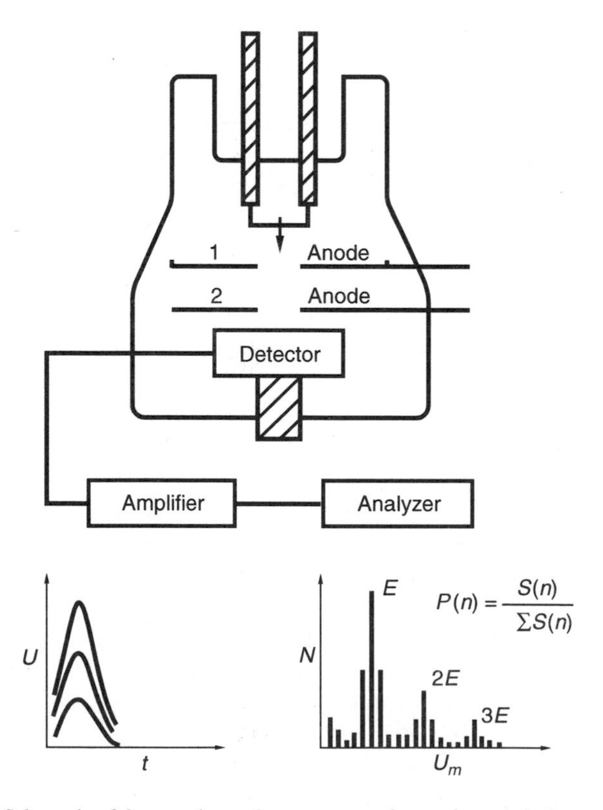

Figure 6.1. Schematic of the experimental apparatus used to study statistical processes in FE.

Measurements of FE statistics from atomically clean surface of polycrystalline tungsten with a reproducible current–voltage dependence in ultra-high vacuum have been conducted.[372, 373] This research has shown that FE from W is to a considerable extent a single electron emission process. However, systematic investigations of FE statistics became possible only after improvements in the experimental device[373] wherein surface cleanliness and perfection were controlled by emission patterns. FE from different crystallographic planes was studied using a rotating cathode assembly and special measures were taken to suppress emission from intermediate electrodes.

Figure 6.2 is a schematic of this improved experimental arrangement.[374] The field emitter (1) is fixed in the rotating system. Motion of the cathode is accomplished with bellows (2) which enables one to observe any region of the FE image by the probe hole in the screen (3). The electrons are registered by the semiconductor detector (4). The cathode may be heated by passing current through the support loop and cooled by liquid nitrogen.

Special attention was paid to make the apparatus highly sensitive to detecting groups of correlated electrons.. The counting rate was 500 electrons/sec. This value was considered as the optimal rate because it enabled us to establish the appearance of electron pairs when they comprised more than 0.1 percent of the total emission current.

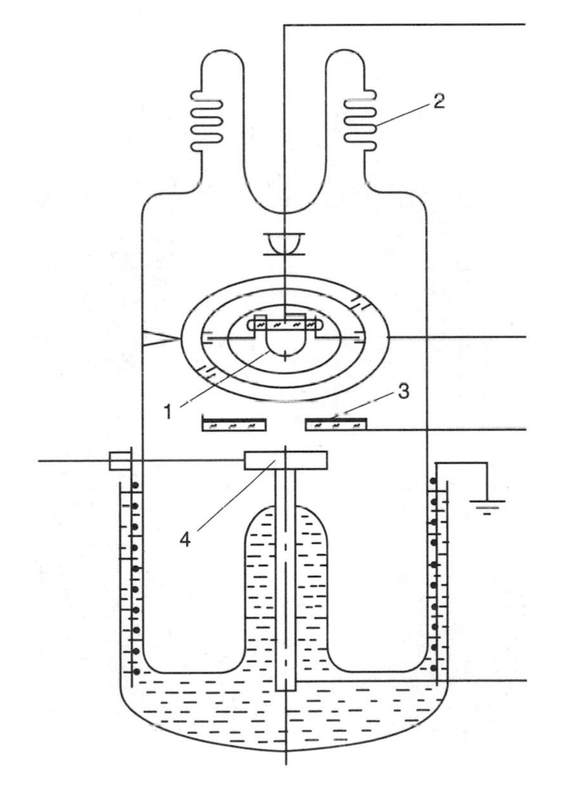

Figure 6.2. Device for the investigation of FE statistics.

6.3. FIELD EMISSION STATISTICS FROM METALS

The first complete investigation of the statistics of FE from W is described by Fursey et al.[375] Here emission processes from various facets; (001), (111), (110), and (112) (Fig. 6.3), were investigated for which anomalies in energy distributions of field electrons have been observed.[122] In addition, emission processes were studied from all other regions of the emission pattern, in particular, the (120), (130), (113), (115), (116), and (117) facets. Pair tunneling was not observed from any crystallographic direction. Some of the spectra obtained are shown in Fig. 6.4 and analysis of some of the spectra in Table 6.1.

The role of thermal fluctuations in the FE statistics of W has also been studied[375] in the temperature range from 77 to 1000 K. For the temperatures investigated, FE had a single electron character. The influence of the adsorption of residual gases on the FE statistics was also investigated. In the first of two experiments, the investigations were performed with different degrees of residual gas coverage on the W surface at a background pressure of $5 \cdot 10^{-10}$ Torr. In the second experiment the statistics were measured at a pressure of 10^{-6} Torr, as in Gazier.[370] Again, many-particle effects were not observed (Figs. 6.4e, 6.4f). These results enable Fursey et al.[375] to infer that the multipeak spectra observed in Gazier[370] are connected with parasitic secondary emission from intermediate electrodes. In measurements carried out with an intermediate accelerating electrode, a secondary emission multipeak spectrum (Fig. 6.4g) was obtained. Later, a detailed investigation of the FE statistics for tantalum (Ta), molybdenum (Mo), and niobium (Nb)[378] was conducted.

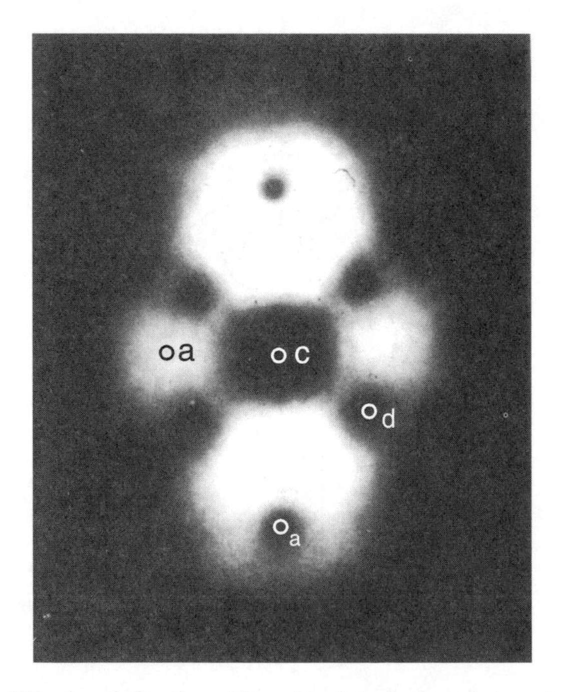

Figure 6.3. FE pattern of W: a, b, c, d – locations of the probing holes for the emission patterns of W corresponding to crystallographic directions ⟨010⟩, ⟨122⟩, ⟨011⟩, and $\langle\bar{1}12\rangle$ respectively.

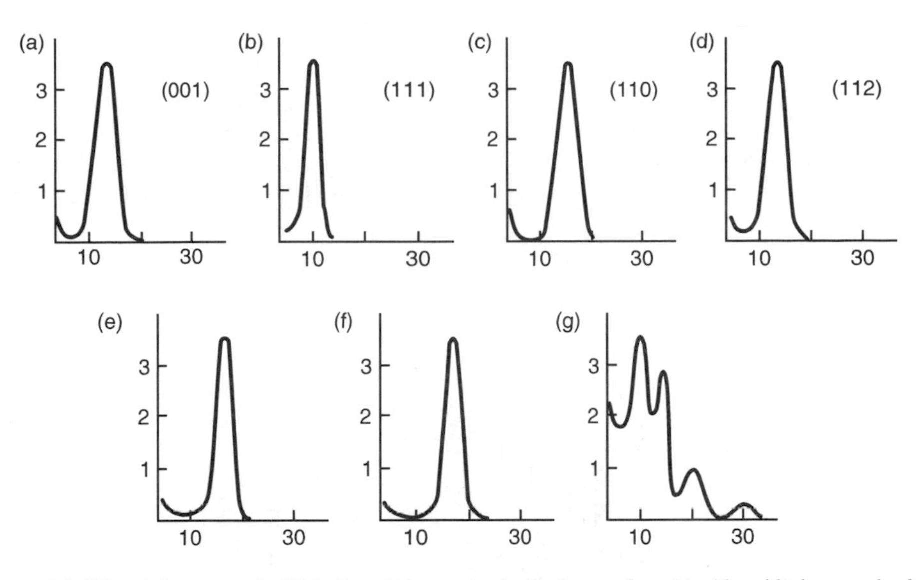

Figure 6.4. FE statistics spectra for W: (a, b, c, d) for an atomically clean surface; (e) with residual gases adsorbed in a vacuum of 5×10^{-10} Torr; (f) in a vacuum of 10^{-6} Torr; (g) with additional acceleration. Vertical scale: pulses ($N \times 10^{-3}$); horizontal scale: energy (keV).[375]

Table 6.1. Relative height of the second peak in spectra from metals

	$[S(2)/S(1)] \times 10^4$			
Facet (hkl)	W	Mo	Ta	Nb
111	8	8	8.5	9
112	9	8.5	9	9
110	8.5	8	9	8
120	9	8	8	9
116	9	9	9	8
001	9	9	8	8.5
130	8.5	—	—	—
332	9	—	—	—

The results presented in Table 6.1 show, that in the temperature range of 77–1000 K, FE from the metals is single particle to within an accuracy of 0.1 percent.

6.4. FIELD EMISSION STATISTICS AT CRYOGENIC TEMPERATURES

The experimental apparatus for the investigation of FE statistics at liquid helium (He) temperature (4.2 K) is described in Fursey et al.[376] A schematic of the experimental arrangement used to investigate FE statistics at liquid He temperature is shown in Fig. 6.5. The

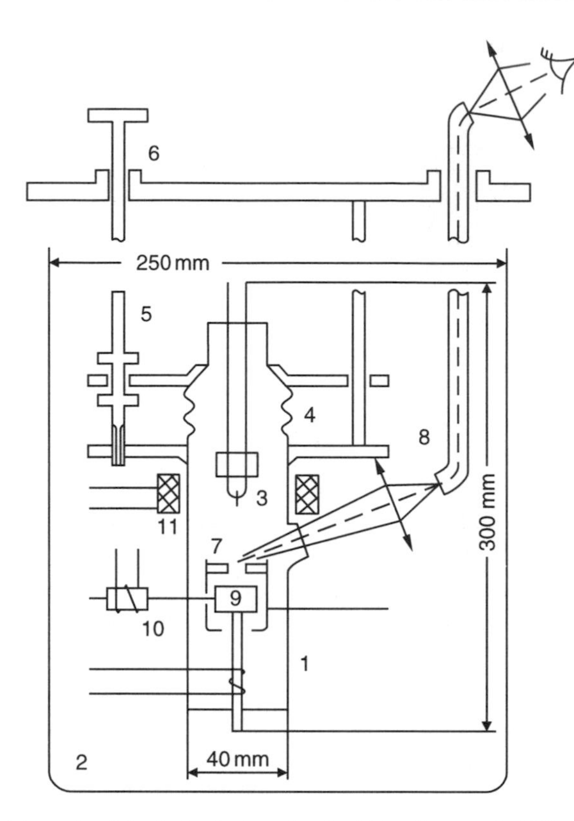

Figure 6.5. Schematic of the device for investigating FE statistics at liquid He temperature.[376]

glass vacuum tube (1) is placed into a cryostat (2). Transfer of the cathode (3) is carried out by bellows (4) with the help of a mechanical manipulator (5, 6). The FE image on the screen (7) is observed by a fiber optic assembly (8). The semiconductor detector (9) and the amplifier (10) are maintained at their optimum operational temperatures by heaters. In this device it is also possible to investigate the electron emission characteristics of superconductors in high magnetic fields. The superconducting electromagnet (11) can create a magnetic field magnitude of 1 T in the region of the emitter.

A comparative investigation of W and Nb at 4.2 K has been reported.[377, 378] It is known that W remains in the normal (nonsuperconducting) state at 4.2 K. Spectra of the FE statistics have been obtained by probing individual facets (low index crystal planes) such as the (110), (112), and (120) planes. For all the W facets investigated, FE had a single-electron character within the accuracy mentioned above ($P(2) < 0.1$ percent). We also investigated superconduction single crystals of Nb. The measurements of the FE statistics were conducted on different crystallographic facets of an atomically clean Nb surface. For all values of current density (10^{-9}–10^{-4} A/cm^2) and field strength investigated, FE from Nb in the superconducting state had a single-electron character (Fig. 6.6). The presence of a magnetic field did not induce any changes in the character of the electron emission from superconductiong Nb, that is, the spectrum of the FE statistics remained a single-electron for all values of magnetic field up to 1 T.

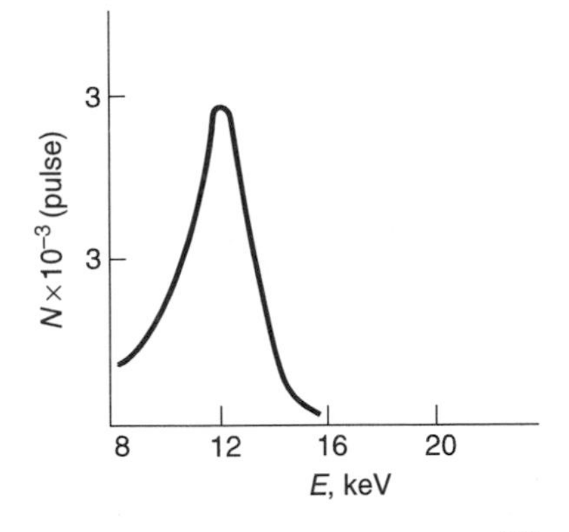

Figure 6.6. FE spectrum for Nb at 4.2 K.[378]

The absence of electron pairs in these experiments may be related to the dispersion of the electrons and their low collection efficiency by the probe hole in front of the detector. The space between electron groups on the emitter surface is limited by their correlation length, which for Nb is equal to 2000 Å. Such an area corresponds to a part of the screen of the FE projector exceeding the dimensions of the usual probe window. For this reason, it is necessary to study superconductors with a comparatively small correlation length; Nb–Ti (300 Å), Nb–Sn (70 Å), and ceramics (20 Å).[147]

FE statistics from a Nb–Ti alloy were measured in the temperature range of 4.2–300 K.[147] This superconductor has a critical temperature of 18 K and a correlation length of 300 Å. Emitter tips were prepared by electrochemical etching. Prior to experiments we first measured the current–voltage characteristics of the Nb–Ti emitters and then observed the FE pattern. The current–voltage dependence exhibited linear Fowler–Nordheim (FN) characteristics. The emission pattern was symmetric after thermal cleaning of the crystal emitter. FE statistics of the Nb–Ti alloy were investigated for current densities of 10^{-6}–10^{-4} A/cm^2 from various crystallographic directions. Measurements at He temperature were carried out in a cryostat where a fiber-optic bundle allowed observation of the FE pattern to check the degree of cleanliness of the emitter surface. Measurements in the cryostat were carried out at a vacuum of 10^{-12} Torr. Analysis of the spectra of the FE statistics obtained for Nb–Ti revealed that in the temperature region 4.2–300 K, FE has a single-electron character to an accuracy of 99.9 percent (Fig. 6.7).

6.5. MULTIELECTRON FIELD EMISSION FROM HIGH TEMPERATURE SUPERCONDUCTING CERAMICS

For high temperature superconductors FE statistics were reported first in Fursey et al.[148] In this paper many-particle tunneling was observed from yttrium (Y) ceramics (Fig. 6.8). The maximum fraction of paired events was 10 percent. The fraction of three electron and four electron groups was roughly 1 and 0.2 percent, respectively.

The sample was sharpened mechanically by diamond polishing. Emission was observed to occur from separate microtips (protuberances) on the surface of the sample yielding

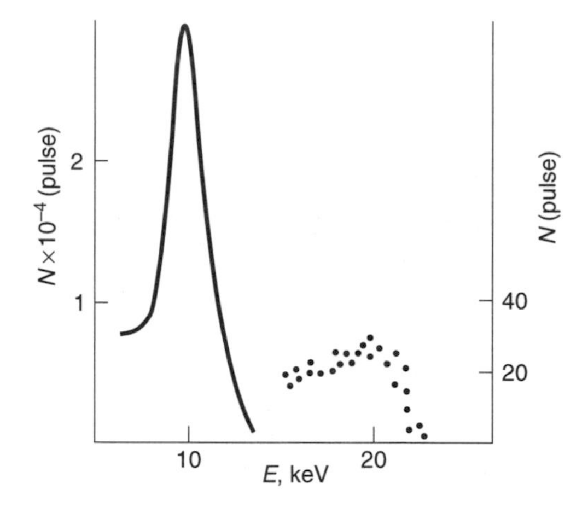

Figure 6.7. FE spectrum for Nb–Ti at 4.2 K.

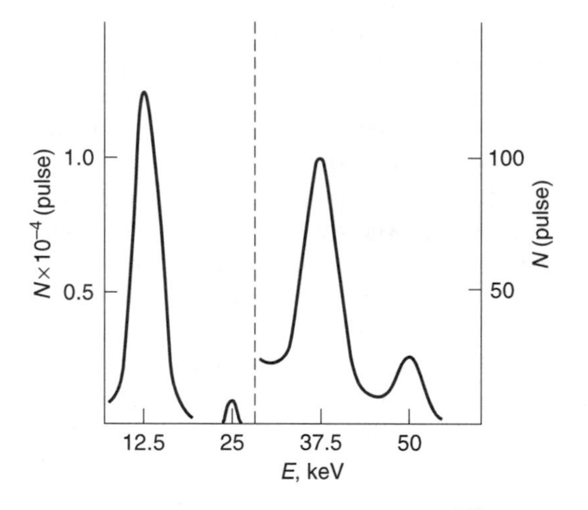

Figure 6.8. FE spectrum for Y ceramics.[148]

an emission pattern with an irregular structure in the form of separate bright emission regions. The superconducting properties of the ceramics were checked before and after the experiments by observing the Weissner effect. The vacuum in the device at 300 K was 10^{-9} Torr, and at 4.2 K, better than 10^{-12} Torr.

Up to the present, current–voltage characteristics, the emission images, and the FE statistics for Y high temperature superconductor single crystals and for Y and Yi high temperature superconductor ceramics have been investigated.[379] It has been established that multielectron FE takes place only from separate, weakly emitting regions of Y-based ceramics. The current–voltage dependence of such samples (Fig. 6.9) has some unusual features also found with semiconducting field emitters: Both have nonlinear FN characteristics that are temperature sensitive.

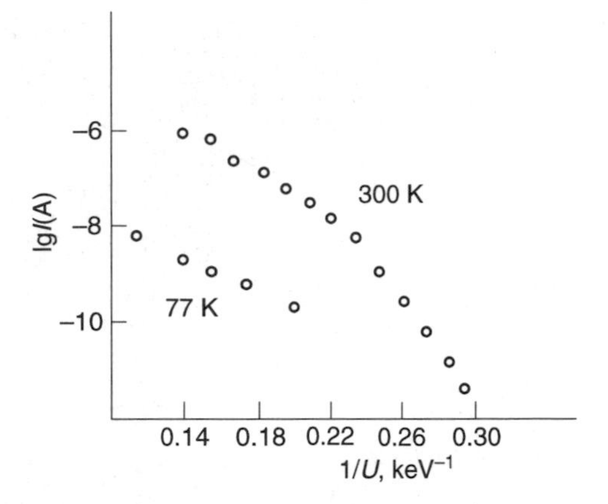

Figure 6.9. Current–voltage dependence of Y ceramics.

It is important to note that multielectron FE was observed from 4.2 to 300 K, the critical temperature being equal to 90 K. It is clear that the effect observed is not related directly to the superconducting state. This observation cannot be explained at present and requires additional investigation.

6.6. INVESTIGATIONS OF FIELD EMISSION STATISTICS FOR HIGHLY TRANSPARENT BARRIERS

The probability of multiparticle tunneling essentially depends on the transparency of the potential barrier.[121] A considerable part (several percent) of the tunneling current from W into vacuum involves electron pairs at FE current densities from 10^1 to 10^4 A/cm^2.[121] This range of current density was necessary to maintain the number of random double superpositions at a level of 0.1 percent as the resolution of the amplitude analyzer was $\sim 10^{-6}$ sec.

The only practical way to investigate FE statistics at current densities in the range of 10^1–10^4 A/cm^2 is to use very short detection time intervals (less than 10^{-10} sec). Such studies have been conducted.[380] The idea is based upon quickly deflecting a narrowly focused electron beam with the leading edge of a nanosecond duration voltage pulse. The time it takes for the beam to pass through a probe hole placed in front of the detector is the detection time. The minimum time between the pulses should be larger than delay time associated with the total detection system (detector and associated electronics). For conventional devices this time is of the order of 10^{-5} sec. The experimental apparatus (Fig. 6.10) includes a FE projector with a probe window, an electron focusing and deflection system, and a semiconductor detector.

The emission statistics for W was investigated up to current densities of 10 A/cm^2. The spectra obtained (Fig. 6.11) show that the second peak corresponds only to occasional superpositions; FE has a single particle character. It should be mentioned that the full capacity of this device was not completely utilized; it is possible to make the time gate smaller than 10^{-12} sec and thereby increase the FE current density to 10^4 A/cm^2.

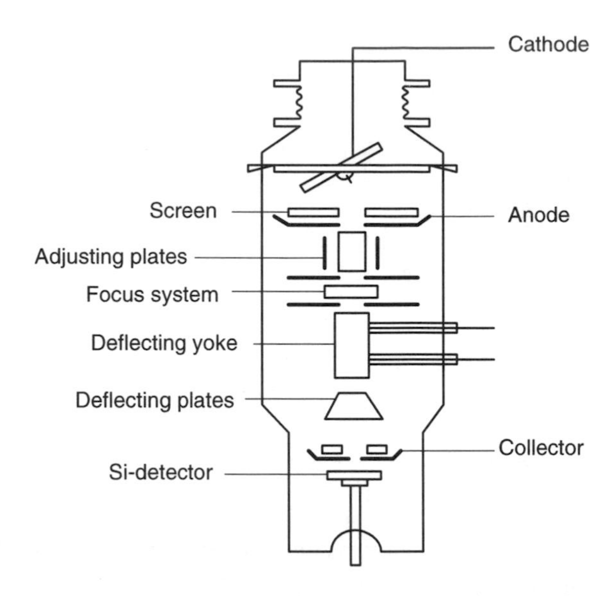

Figure 6.10. Schematic of the experimental apparatus for the investigation of FE statistics at high current densities.

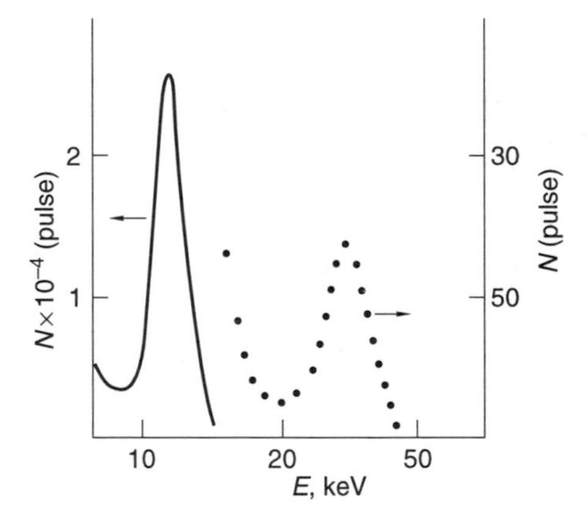

Figure 6.11. FE statistics spectrum for W at a current density of 10 A/cm^2.

6.7. RESUME

1. It is to be noted that a correct method to determine the statistical distribution of "elementary emission events" has been developed by now.
2. Studies of the rate of elementary events and statistics of the field electron emission in direct experiments confirm the adequacy of the model of free electrons for the field electron emission from metals with an accuracy of not worse than 0.1 percent.

In some materials, in particular, superconducting Y ceramics, multiple-particle tunneling has been observed. Explanation of this effect has not yet been given.

7

THE USE OF FIELD EMISSION CATHODES IN ELECTRON OPTICAL SYSTEMS: EMISSION LOCALIZATION TO SMALL SOLID ANGLES

7.1. INTRODUCTION

In this chapter several approaches to the fabrication of high-brightness field emission (FE) electron sources are discussed. Many electron optical systems require that the electron beam be formed into a very small spot, of 10 nm or less. The current density per unit solid angle, or brightness β, of the electron source, is limited by the Langmuir equation to a value $B_{\max}$.

$$B_{\max} = jeV/(\pi kT), \tag{7.1}$$

where j is the current density at the cathode, T is the cathode temperature, k is the Boltzmann's constant, and e is the electron charge. To exceed this Langmuir limit, the size of the cathode must be reduced, and the current density increased. Field electron emission meets these requirements. Unique features of FE electron sources are:

1. very high-energy efficiency;
2. available steady-state FE current densities approaching 10^5–10^6 A/cm^2;
3. a very small ($\sim$0.4 eV) electron energy spread;
4. a high beam coherence due to the narrow energy spread and the use of small (atomic dimension) emission areas.

Conventional tungsten (W) FE emitters have been utilized in a new generation of high-resolution electron probe devices, such as: scanning electron microscopes having resolutions approaching several Angstroms; transmission electron microscopes with resolutions approaching one Angstrom over a wide range of electron beam energies; electron lithography machines; Auger spectrometers; X-ray micro-analyzers; etc.

FE can be further exploited in such devices if the emission process can be localized to a small area on the cathode surface. Several procedures leading to localization of the emission process are reviewed in this chapter:

1. Thermally activated, field-assisted, surface self-diffusion, or "thermal-field build-up" of the emitter surface[71–74, 381, 382] can result in localization of the electron emission by the formation of sharp ridges at the edges of crystal facets in

the vicinity of the emitter tip's apex. These ridges serve as regions of local field enhancement, and thus preferentially emit electrons. Thermal-field build-up using pulsed fields has recently been shown to produce ridges of atomic sharpness.[247]

2. Adsorption on the emitter tip surface can result in a decrease in the surface work function of certain crystal planes, thereby localizing the FE to those planes.[55, 382] It has been shown that emission regions having smaller dimensions can be obtained if high electric fields are applied while heating the adsorbed layer.[253] The example of ZrO on W will be discussed.
3. Formation of very sharp points having a single atom at the apex of a tip effectively localizes the emission to that atom by field enhancement. This method of FE localization offers many opportunities for electron holography.[383, 384] The possibility of developing high-resolution electron transmission microscopy and electron holography without utilizing focusing electron beam lenses is also discussed.

7.2. THE OPTIMUM CRYSTALLOGRAPHIC ORIENTATION OF THE FIELD EMISSION CATHODE

A conventional FE cathode whose surface has been cleaned, smoothed, and rounded by high temperature thermal flashing serves as a virtual point electron source having dimensions of the order of 10^{-5} cm. The cathode's surface is comprised of different crystallographic planes having different work functions and local radii of curvature. By the Fowler–Nordheim (FN) equation, one sees that these distinctions in geometry and work functions result in a variation in the emitted electron current density over apex of the cathode tip. If the cathode has been formed from extruded polycrystalline wire of a material belonging to the body-centered-cubic (bcc) class, the apex of the tip is typically a $\langle 110 \rangle$ oriented crystal grain. Unfortunately, the $\langle 110 \rangle$ direction has a relatively high work function in addition to being a relatively large and flat crystal plane on the tip's surface, resulting in the emission of a small electron current density along the tip axis (see Fig. 7.1b). It is preferred in electron optical applications that the most intense electron emission be directed along the optical axis of the device. To increase the current density emitted on axis using bcc materials such as W, the emitter tip cathode is typically fabricated from zone-refined, single crystal wire. With tungsten, orientations of choice have been $\langle 111 \rangle$, $\langle 100 \rangle$, and $\langle 310 \rangle$ (see Fig. 7.1a,c).

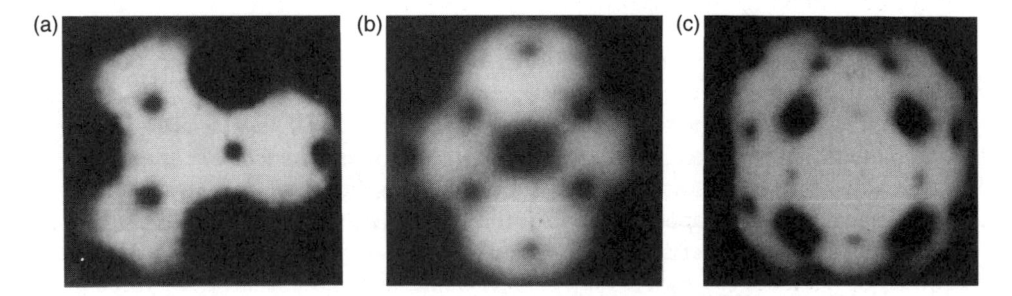

Figure 7.1. Distribution of field electron emission over the cathode surface for different crystallographic orientations of a W tip: (a) $\langle 111 \rangle$; (b) $\langle 011 \rangle$; (c) $\langle 001 \rangle$.

Table 7.1. Comparison of FE electron sources with conventional ones

Parameters of electron source	FE (cold cathode)	LaB_6 (thermocathode)	Tungsten filament
Brightness, $A \cdot cm^{-2} \cdot sr^{-1}$	10^9–10^{10}	10^7	10^6
Dimension, Å	$<10^2$	10^5	$>10^5$
Energy spread, eV	0.22	2.0	2.0
Service life, hr	>4000	1000	40
Vacuum conditions, Torr	10^{-9}–10^{-10}	10^{-7}	10^{-5}

Devices utilizing such FE cathodes are already designed, and outstanding results have been obtained, for example, in electron microscopes developed by Hitachi and Amray. The advantages of these electron sources, over conventional thermionic cathodes, have been demonstrated quite unequivocally with several parameters: the electron source brightness and coherency, the electron beam energy distribution, noise characteristics, and service life. (Table 7.1).

7.3. FIELD EMISSION LOCALIZATION BY THERMAL-FIELD SURFACE SELF-DIFFUSION

7.3.1. Localization by Build-Up Processes

Emission into angles approaching 100° at a current density of $\sim 10^5$ A/cm^2 is typical for thermally annealed FE cathodes. Such large beam divergences are undesirable in an electron source since they are associated with low source brightnesses. In addition, a number of technological difficulties arise, for example, since the majority of the beam current is intercepted by apertures electron stimulated desorption impedes the maintenance of high-vacuum conditions in the electron gun. These problems are somewhat alleviated by selection of the proper emitter tip orientation; however, much more significant improvements in the source brightness can be achieved through emission localization techniques, such as those involving build-up phenomena.

It has been shown[385–390] that micro-ridges and crests can be formed in the apex region of a W emitter by heating the tip to temperatures of 1800–2200 K in the presence of an applied electric field F_2 at the apex of the emitter, where

$$F_2 > (16\pi\gamma/r)^{1/2}, \tag{7.2}$$

where γ is the surface tension of the emitter tip material, and r is the radius of the tip endcap. The micro-ridges are formed in a few tens of seconds under these conditions. It was later shown[391] that these ridges also form over a time period of several tens of hours at similar temperatures if the applied field is of the order of $F_1 = (8\pi\gamma/r)^{1/2}$.

The effects of such thermal-field processes on the electron emission distribution of the emitter tip cathodes is shown in Fig. 7.2. It is clear that emission can be enhanced in the $\langle 100\rangle$, $\langle 111\rangle$, and $\langle 013\rangle$ directions. Note in this figure that the low index planes such as the (110) and (211) planes grow in size during this process, to keep the surface free energy a

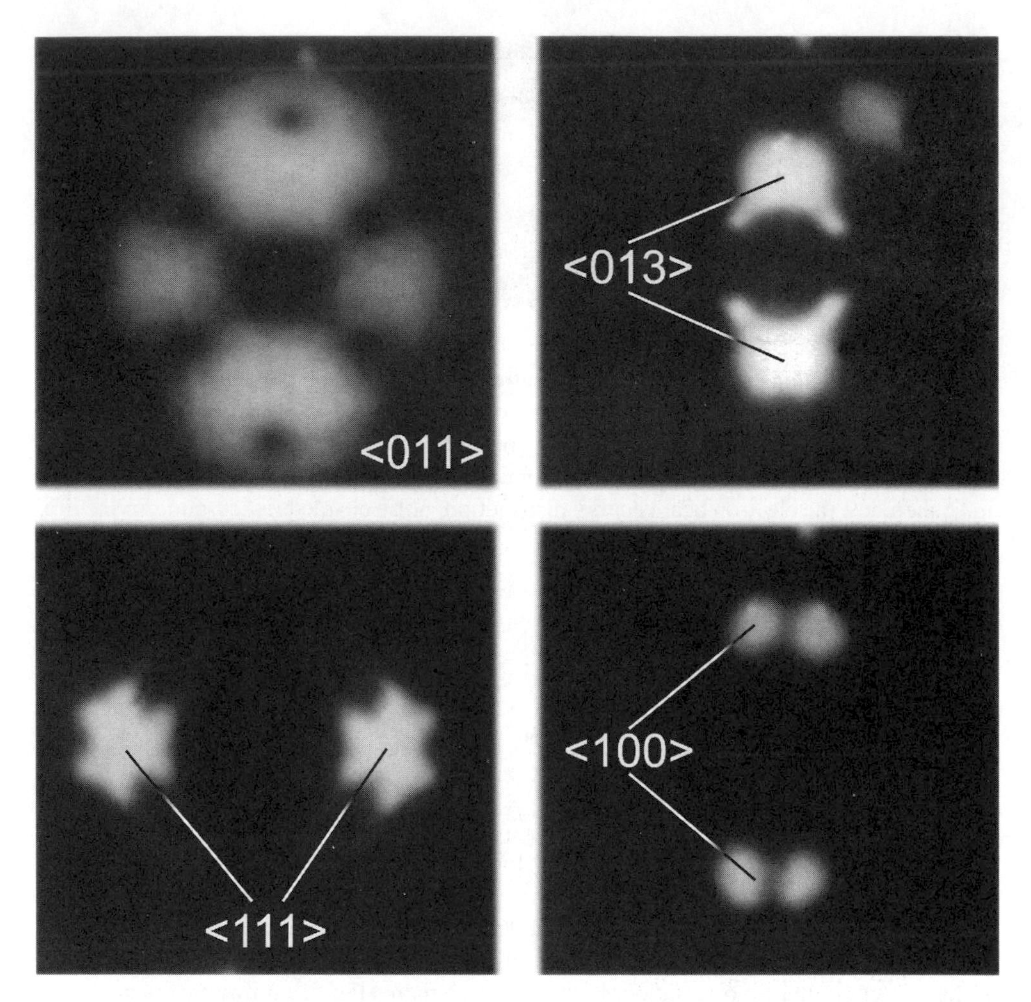

Figure 7.2. Localization of FE to several crystallographic planes by thermal-field build-up.

minimum. The edges formed by these planes are the most highly emitting, due to the large local field enhancement factor $\beta = F/U$, where F is the magnitude of the applied electric field and U the applied voltage. Changes in the electron emission spatial distribution that occur during such build-up process are shown for the case of $\langle 100 \rangle$ buildup in Fig. 7.3.

7.3.2. Localization by Micro-protrusion Growth

The FE current can be even more highly confined if a sharp micro-protrusion is formed at the apex of the emitter tip. The formation of such micro-protrusions in a high electric field was observed in early studies by Becker,[393] Sokolskaya,[385] Drechsler,[392] and others. To date, the mechanism of protrusion formation, as well as their shape and crystallographic structure, have been only partly established.[387–390, 397, 398]

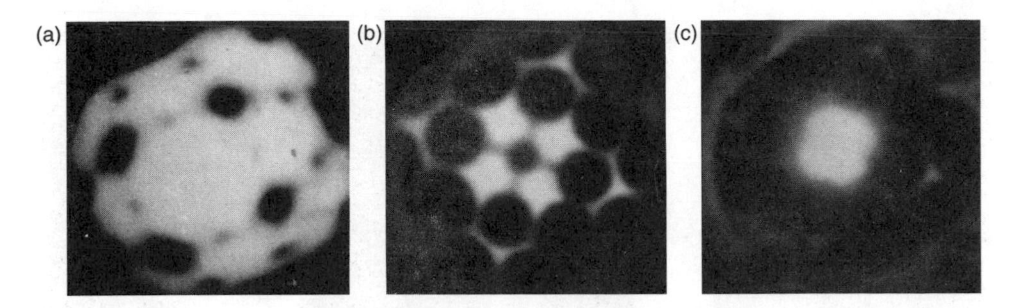

Figure 7.3. FE localization on a built-up W ⟨100⟩ tip: (a) the initial surface; (b) after some build-up; (c) after further build-up.

At present, three general kinds of such micro-protrusions have been identified: Type I: comparatively rough structures having dimensions comparable to those of basic grains (for a typical tip with a radius of 10^{-5} cm, the protrusion's dimensions are of the order of $3–5 \times 10^{-6}$ cm); Type II: fine micro-protrusions having nanometer dimensions (10^{-6}–10^{-7} cm); Type III: atomically sharp ridges and crests. The formation of atomically sharp structures (Type III) was recently observed when very high electric field pulses were applied to (and the resulting large emitted current densities were extracted from) emitter tips.[247]

Several experimental difficulties are involved in the investigation of micro-protrusion growth: (1) All three type of protrusions form simultaneously making it difficult to determine the type of protrusion which corresponds to a particular emission site by conventional FE microscopy; (2) High spatial resolution is required for studying the second and third types due to their proximity to one another. The combination of two microscopy based approaches must be applied if a better understanding of these growth phenomena is to be achieved: (a) FE and the high-resolution transmission/scanning electron microscopy[387, 395, 399]; (b) FE and field ion microscopy.[389, 398]

7.3.3. Large Micro-protrusions

It has been shown that large micro-protrusions are characterized by a slightly truncated, pyramidal shape. Some typical shapes of this type of outgrowth on a W emitter are shown in Fig. 7.4. Figure 7.4c shows transmission electron microscope images of the emitter shape,[388] and Fig. 7.4a are schematics of these images.[390] Figure 7.4b shows the corresponding field electron emission images.[388, 389] The FE micrographs were taken after brief heating cycles to slightly smooth the surface. The fact that the outgrowths are a single crystalline and oriented normal to the corresponding close-packed crystal facets was confirmed by subsequent field ion imaging and field evaporation of the protuberances (see Fig. 7.5).[389, 400] It has been proposed that these outgrowths are the result of "auto-epitaxial" growth of the close-packed facets, where the source material is comprised of atoms migrating from the sides, i.e. shank, of the emitter tip.[388–390, 400, 401]

The process of micro-protrusion formation is typically performed with the tip held at a positive potential relative to its counter-electrode (negative field). In this case micro-protrusion growth occurs simultaneously with field evaporation of the emitter tip, resulting in an overall shortening of the emitter tip.[388] This fact was misinterpreted is several

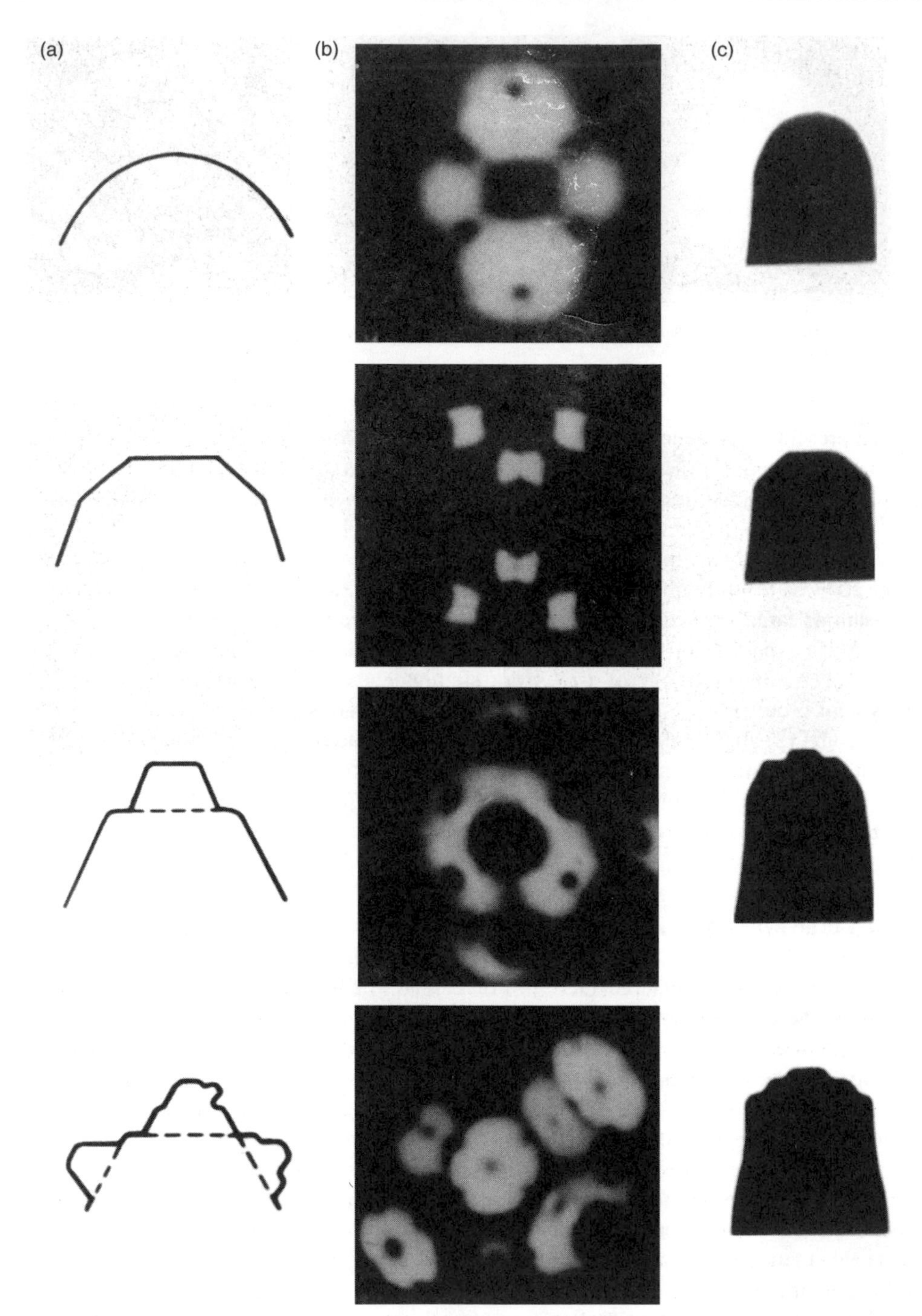

Figure 7.4. Typical shapes of micro-protrusions on a W tip: (a) schematic representation; (b) FE micrographs; (c) transmission electron microscope shadow graphs.

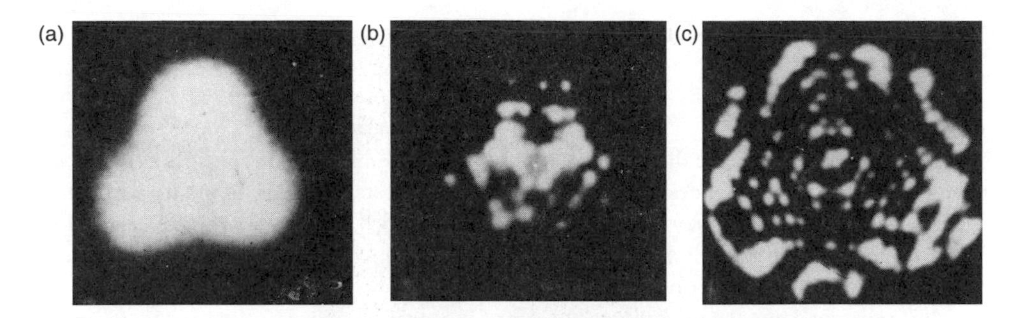

Figure 7.5. Micro-protrusions observed in the field ion microscope.[389]

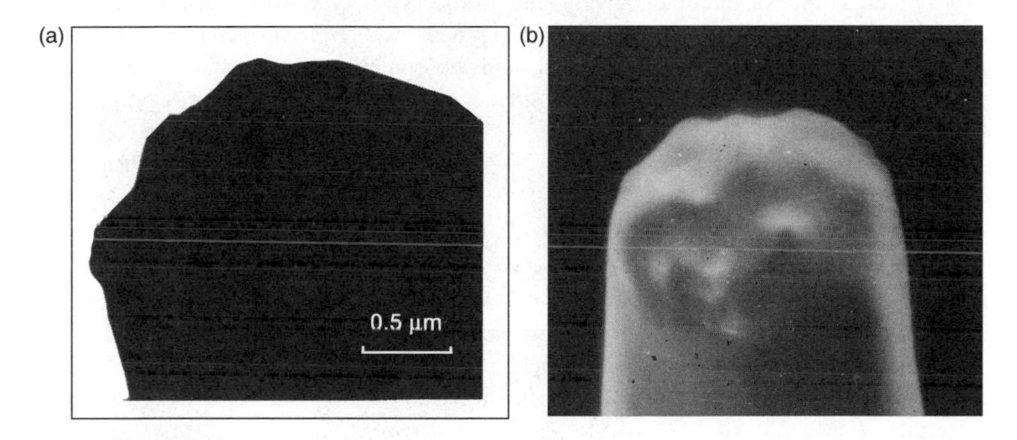

Figure 7.6. Field outgrowths at a W surface: (a) positive (negative potential on tip) field growth, observed in a transmission electron microscope; (b) negative field growth, observed in a scanning electron microscope.

instances and used as an argument against the formation of micro-protrusions by a growth process.[391, 402] These arguments were countered by studies of micro-protrusion formation in a reversed field, where the emitter tip is held at a negative potential relative to its counter-electrode (positive field), thereby eliminating the effects of field evaporation.[394–396]

Direct evidence of micro-protrusion outgrowth in high electric fields was obtained by conventional electron microscopy.[394, 395] A transmission electron micrograph of a built-up tip, at a resolution of $\sim$7 nm, is shown in Fig. 7.6a. The pattern of the surface formed by build-up in a negative field (Fig. 7.6b) is similar to that obtained with a positive field, although the surface structure formed during build-up in a negative field shows effects of partial field evaporation of the pyramidal protrusion. The tip profile obtained during build-up in a positive field is similar to the profiles observed earlier with build-up in a negative field only at lower imaging resolution.[388, 391]

The similarity in the shape of the tips as the result of build-up in both positive and negative fields is consistent with the formation of micro-protrusions by a growth process. The suggestion that micro-protrusions are formed due to the removal of surrounding emitter tip material by field or thermal evaporation[391, 402] is not consistent with the experimental results.[395]

7.3.4. Fine Micro-protrusions

Localized electron emission due to build-up processes was observed in early work with the FE microscope.[385, 386, 403] It was immediately realized that this localized emission was a result of the presence of micro-protrusions on the emitter tip surface. Initially it was thought that the protuberances were merely random clusters of atoms on the tip surface.[385] Later it was realized that they were actually crystalline.[398, 404] Typically these protrusions had a ridge structure with dimensions on the order of 50 Å with the radius of curvature of the ridge's edge of the order of a few angstroms.

It is presumed that the growth mechanism of these small micro-protuberances is similar to the larger outgrowth discussed previously. These protuberances typically take the shape of short, pointed hillocks, or comparatively long needles. Both types form most readily: (a) in the neighborhood of the (110), (111), and (211) facets where the surface is comparatively rough, where atom migration is hindered, and the conditions for crystal nucleation are favorable,[388, 400] and (b) in the regions of greatest field strength, that is, in regions where facets meet (see Fig. 7.4).[397]

Comparatively high temperatures ($T = 1700$–2000 K)[388, 397] and high electric fields, F_3,[388] where

$$F_3 = 2(16\pi\gamma/r)^{1/2} \tag{7.3}$$

are the most favorable for the formation of these micro-protrusions. A comparison between experimental results[399, 401] and Eq. (7.3) is shown in Fig. 7.7. Additional studies of the conditions leading to the formation of these protuberances were made in work.[404] The process of protuberance formation has been shown to be the same for a wide variety of materials (W, Mo, Ta, Nb, Ir, Re).[390]

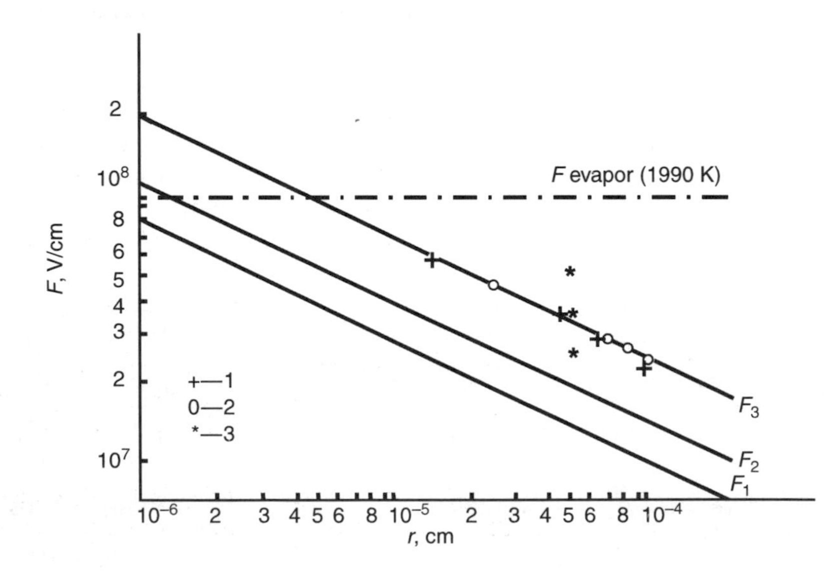

Figure 7.7. Dependence of F_1, F_2, and F_3 on the tip radius r and experimental values of the fields: (1) FE measurements;[388, 399] (2) electron microscopy investigations;[399] (3) Drechsler's results.[401]

The presence of small protuberances allows for the confinement of FE to very small regions on the emitter tip surface. Several desired characteristics of an emission localization process allow for: (1) the ability to periodically heat the emitter tip for cleaning; (2) the ability to reproducibly localize the emission to a given region of the tip surface; (3) high current stability during operation; and (4) the generation of one emission site (for beam applications).

It appears that the work of Crewe and coworkers[71–74] is among the first successful attempts to use built-up emitter tips in a technological application. Such tips, following build-up in the ⟨310⟩ direction on W, were used in a scanning electron microscope to obtain a record resolution at that time (~5 Å at 25 kV). Single uranium and thorium atoms, suspended in a specially fabricated polymer compound, were resolved. The maximum FE current was 500 μA, with the current stability being better than 4 percent over time periods of ~90 min.[71,73]

Swanson and Grouser[381] localized the emission by ⟨100⟩ build-up on W, and thereby reduced the beam divergence angle to 18°. Pautov and Sokolskaya[405] obtained a beam divergence angle of 12° on tantalum (Ta) in the ⟨111⟩ direction and noted that similar results would be expected with niobium (Nb). In this case of ⟨111⟩ build-up, a ⟨111⟩ oriented tip was used, resulting in confinement of the emission on the tip axis. Single micro-ridges emerged as a result of the enlargement of the {110} and {112} planes.

Further studies of build-up on ⟨111⟩ oriented tips have been conducted.[398] By varying both the field and temperature, it was possible to form a single micro-ridge and remove the satellites (see Fig. 7.8). The authors also proposed a method of fine micro-ridge formation in the ⟨011⟩ direction by the adsorption of silicon (Si) atoms on a W surface. The possibility of the formation of a ridge at the (011) face is increased significantly in this case and the ridge has a crystalline structure; FE and field ion micrographs are shown in Fig. 7.9. In these micrographs the crystallinity of the structure is evident with the lattice constant appearing to be close to that of a pure W lattice. Experiments with an atom probe have shown that the micro-ridge is composed of a mixture of Si and W atoms.[398]

Finally, it should be noted that the micro-protrusion formation process is very sensitive to the presence of adsorbates. The main drawbacks in the use of micro-protrusions for emission localization in the build-up process are: (a) difficulties in the controlled generation of micro-protrusions of a pre-set shape in a fixed place at the cathode surface; (b) poor beam current stability over extended periods of cathode operation, and (c) the difficulty in restoring the emitter tip emission characteristics following degradation.

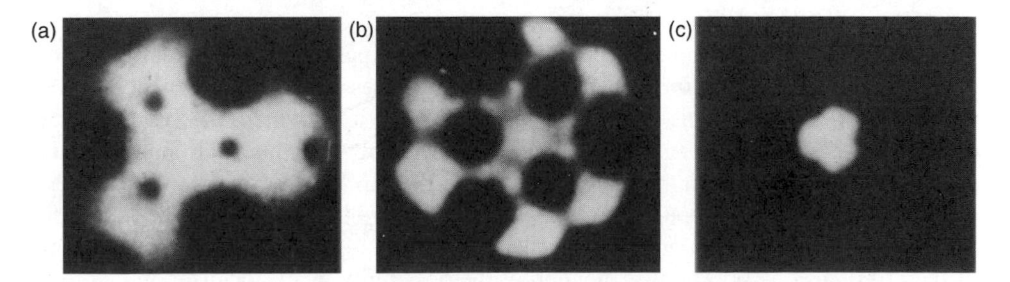

Figure 7.8. Field electron images of a W ⟨111⟩ tip with ⟨111⟩ during thermal-field treatment. (a) the initial W tip; (b) the built-up W tip; (c) localization of the FE by the formation of a single emission site.

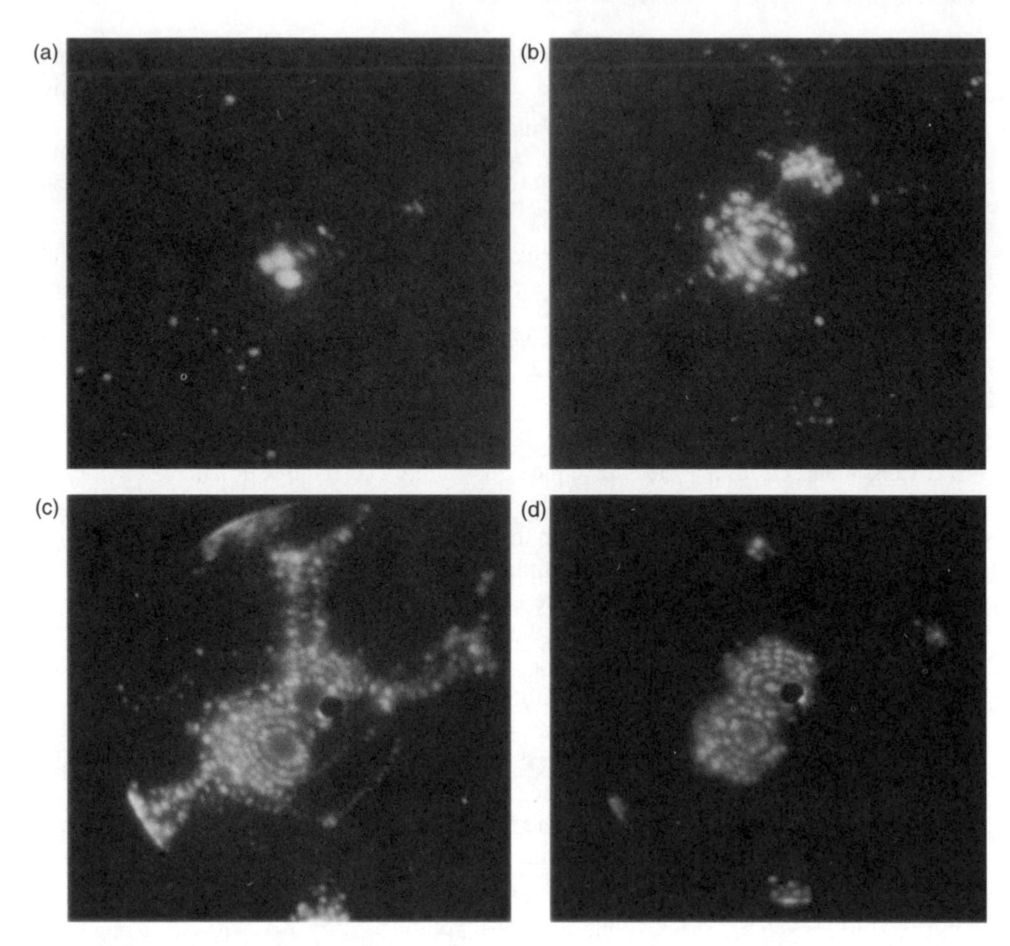

Figure 7.9. FE (a) and field ion micrographs (b, c, d) of micro-ridges formed at W (011) crystal planes.[404]

7.4. FIELD EMISSION LOCALIZATION BY A LOCAL WORK FUNCTION DECREASE

7.4.1. Basic Principles of Localization by Decreasing the Work Function

The selective adsorption of electropositive atoms on particular crystal planes, thereby resulting in the reduction of the work function of these planes, is one of the most elegant ways to localize FE on the cathode surface. If the crystal plane upon which selective adsorption occurs is at the apex of the emitter tip cathode a reasonably paraxial electron beam can be formed from a conventional thermally annealed emitter tip.

The adsorption of zirconium (Zr) on W is the most illustrative example of selective work function reduction by adsorption. The absorbed Zr forms small islands in a vicinity of the (100) facet of W. The first mention of the possibility of FE localization due to the deposition of Zr atoms on W appeared in a review by Dyke.[39] The emission characteristics from the adsorbed Zr islands have been investigated under various vacuum conditions and at maximum current densities.[55, 253, 406–408] Our research group suggested the use of W tips oriented in the ⟨100⟩ direction.[253, 406, 407] Comprehensive investigations of the Zr/W system were also carried out by Swanson and Martin and Tuggle et al.[409, 410]

A number of other systems, similar to the Zr/W system have been investigated. Hf on W and Mo was investigated by Shrednik[411] who demonstrated that emission localization similar to that observed with Zr on W occurs. A series of adsorbates (Al, Cr, Ge, Mb, Ti, Si, Zr, Hf) on W and Mo have also been suggested.[412] However, the most thoroughly studied system of Zr on W offers the greatest potential. It is therefore considered in greater detail below.

7.4.2. Zr on W

The first thorough investigation of the topology of Zr layers adsorbed on W was reported by Shrednik.[413] Through the use of field ion microscopy, he was able to show that following the deposition of monolayer coverages of Zr, the bright emission regions in the vicinity of W (100) planes were due to a local decrease in the work function and not due to local field enhancement. The minimum work function measured for the W/Zr system[413] was 2.62 eV. It was also shown[413] that with thicker Zr coverages, the contrast in the emission micrographs is not only due to the lower work function, but also due to field enhancement by roughness at an atomic level. It was suggested that the formation of close-packed Zr layers may be due to the small lattice mismatch between Zr and W. These close packed layers were also reported by other researchers.[55, 253]

Actually, two types of structures can be assigned to the Zr overlayers: (1) close-packed layers yielding fairly homogeneous emission throughout the adsorption region (Fig. 7.10a); and (2) more loosely packed layers that form a series of concentric rings with a dark region in the center (Fig. 7.10b). This ring structure (Fig. 7.10c) is clearly visible in the field ion microscope (see Fig. 7.10d). With the close-packed layers rapid surface migration is visible at 1400 1500 K. The concentric ring shape of the Zr layers appears to be more stable, possibly due to the fact that in this configuration, the Zr atoms reside on the terraces surrounding the {100} facet, which may hinder surface migration. The presence of the low emission region located in the center of the emission distribution is not convenient for electron optical device applications.

Illustration of emission localization to a small solid angle using ZrO on W ⟨100⟩ is presented in Fig. 7.11.

7.4.3. The Influence of the Vacuum Conditions

Several important considerations involved in the use of adsorbates for selective work function reduction and localization of the emission are: (1) The lifetime of the adsorbate's properties that lead to emission localization, (2) Reactivation of the localization mechanism, and (3) Replacement of the adsorbate. A number of experiments have been conducted in which these issues have been investigated.[55] At vacuums of 10^{-8} Torr, it was found that the emission current was stable for ~1 hr with the Zr overcoating, whereas for clean W under the same conditions, decay in the emission current occurred in 10–15 min.

Cathode stability, as well as the reproducibility of the current–voltage characteristics, over time periods of 136 hr were also investigated. Curve I, in Fig. 7.12 is data accumulated

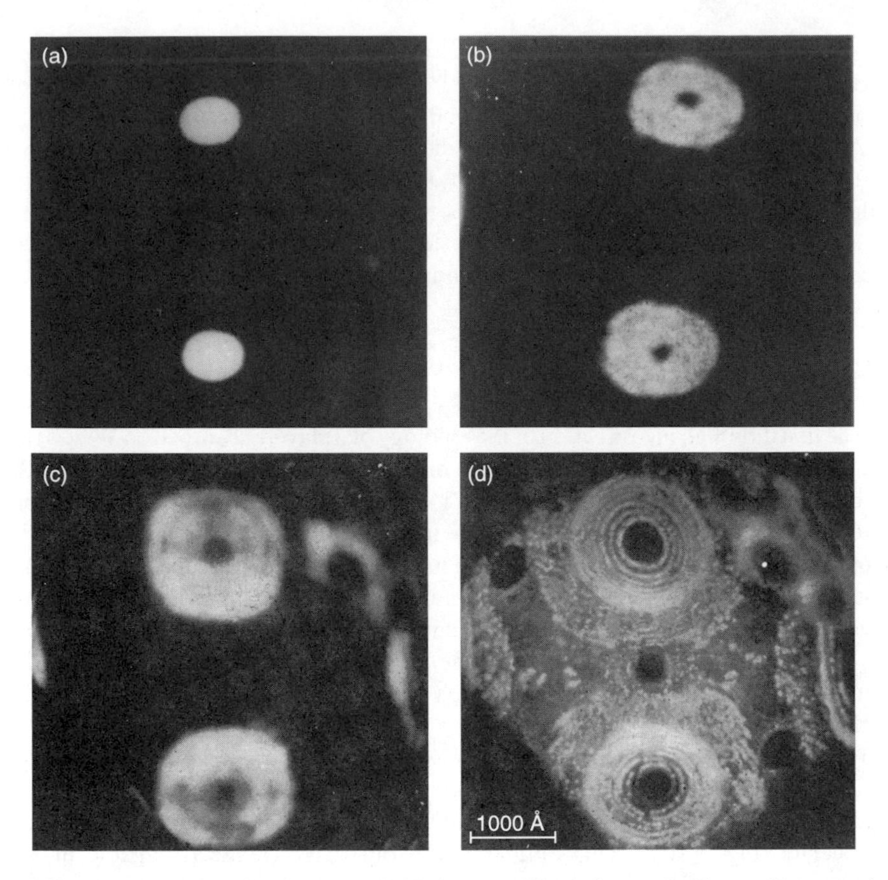

Figure 7.10. Field electron and ion micrographs of Zr layers on W: (a) close-packed layers; (b) concentric ring layers;[55] (c) concentric ring layers;[413] (d) hydrogen field ion pattern.[413]

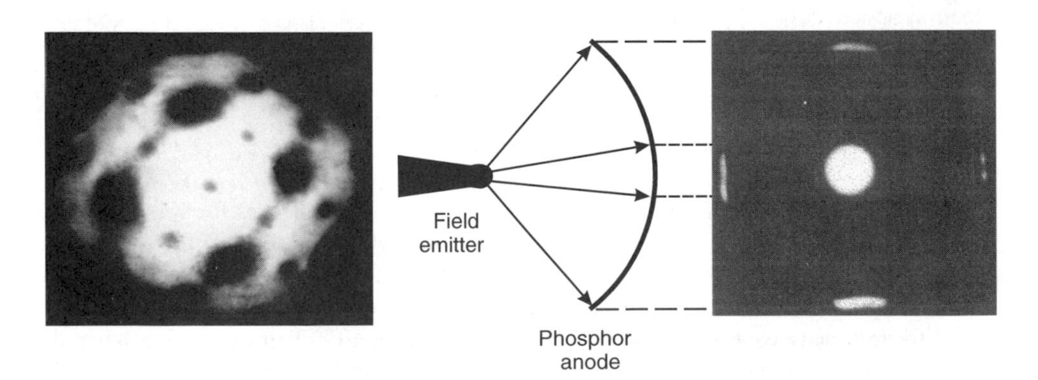

Figure 7.11. Emission localization to a small solid angle using ZrO on W ⟨100⟩.

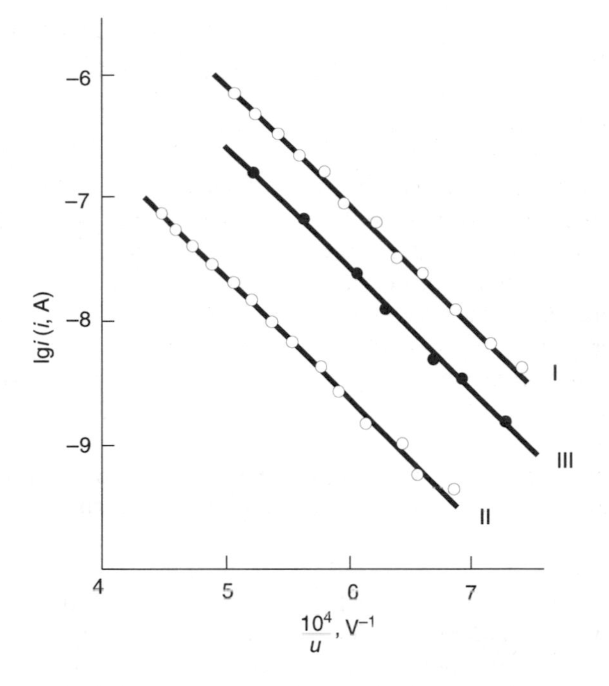

Figure 7.12. Restoration of emission characteristics after operating Zr coated W in a vacuum of 10^{-8} Torr;[55] (I) immediately after Zr deposition; (II) after 136 hr of operation ($P = 10^{-8}$ Torr); (III) after annealing at $T = 1300$ K.

immediately following deposition of the Zr overlayer, and Curve II subsequent to 136 hr of cathode operation. It is apparent that the emission characteristics have changed significantly during this time period, presumably due to adsorption processes. Note however, that the original current–voltage characteristic can nearly be restored following heating of the cathode to 1200–1300 K for 30 sec. as shown in Curve III. This process can be repeatably reproduced over several degradation/reactivation cycles. Similarly the emission characteristics following deposition of additional Zr are repeatable if heating times and temperatures are accurately reproduced.

Several experiments have been conducted in order to examine the possibility of employing the Zr/W system under less stringent vacuum conditions: (a) the effects of admitting laboratory air; (b) restoration of current–voltage characteristics in vacuums of 5×10^{-6}–10^{-5} Torr, and (c) the effects of venting to laboratory air to 10^{-3} Torr followed by reevacuation to 5×10^{-6} Torr.

The results of these experiments lead to the following conclusions.

(1) The emission characteristics can only be restored by the redeposition of Zr following venting to atmospheric pressure of laboratory air.
(2) Gradual venting to laboratory air up to pressures of 5×10^{-6}–10^{-5} Torr results in degradation of the emission confinement almost to the point of the predeposition characteristics. The emission localization is partly restored following heat treatment to 1200–1300 K, presumably by contaminant desorption.

(3) An abrupt pressure increase to 10^{-3} Torr substantially reduces the emission, and localization totally disappears. Localization cannot be restored by heating.

7.4.4. The Dispenser Cathode Approach

A major practical problem associated with the localization of FE by the selective adsorption of work function reducing materials is the sensitivity of the resulting surface to cathode surface heating, ion bombardment, adsorption of residual gases, etc. As noted above, the localizing effect of a monolayer, or nearly a monolayer, coverage is destroyed in poor vacuum conditions, including of course exposure to the atmosphere. These characteristics show that regeneration of the cathode is necessary at periodic intervals, a problematic issue to their use in practical electron optical devices. If restoration of the cathode characteristics is possible by straight forward techniques, one can consider this type of source for use in commercial devices.

It has been observed[412, 415, 416] that Zr may be diffused into the cathode by heating to form a solid solution. In this case, degraded emission characteristics can be restored simply by heating the cathode.

Special placement of an external source of Zr near the apex of the tip (see Fig. 7.13) facilitates the impregnation of the tip's bulk and the transport of Zr atoms toward the emitter tip by surface migration. Experiments with an electron source of the type shown in Fig. 7.13 were performed.[408, 416, 417] The FE cathodes could be stored in a closed container under normal pressure for several months without degradation. Following subsequent installation into the vacuum system, and heat treatment at 1300–1500 K, the original localized emission current–voltage characteristics, the shape of the emission site, and the current density distribution within the emission site were restored. Such results make the Zr/W system promising for use in commercial electron devices.

7.4.5. Stabilization by Oxygen Treatment

Swanson[381, 382, 418] was the first to notice that localization is facilitated if Zr is deposited on a W surface in the presence of oxygen. In fact, the lowest work functions, highest current stability, and highest thermal stability (to $\sim$1800 K) are obtained by the adsorption of ZrO molecules.[412, 415, 419] The thermal stability is very important if the emitters are to be operated in the Schottky emission regime.[382, 408–410, 420]

Several other characteristics of the emission properties of the coadsorbed Zr/O system are beneficial: (1) The electron current density throughout the emission region is very uniform, that is, it is possible to obtain an emission site without a dark area in the center;[415] (2) The ZrO/W layers decrease the level of low-frequency noises in comparison to that observed with Zr/W layers, a result that is particularly important for the thermal-field emission regime; and (3) The stability of the average current is higher for longer time periods. The decrease in low frequency noise is apparently the result of the formation of a strongly bonded chemisorbed compound that decreases the mobility of the layer.

It is possible to confine the adsorption of Zr (ZrO) to one small $\{100\}$ plane on the emitter tip if the emitter tip build-up process is performed in a background pressure of oxygen prior to deposition. The terraces surrounding the $\{100\}$ plane are decreased in size and the subsequent adsorption of Zr (ZrO) does not result in a decreased work function in these

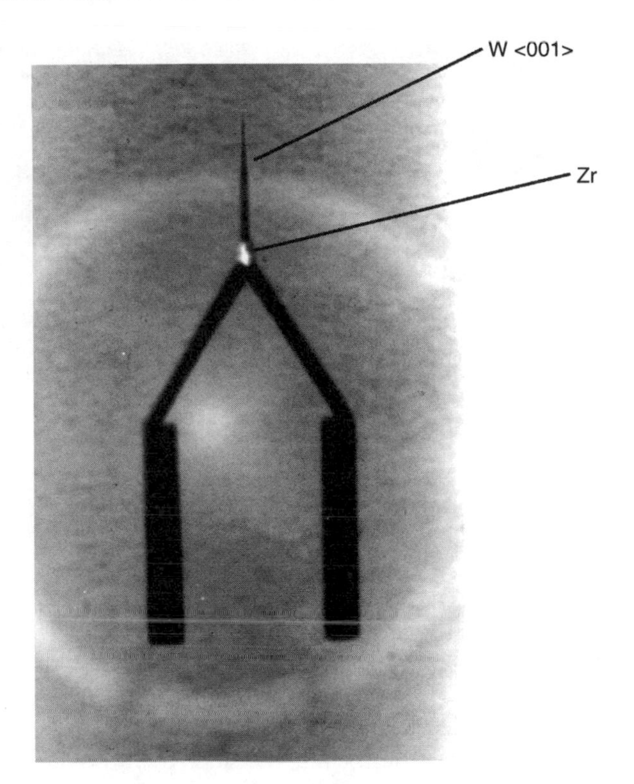

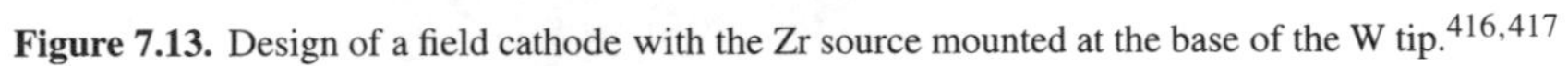

Figure 7.13. Design of a field cathode with the Zr source mounted at the base of the W tip.[416,417]

regions, presumably as a result of an average change in distance between surface atoms.[415] The procedure for the formation of ZrO/W point cathodes has been described.[412, 415] It has also been determined that reproducible results can be obtained when oxygen impregnates the whole volume of a tip near its apex. This requires an oxygen exposure of tens of hours at a pressure of $\sim 10^{-8}$ Torr.

Investigations[408, 416, 417] confirm these data and stable layers were formed in one to two minutes with residual gas pressures of 10^{-2}–10^{-3} Torr.[417] The best operational parameters for the cathode are obtained with this treatment: a brightness of 10^9 A · str^{-1} · cm^{-2}, a current drift of 1 percent, and a noise level of 1 percent. The probe currents were $\leq 1\ \mu$A. The maximum probe current obtained in our experiments was 10 μA at the total current of several mA.[417] The current drift at these currents was 2–3 percent with a noise level of 3 percent at 1800 K.

7.4.6. Shaping of the Zr/W Emission Sites in an Applied Electric Field

It has been shown that the direction of Zr migration on W depends on the sign of electric field.[55, 253] With the emitter tip serving as the cathode Zr migrates from the apex toward the emitter tip shank. With the field direction reversed, the Zr is drawn toward the tip's apex. By using this reversed emitter tip polarity and heating the emitter tip it is possible

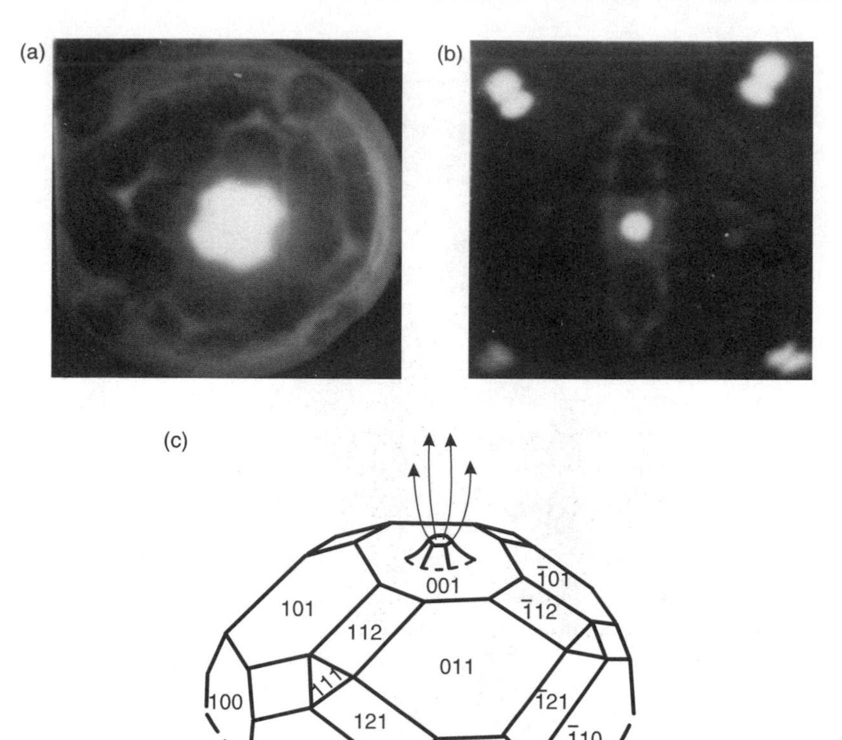

Figure 7.14. (a) Initial micrograph of a built-up W ⟨100⟩ tip; (b) FE localization at the surface of a built-up Zr/W ⟨100⟩ tip; (c) scheme of microtip formation at W ⟨100⟩.

to decrease the beam divergence from 15–20° to 5–10° (see Fig. 7.14). The field strength employed was chosen to be equal in magnitude to that required to draw an electron current density of $\sim 10^5$ A/cm^2 in the FE mode of operation. The temperature at which Zr migration is observed for the initial deposit is 1300–1500 K. The work function determined from the relative slopes of FN plots for these layers (Fig. 7.14b) was 2.9 eV.

Recent experiments[416, 421] confirm the possibility of obtaining such narrowly collimated FE beams. A thorough analysis of the FE images indicates that tip build-up is the process leading to such localization. A region of enhanced field strength emerges at the apex with the orientation of the tip in the ⟨001⟩ direction, and a micro-ridge appears during more advanced stages of build-up (see Fig. 7.14c). The area of the {100} facets decreases, thereby leading to further emission confinement (Fig. 7.14b). The reduced size of the terraces surrounding the (100) plane prevent the appearance of the emission zones of concentric rings.

Direct confirmation of the important role played by the build-up in this localization process is the emergence of a rebuilt W surface coming out from beneath the Zr layer (Fig. 7.14c). The size of the emission site is of the order of ten's of angstroms. Thus, the two emission confinement techniques can be combined for "double" localization: forming a build-up surface and generating an ultrasmall emission site at (100) plane.

7.5. FIELD EMISSION FROM ATOMICALLY SHARP PROTUBERANCES

Electron holography, one of most promising methods for investigating nanostructures and atomically sized objects, was proposed by Gabor in 1948.[422] The key to practical implementation of this idea is the development of an electron source having sufficient coherence and brightness. FE cathodes offer new opportunities for the development of electron holography due to their very high brightness (two orders of magnitude higher than thermionic cathodes (see Table 7.1)[423] and high coherence. Excellent interference patterns using an FE cathode were obtained by Zeitler and Ohtsuki[424] on a gold film having 10^{-4} cm diameter holes. Recently, FE microtips (Spindt-type Mo tips with a gate diameter aperture of $\sim$1 mm)[425] were used for the observation of interference fringes with low-energy (140 eV) electrons.

Fink[80,82,83] suggested and investigated ultrasharp point emitters having a single atom at the apex. The emitters were fabricated by field ion techniques, where in the course of controlled field evaporation of a $\langle 111 \rangle$ oriented W tip (Fig. 7.15a). The W trimer is formed on the (111) plane (see Fig. 7.15b). W atoms are deposited from the vapor phase until one fills the threefold hollow site formed by the trimer (Fig. 7.15c). The procedure results in an angular divergence in field ion mode of operation of less than 0.5°.

The formation of atomically sharp outgrowths was also demonstrated by Bihn and Garcia by a build-up process called the "field-surface melting" technique.[84, 426] This process involves the application of an electric field to a W $\langle 100 \rangle$ tip coated with gold. The tip's radius of curvature was 70–100 nm. The emitter tip is heated by drawing very high current densities in the pulsed mode of operation. If the surface is rough due to the presence of adatoms, vacancies, kinks, etc., the action of the electric field and the surface temperature leads to so-called "field-surface melting" and the formation of nanoprotrusions, that can terminate in one atom. By rapid cooling of the nanoprotrusion in the applied field, one obtains highly localized, and thereby coherent, electron emission. Measurements have shown that such a source, formed by a protrusion of 20–30 nm in height and terminated by one atom provides emission angles for ions and electrons of $\sim 2°$ and 4° respectively. The electron beam provides currents of $<1\ \mu$A and an ion beam currents of $\sim 10^{-14}$ A.

Atomically sharp emission sites can be also formed by this build-up process without overcoating the emitter with a low melting point material[247] as has been demonstrated by the formation of atomically sharp protuberances and ridges on W, Mo, Ta, Nb surfaces.

7.6. APPLICATIONS OF FIELD EMISSION CATHODES IN ELECTRON OPTICAL DEVICES

In conclusion we summarize some important parameters of FE cathodes obtained by different investigators. Maximum brightness measured using built-up tips are of the order of 10^{10} A $\cdot$ sr^{-1} $\cdot$ cm^{-2} and the smallest energy spreads are $\sim$0.2 eV using the cathodes at low temperatures. Substantial interest exists in using ZrO coated W $\langle 100 \rangle$ cathodes in

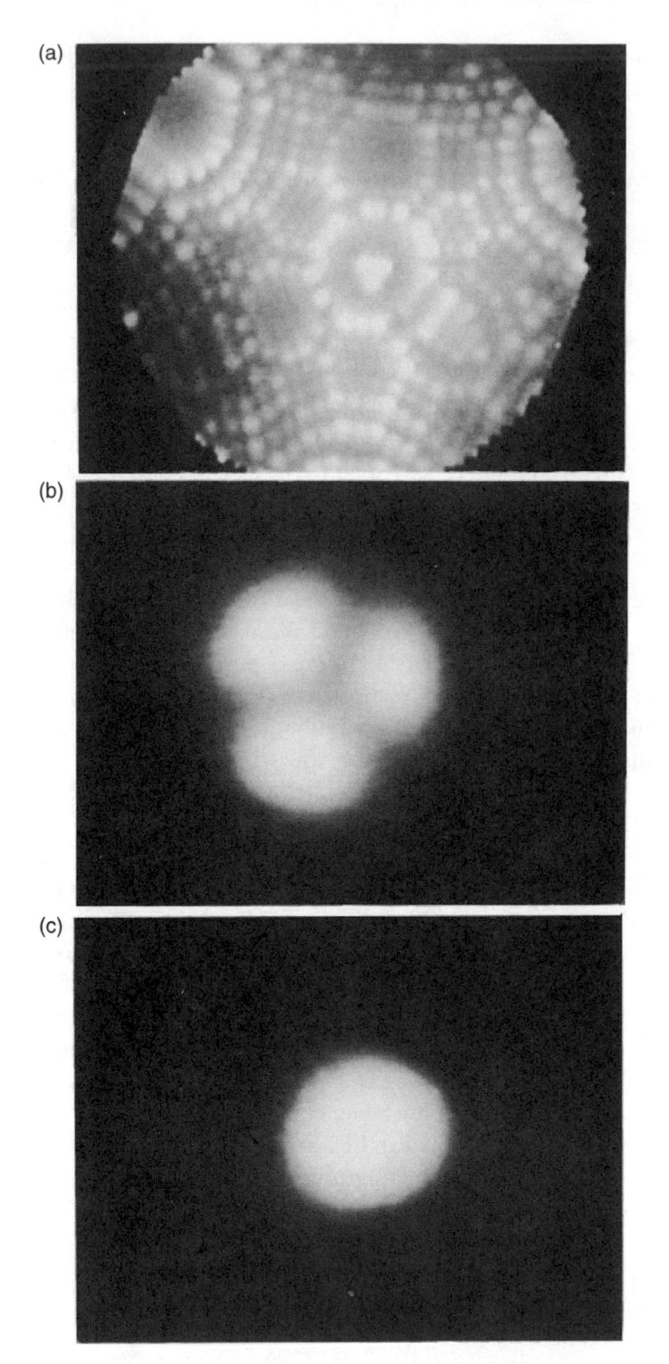

Figure 7.15. Preparation of a single-atom tip of W [111]-oriented single-crystal wires: (a) common field ion image showing the (111) apex plane with a trimer as well as vicinal planes; point source tips with a significantly high electric field confined to just (b) three atoms and (c) one atom.[221]

Table 7.2. Characteristics of scanning electron microscopes (SEM) and transmission electron microscopes (TEM)

Model of microscope	Resolution, Å	Magnification	Max brightness, $A \cdot cm^{-2} \cdot sr^{-1}$	Accelerating voltage, kV	Vacuum conditions, Torr	Service life, hr
1–3 SEM, W ⟨310⟩	5			25	10^{-9}	
SEM S-800 Hitachi	20	20×–200,000×	10^9	1–30	10^{-9}	1000
SEM S-900 Hitachi	7	100×–800,000×	10^9	1–30	10^{-9}	>2000
SEM S-5000 Hitachi	6 at 30 kV 3.5 at 1 kV	30×–800,000×		0.5–30	10^{-9}	>1000
TEM HF-2000 Hitachi	2.3 point 1 lattice 1.6 inf. lim.	200×–1,500,000×		200		>4000
TEM HF-3000 Hitachi	2 point 0.7 lattice 1.1 inf. lim.			300	10^{-10}	>4000
SEM 1860 FE Amray	20 at 10 kV 80 at 1 kV	25×–400,000×		0.5–10	10^{-10}	
SEM 1860 CFE Amray	20 at 5 kV 80 at 1 kV	25×–400,000×		0.5–5	10^{-12}	

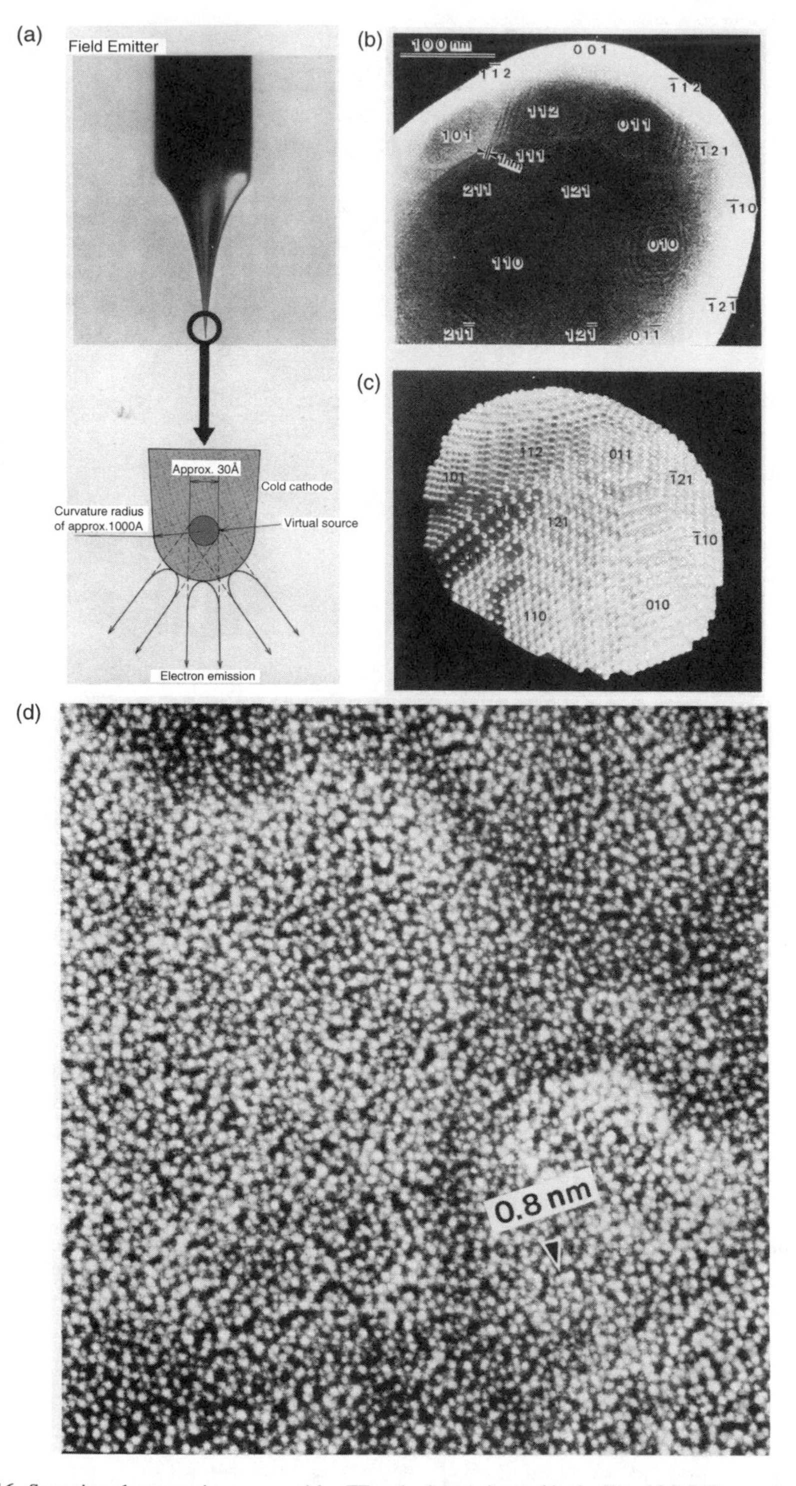

Figure 7.16. Scanning electron microscopy with a FE cathode: (a) tip used in the Hitachi S-900 scanning electron microscope; (b) scanning electron microscopy of the tip apex; (c) ball model of the tip apex; (d) high resolution image of Au–Pd particles in a carbon substrate.

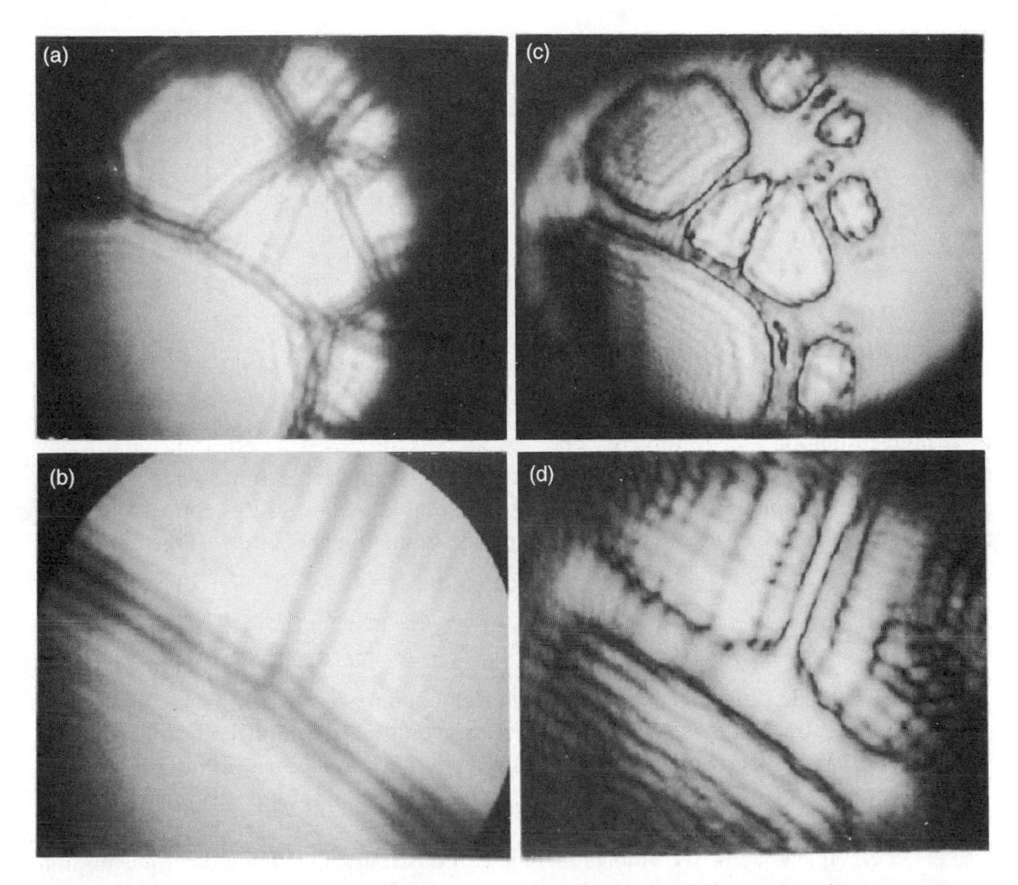

Figure 7.17 (a) Holograms taken with 80 eV electrons of carbon fibers with the electron point source 2400 nm away from the carbon sample; (b) closeup of the central region with the source just 550 nm away from the sample-emitting electrons at 65 eV (bottom); (c and d) results of the numerical reconstruction of the above holograms.[221]

the thermal-field emission regime. Maximum current densities for localized emission in this case were roughly 10^9 A/cm^2 with minimum electron beam devergencies of ~5°. The cathode service life in electron optical devices is several thousand hours. Maximum total current in the beam probe is 10 μA.

The first attempts at employing FE cathodes in electron microscopy were made by Crewe et al.[71–74] who have succeeded in showing the potential advantages in the use of such cathodes. They attained a 5 Å resolution with a scanning electron microscope at 25 kV. At present, the most significant applications of FE cathodes in electron microscopes have been made by Hitachi (see in 423) and Amray firms. The basic operational parameters of their microscopes are listed in Table 7.2. Figure 7.16 shows several important features and results of applying FE to scanning electron microscope cathodes.

There have been only a few attempts at applying electron holography using FE.[82, 221] Interferograms obtained by Fink et al.[221] on thin Au films are shown in Fig. 7.17. Electron holography was demonstrated using a well-collimated, high-energy electron beam produced

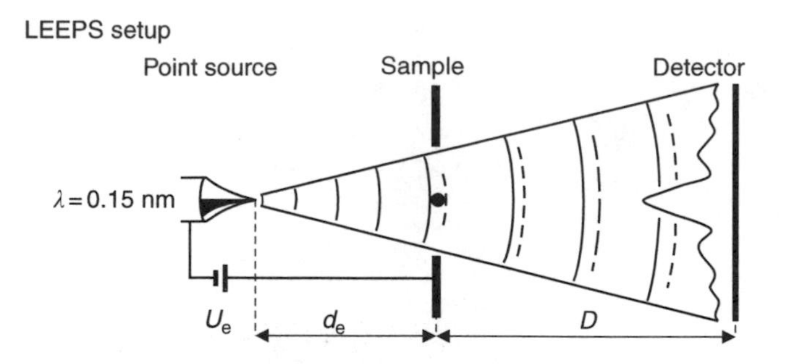

Figure 7.18. Setup for the holography experiments with low-energy electrons.[221]

by a high-resolution electron microscope.[383, 384] Fink proposed an electron microscopy method to investigate objects only several micrometers from the electron source.[82] A magnification of more than 10^5 is expected. One remarkable feature of this form of electron microscopy is that no lenses are required, thereby eliminating aberrations (Fig. 7.18).

7.7. RESUME

1. FE cathode is the most pointed, most bright, and most monochromatic of electron emitter. Usual dimensions of a field electron emitter lie within tenths of a micrometer. With special methods, one can reduce the linear dimension of an emitting spot down to 10^6–10^{-7} cm, or even make the emitter atomically sharp in some particular cases.
2. Tip FE cathodes are employed in the industrially manufactured transmission and scanning electron microscopes of superhigh resolution, in Auger-electron spectrometers, and in the electron holography.
3. A new approach to the development of electron optical systems is being developed based on the vacuum microelectronics techniques, which permits forming of the cathode and of electron optics elements in a single block.

8

ADVANCES IN APPLICATIONS

8.1. INTRODUCTION

Vacuum microelectronics (VME) is a new field in micro- and nanoelectronics that has been developed during the recent few decades. VME is based on the employment of electrons in vacuum at the dimensions of active elements in tenths and hundredths of a micrometer. Practically, a field electron emitter (FEE) is used as an active element in VME systems. By employing microscopic anode-to-cathode gaps, one can build devices with a field emission (FE) cathode controlled by a voltage in tens of volts.

VME comprises of devices and active system components of micrometer size commensurable with nanometer-scale dimensions. VME presents a sort of an alternative to the solid-state electronics, employing the ballistic motion of electrons in vacuum instead of the transport of current carriers (electrons and holes) in semiconductors.

Small dimensions of a FEE provide an opportunity of bringing the size of active elements down to a few tens of atomic dimensions (i.e., to 10^{-6}–10^{-7} cm). Some up-to-date experimental data indicate the existence of elective elements of a size on the atomic scale (1–3 nm) on such special materials as, for example, carbon nanotubes.

The small dimensions of FEEs also permit achieving of a density of active elements of up to 10^{8}–10^{10} cm^{-2}, and even of up to 10^{12} cm^{-2}, if one employs such "self-organizing" systems as fullerenes, nanotubes, etc.

The advantages of VME as compared to the solid-state electronics are as follows:

1. no dissipation of energy at electron transport in the medium, that is, in vacuum. That gives the VME devices advantages for certain classes of devices such as those generating high power at high frequencies;
2. the VME devices are acting with a high speed due to the virtual lack of inertia in the FE process and a very small electron transport time in the vacuum gap, which permits designing of fast-acting high-frequency devices;
3. a marked nonlinearity of the FE current–voltage characteristics, which permits designing of frequency converters and multipliers;
4. no energy expenses on the act of electron emission to quantum origin of the tunneling process;
5. a high radiation tolerance and heat resistance of the emitters.

Main VME applications

1. the most outstanding and widespread application is the production of flat low-voltage displays of high brightness and high resolution based on field emission arrays (FEAs);
2. employment of nano–FE cathodes in electron optic systems of superhigh resolution (in scanning and transmission electron microscopes, electron lithography systems, Auger electron spectrometers, and in tunnel microscopy);
3. production of active elements for integrated circuits (diodes, transistors, etc.);
4. employment in microwave devices;
5. devising of different types of pressure gauges, magnetic field gauges, etc.;
6. devising of a new type of ion source;
7. Realization of new very efficient production technologies for microelectronics (lithography, etc.).

8.2. SHORT HISTORICAL REVIEW AND MAIN DEVELOPMENT STAGES

A fairly comprehensive history of the subject has been presented in the last few years in reviews by Brodie and C. A. Spindt,[87] I. Brodie and P. R. Schwoebel,[428] H. H. Busta,[88] R. Baptist,[105] and H. F. Gray.[104, 429]

VME had began from the idea of a first vacuum transistor resistant to high levels of radiation and high ambient temperatures and having, as does a semiconductor transistor, a nonlinear current–voltage characteristic. In a vacuum microtransistor, in contrast to a solid-state one, transport of the charge carriers takes place in vacuum and by a ballistic method, in structures of micron and submicron dimensions.

The idea of a vacuum transistor had first been suggested by Shoulders in 1961[102] in his designs of vertical and horizontal microtriodes. Shoulders proposed a technology of active emitter elements 0.1 μm in size, applied FE microscopy for visualizing the emitter surface microstructure and was one of the first to formulate a concept of microminiaturization of vacuum electronics devices as a way to achieve such new features as high accuracy of parameter reproduction, stability, low noise level, and others.

The next important phase in the growth of VME was the development of a new class of cathodes, namely, the FEAs. An important achievement in the technology of FEAs was integration of the electron collector (anode) with the FE matrix and the "honeycomb" control electrode (gate) separated by a gap of a few microns, which made possible the addressing of individual cathode elements and thus design amplifiers, frequency multipliers, current converters, and elements of display panels. Because of the small separation between the cathode and the gate these FEAs can be operated at low voltages (hundreds of volts). The first report of a FEA with a multi-tip system with developed working area was made by Spindt (1968,[214]). The FEA proposed by Spindt (Spindt cathode) consisted of a large number of individual molybdenum emitter tips arranged with a density of more than 10^6 cm^2 at micron distances from one another. One more direction of the development of FEAs was the work on increasing the operating area of cathodes and density of the elements in order to reduce spatial fluctuations of the emission current.

Work on fabrication of FEAs of an area greater than 140 cm^2 was done by Spindt and Brodie with co-workers.[87,428]

Due to modern technological achievements in preparing thin films of submicron dimensions and advances in electron beam lithography, emitter packing densities of up to 10^7–10^8 cm^2 were reached and matrix currents averaged over cathode surface raised to over 1000 A/cm^2.[445] In 1989, Makhov[442] had demonstrated that FEAs of silicon pyramids coated with a thin (30 Å) film of dielectric (silicon dioxide) can be operated at voltages as low as 10 V due to lower Schottky barrier at the semiconductor/dielectric boundary and low electron affinity of the surface of the dielectric.

In 1970, Crost et al.[431] proposed a technology of flat display panels based on Spindt's FEAs.

VME received further impetus when FEAs were developed based on tip, blade, conical, and pyramidal structures of heavily doped silicon. An advantage of amorphous silicon as a material for FEAs is the highly developed technology of forming microstructures on the surface of semiconductors relying on the vast experience of mass production of silicon integrated circuits. The traditional methods of fabricating microcomponents of a silicon FE cathode are anisotropic selective chemical etching, ion beam lithography, reactive ion etching, and others. Success in obtaining stable and uniform emission from wide band semiconductors made them competitive with metals owing to their superior chemical surface stability and low operating voltages.

In 1986, Gray et al. built on the basis of silicon FEAs, the first efficient vacuum transistor or a miniature FE triode,[440] which in essence followed the concept of Shoulders. This vacuum microtransistor has a number of advantages over solid-state semiconductor transistors: transport of electrons is not affected by properties of the semiconductor making unnecessary the control of crystallinity and impurity content; due to its micrometer-scale dimensions, the packing density of integrated circuit components can be more than 10^7 per cm^2; and because of small cathode–anode separation (a few microns), the electron transit time is short making possible the generation of high-frequency currents. Fabrication of the vacuum microtransistor, stimulated the development of vacuum integrated circuit technology.

In the years to follow, field sources of electrons based on stable FEAs have been intensively studied as candidates for application in flat display panels of high-brightness, nanometer-scale electron beam lithography, electro-optical devices of high electronic brightness and small spot size, microwave amplifiers and generators for frequencies from 100 to 100 GHz, and high-speed digital switching devices of the subpicosecond range.

Work on microwave vacuum electronic devices with high-current FE cathodes was begun by Brodie and Spindt (1979,[92] 1976[91]; Lally et al. 1986[439]). Spindt initiated work on frequency generators for a range of hundreds of GHz based on FEAs of submicron dimensions with periodic structure (Spindt, 1992[448]).

First color display of high resolution (300 pixels per in.) with individually addressed FEAs has been demonstrated in 1987 (Holland et al.[441]). Work on efficient commercial low-voltage FE displays has been started at *PixTech, Motorola, Micron Display Technologies, FED Corporation, SIDT, Honeywell, Silicon Video, Raytheon, Futaba, and Samsung*.

FE from ultrasmall sites at the tip cathode surface was utilized in designing various surface diagnostics systems of superhigh resolution, in electron microscopy,[473, 474] and in atomic resolution electron holography.[82, 475] In these applications, advantage was taken of high brightness and spatial coherence of the electron beam emitted by a point FE cathode.

Discovery of the scanning tunneling microscopy[438] stimulated the development of large number of techniques for visualizing the surface topography with atomic resolution.

Microminiaturization of the electron optics made possible by the use of multielectrode systems of cylindrical microelectrodes as electrostatic lens coaxial with individual tips in FEAs paved the way for multiple-beam electron lithography of nanometer range.[476] This technique is remarkable for its high efficiency and output in producing matrices of silicon integrated circuits. Efficient fabrication of individual nanostructures is achieved with a probing "resist-free" lithography in a scanning electron microscope, where at a potential difference between the probe and the substrate of several volts a certain composition of the ambient modification of the substrate surface by the probe is produced as a result of localized chemical reaction (e.g., of oxidation).[477]

In the works of Baptist, a possibility, in principle, of producing high-power ion beams and building mass-spectrometers with the use of FEAs has been demonstrated.[105]

In 1995, Denis Herve built a compact blue semiconductor laser excited by an electron source based on an FEA (see Gray[104]).

In the last decade the technologies of producing efficient FEAs has been actively developed. Using the method of optical interference lithography, FEAs of large operating area have been produced having a cell area of 0.1–0.2 μm.[104]

Fabrication of large-area multi-tip FEAs with a control electrode having small grid cells became possible with the use of chemical surface polishing[478] eliminating the need for lithographic methods.

Molecular beam and gas phase epitaxy provided means of dorming nanometer tips of silicon,[444] as well as heterostructures, for example, Si–Ge[104] and nanocrystals of diamond[479] on the surface of silicon tips.

A considerable contribution to the development of FEA technology was made by research groups at LETI and SRI.[105]

A new field of VME originated from the work on FEAs made of materials with low electron affinity or highly nonuniform nanostructure, from which emission currents can be obtained at relatively low electric fields (of the order of 10^5 V/cm). Of the materials known to possess such properties and likely to have practical applications the most intensively studied are films and coats based on nanoclusters, namely diamond-like structures, fullerenes, tubelenes (nanotubes), and their derivatives.[449, 480]

8.3. FIELD ELECTRON MICROSCOPY

An important development in the FE investigations had been the invention by Muller in 1936 of the FE microscope—the projector.[21] Practically from this moment a systematic accumulation of the data on surface properties of the field emitters started. This device made it possible to identify many factors causing instability of the FE process, to study the types of changes induced in the shape of a field emitter by the electric field, temperature adsorption of foreign atoms, electron, and ion bombardment. High magnification (10^5–10^6) and resolving power (10–30 Å) of the field projector together with the possibility of active action on the object of study (in situ) made it an indispensable tool in the studies of adsorption, desorption, epitaxy, surface diffusion, phase transitions, and so on.[21, 23, 24, 29]

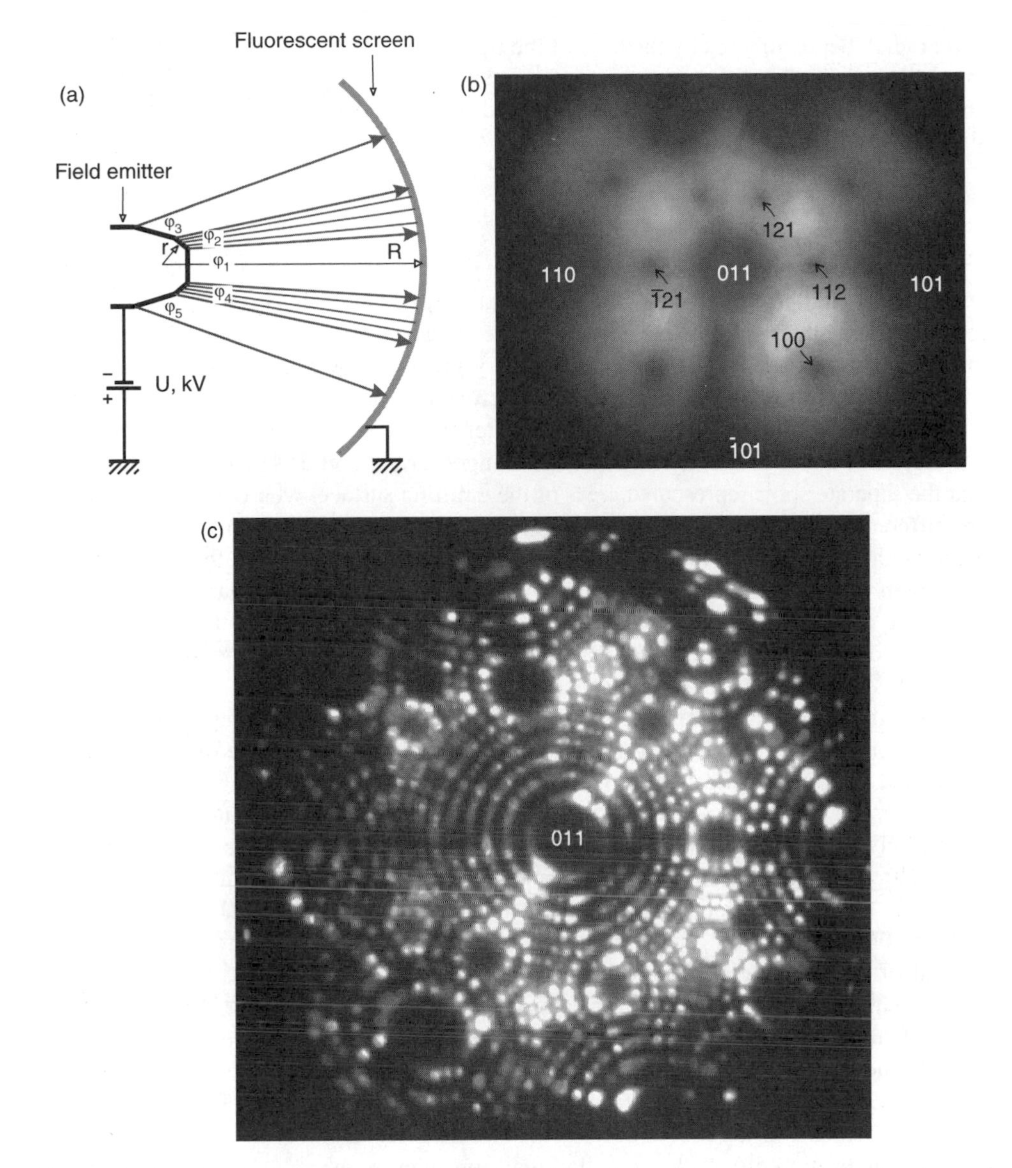

Figure 8.1. The principle of FE microscopy. (a) Setup; (b) field emission picture of w tip; (c) field ion picture of w tip.

The FE microscopy is based on the fact that if a fluorescent screen—anode is placed at a macroscopic distance (a few centimeters) across the path of electrons emitted from a sharp tip (a few tenths of a micron), then the electron rays will produce on this screen a projection of the tip apex with a very high magnification (Fig. 8.1a).

Because the electrons emitted from the tip surface scatter almost radially, the magnification of such a microscope–projector is equal to the ratio of the distance between the screen and the tip to the radius of the tip apex. Actually, the trajectories of electrons are not quite radial, being affected by the base of the tip and the electrodes holding it; so the electrons move along slightly sloping parabolas and the image on the screen appears somewhat compressed. Taking this into consideration, the magnification M can be represented by a simple formula

$$M = \frac{1}{\gamma} \cdot \frac{R}{r}, \tag{8.1}$$

where γ is the compression coefficient ($1.5 < \gamma < 2$); R is the anode–cathode distance, and r is the emitter apex radius. Because the tip apex dimensions are of the order of a few tenths of hundredths of a micron and the distance R can be made (3–10) cm, the magnification of such an instrument can be very great, of power up to 10^5–10^6. Muller[21] had applied this principle of field electron microscopy for the first time in a study of the emission current density at the surface of emitter tip. He noted that the image obtained is not uniform but consists of symmetrically arranged bright and dark spots. He understood that the separate spots represented areas of the emitting surface, where the work functions are different (see Fig. 8.1). He also realized that because the FE current density depends exponentially on the work function ($j \approx \exp(-\varphi^{3/2}/F)$), formula (1.8), the bright spots on the screen corresponded to regions with small work function and the dark spots are areas where the work function is large. He also found out what the tip apex is like.

The following picture can be imagined: put on a cubic crystal (body-centered, face-centered, or hexagonal) is a hemispherical "hat," which cuts in the surface a smooth hemispherical segment composed of sections of different crystallographic planes. It is known from crystallography that different crystallographic planes have different arrangements and packing densities of atoms. It is also known that the packing density of atoms determines the work function. Closely packed planes have large work functions; a "loose," less closely packed ones have smaller work functions. So, at the surface of the fluorescent screen due to different densities of the emission beams the component crystallographic planes are displayed with the high magnification, 10^5–10^6, mentioned above.

The FE image of a tungsten (W) crystal tip is shown in the right-hand side of Fig. 8.1b. Instead of a usual single grain it features two single-crystal grains. Comparison of the emission image with X-ray diffraction data, as well as, comparison of the image parameters with calculated crystallographic projections provided reliable identification of the planes seen on the screen. In the emission image in Fig. 8.1b a closely packed $\{011\}$ facet (in crystallographic Miller indices*), facets of the $\{112\}$ type, cubic $\{100\}$ facet, and so forth.

Also, Muller was the first to develop the atomic-scale microscopy using his discovery of a field ionization effect, the so-called field ion microscope.[285, 492] An ionic image of a W tip oriented in the same crystallographic direction is also shown in Fig. 8.1c. In this image individual atoms can be seen as separate points. It was a great invention, the first in human history, direct observation of the atomic structure of matter. Regrettably, because of the limited size of this article, it is not possible to describe here the principle of field ion microscopy and the outstanding achievements made using this remarkable instrument. This is a subject to be considered separately.

The FE microscope possesses not only large magnification but also high resolution. The resolution is understood as the possibility of separate observation of two close points of an object. In the field electron microscopy the resolution is determined by the wave nature of an electron, that is, in accordance with the de Broglie relation its wavelength is $\lambda = h/p$, where $p = mv$ is the electron momentum. If an accelerating voltage U is applied across the cathode–anode gap of a microscope, the electron wavelength, as can be easily shown, will be $\lambda = 12.3/\sqrt{U}$ Å$(1\,\text{Å} = 10^{-8}\,\text{cm})$, if U is measured in volts. It is found that because of diffraction the spreading at accelerating voltages of several kilovolts usual in emission microscopy is about 8 Å. Besides, the spreading occurs also because of the tangential velocity component of the tunneling electron exiting the surface. The total spreading in usual conditions of the FE microscopy is 20–30 Å. According to estimates in the resolution δ of a FE microscope can be calculated by an approximate formula

$$\delta \approx 2{,}62 \cdot \gamma (r/k\vartheta(Y)\varphi^{1/2})^{1/2} (\text{Å}) \tag{8.2}$$

where γ is the compression coefficient; r is the emitter radius in E; k is a coefficient equal to approximately 5; $\vartheta(Y)$ is the Nordheim function; and φ is the work function in eV.

The emission microscope proved to be a remarkable instrument and made possible studies of a number of important properties of surface in strong electric fields as well as various fine effects on the surface.

First of all, with the emission microscope it has been revealed, what is going on at the surface of the emitter proper, under different conditions: causes of instability in the operation of field cathode established, its emission characteristics specified, local current densities from different areas of the emitting surface measured, and others. With the FE microscope, it is possible to identify, the crystalline structure of materials by comparing the parameters of an emission pattern with calculated crystallographic projections.

With single crystal tips some phenomena can be studied in strong electric fields, 10^7–10^8 V/cm, unattainable with macroscopic objects.

Muller[491] has discovered reconstruction of the tip surface occurring due to migration of the surface atoms in emitters heated in a strong electric field. The fact that the reconstruction is related to changes in the shape of crystal (its cut) has been demonstrated by Benjamin and Jenkins[489] in brilliant works of Sokolskaya.[385]

It has been shown that the built-up phenomenon plays an exceptionally important role in initiation of instability of the emission process due to formation on the emitting surface of sharp "ridges" and small nanometer-size protuberances. Their formation causes local enhancement of the electric field and a drastic increase of the emission current. An example of the reconstruction with formation of nanometer protuberances on the tip apex is shown in Fig. 8.2 (see[388, 389]).

As mentioned above, the resolution of a field microscope is 20–30 Å and it is difficult in principle to further improve it. Yet, with very small protuberances on the surface, the resolution can be still higher.[497] In the studies[396, 247] ridges having a "sharpness" of just 2 or 3 atoms have been observed (Fig. 8.3).

The FE microscope allows one to study extremely slow displacements of the surface, study in detail the surface migration, two-dimensional and three-dimensional evaporation, and various chemical reactions, including reactions in a strong electric field.

Employing FE microscope, major quantitative characteristics of crystal surfaces can be studied: work functions, activation energies for migration, desorption, and evaporation of atoms, the size of two- and three-dimensional formations on the surface, and bonding energies in two-dimensional crystals. The FE microscopy can be applied in studies of high-temperature phase transitions and orientational relationships for transitions between different crystalline modifications.

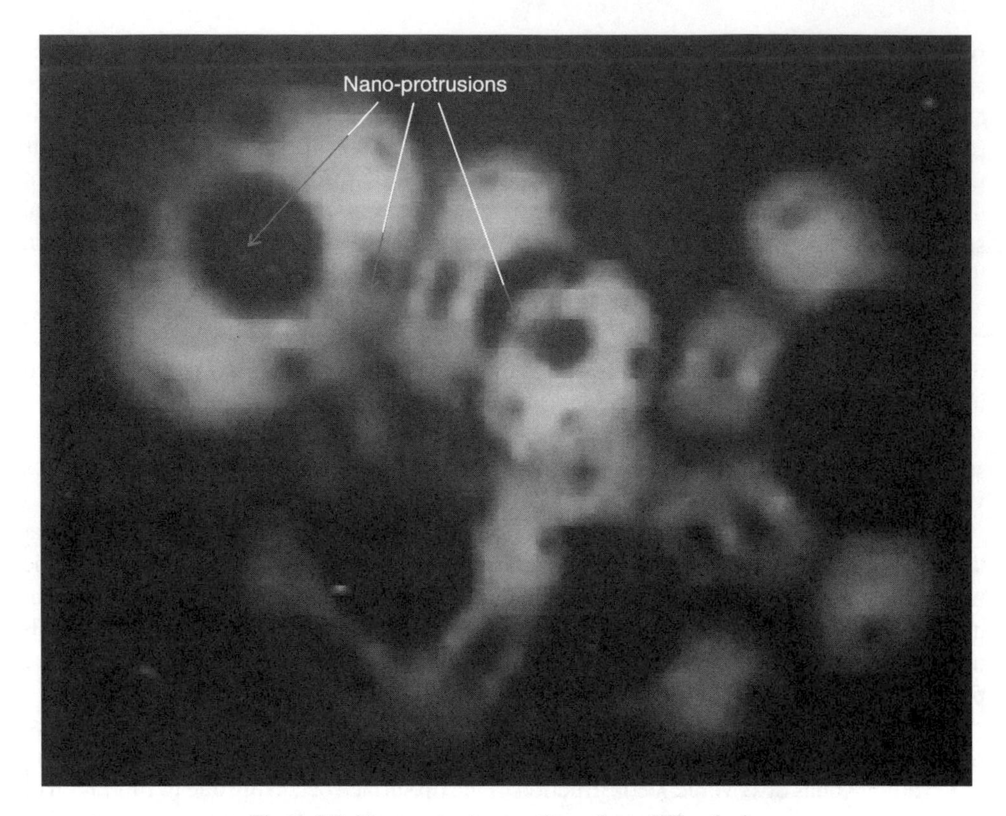

Figure 8.2. Nano-protrusions on the surface of FE cathodes.

With a microcrystal in the form of a tip, a unique method of obtaining atomically clean surfaces can be implemented using desorption and evaporation of atoms by electric field[285] under direct control of the process in the FE microscope. With this method atomically clean surfaces of a number of semiconductor crystals have been obtained, such as Ge, Si, GaAs, and others (see Chapter 5).

The above list does not exhaust possible applications of the field electron emission microscopy. This technique is currently being actively developed and discoveries of new effects and phenomena can be expected.

8.4. FIELD EMISSION DISPLAYS

One of the most promising applications of FEE at the present time is flat FE display panels of high-image quality, components for computing and information processing systems featuring supershort response times (of the order of microseconds) and high radiation and thermal resistance. By replacing thermal cathodes with FE cathodes consisting of multi-emitter arrays considerably higher currents can be attained and the display

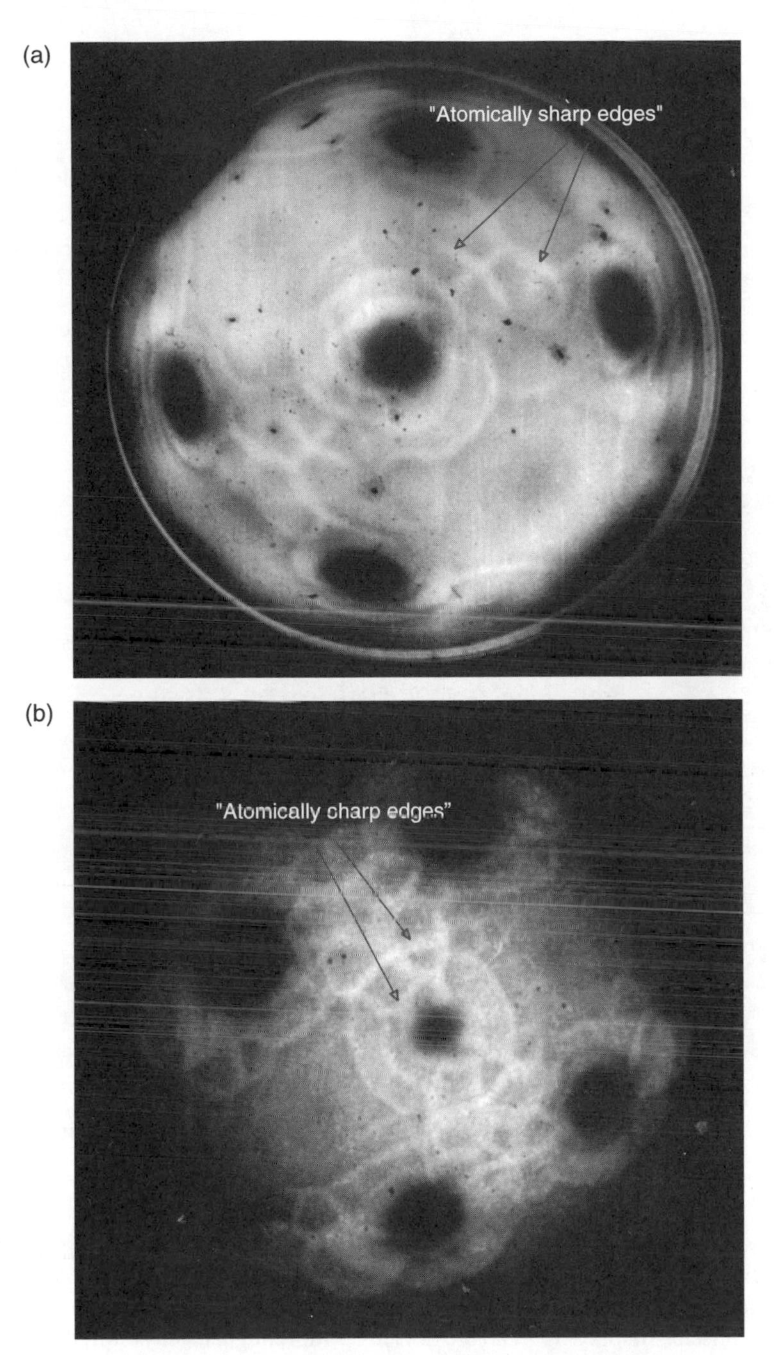

Figure 8.3. "Atomically sharp" edges formed by a strong electric field: (a) on the W surface; (b) on the Mo surface.

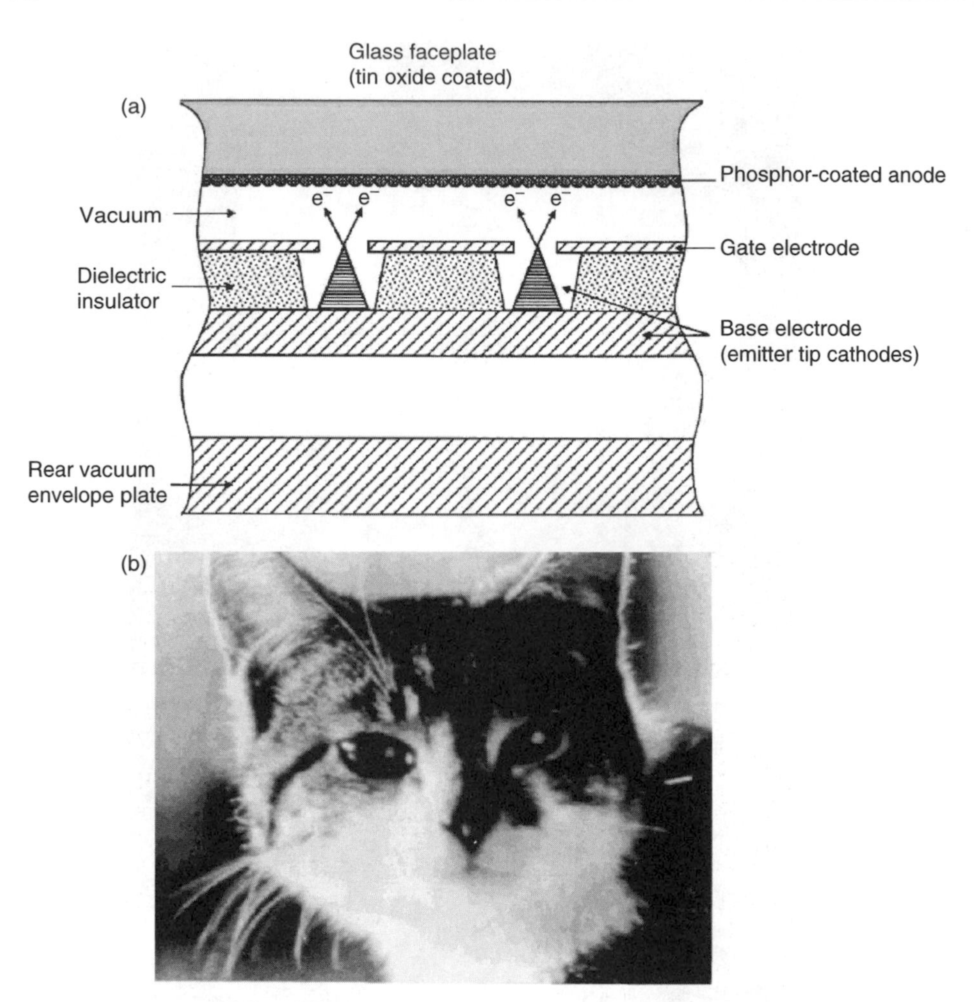

Figure 8.4. The flat-panel display based on microfabricated field-emitter arrays.[87, 428] (a) Schematic diagram of the display. (b) Photograph of the image of a feline on the black-and-white screen of the LETI flat-panel display.

characteristics significantly improved, resulting in greater brightness and spatial resolution (density of pixels), larger viewing angle, and higher speed, as well as lower power consumption.

Field emission displays (FEDs) are devices comprising an array cathode, an extracting electrode, and a luminescent screen, where the image is formed (see Fig. 8.4). A conspicuous achievement of the past years, mainly connected with works by Spindt,[214] is the fabrication of an ordered array of metal or semiconductor tip cathodes integrated with a honeycomb anode (gates) and arranged in such a way that facing each tip is a corresponding anode gate (see Fig. 8.5). The small distance between the emitter and the gate (a few

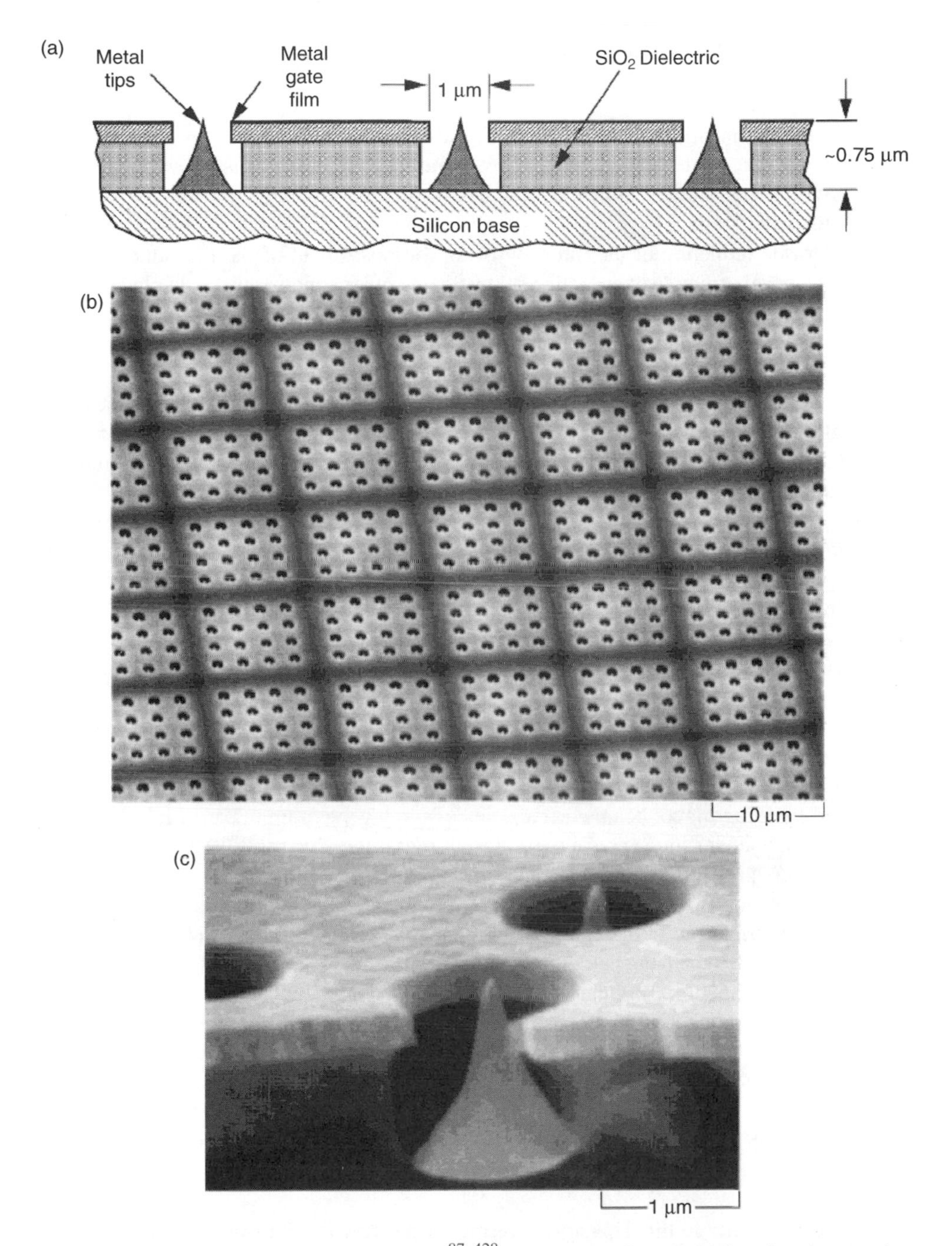

Figure 8.5. Spindt-type field-emitter cathode array.[87, 428] (a) Schematic diagram. (b) Scanning electron microscope micrograph of an array. (c) Scanning electron microscope micrograph of a single emitter in the array shown above.

micrometers), as well as a small (tens of nanometers) tip apex radius solve several important problems:

1. Fabrication of a low-voltage multi-emitter cathode. FE cathodes with a honeycomb gate controlled with a voltage of about a hundred volts are now in operation, with some studies indicating the possibility of reducing the voltage down to ten volts.[442]
2. Drastic reduction of the intensity of ion bombardment of the FE cathode due to the novel principle of operation of the device, which increases the display lifetime, makes the emitter operation more stable and eases the requirements to operational vacuum; a vacuum of the order of 10^{-7} Torr was shown to be sufficient.

Due to smallness of the gap between the FE cathode and the controlling electrode (gate) the signal frequency can be high, covering the range of TV signals, while the high speed of FEDs compared to an ordinary cathode-ray tube ensures high instantaneous brightness of every pixel and, consequently, high average luminance over the screen area.

The field-emitter-array cathode operates cold, needing no warm-up time; the modulation voltages at the gate are in the same range as in conventional cathode-ray tubes (CRT) guns; and the gate-to-cathode capacitance does not impose a limitation at television modulation frequencies (3–30 MHz). The major advantage, however, is that *the much higher currents can be obtained from a similar-sized cathode compared to conventional CRT's.* Higher currents are useful for the high-brightness, large-pixel-count displays required for *high-definition television* (*HDTV*).

For the same frame rate, the dwell time of the beam on a pixel is smaller for HDTV than for a conventional TV display, thus requiring a *higher instantaneous brightness* to maintain a given average brightness for the whole frame. This brightness increase must be attained by using higher beam currents, since X-ray generation makes it impractical to increase beam voltages much above their present maximum levels (30 kV).

One of the most promising FED types are the devices with active matrix addressing, in which every picture element (pixel) has a built-in circuitry facilitating addressing and recording of information, so that the information capacity of a display can be increased for the same or even better image perception characteristics. The idea of a planar addressable matrix as a replacement for ordinary CRTs was first formulated by Crost.[455]

The basic component of such a display is a field electron multi-emitter cathode (array) with gates (Fig. 8.5). The matrix of the multi-emitter cathode is usually fabricated on a dielectric substrate such as glass. The entire cathode surface is divided into individual elements—pixels. Figure 8.6 shows the schematic of an individual pixel forming a three-color image; each pixel consisting of three parts, one for each color.

8.4.1. Requirements to the Tips and Lifetime of Matrix Field Emission Cathodes

A FEE usually has the form of an elongated tip, cone, pyramid, or blade, or it may represent random nonuniformities shaped as sharp angles, ridges, and micro-protrusions, which have close emissive capacities and emit onto the luminescent screen a fairly uniform flow of electrons.

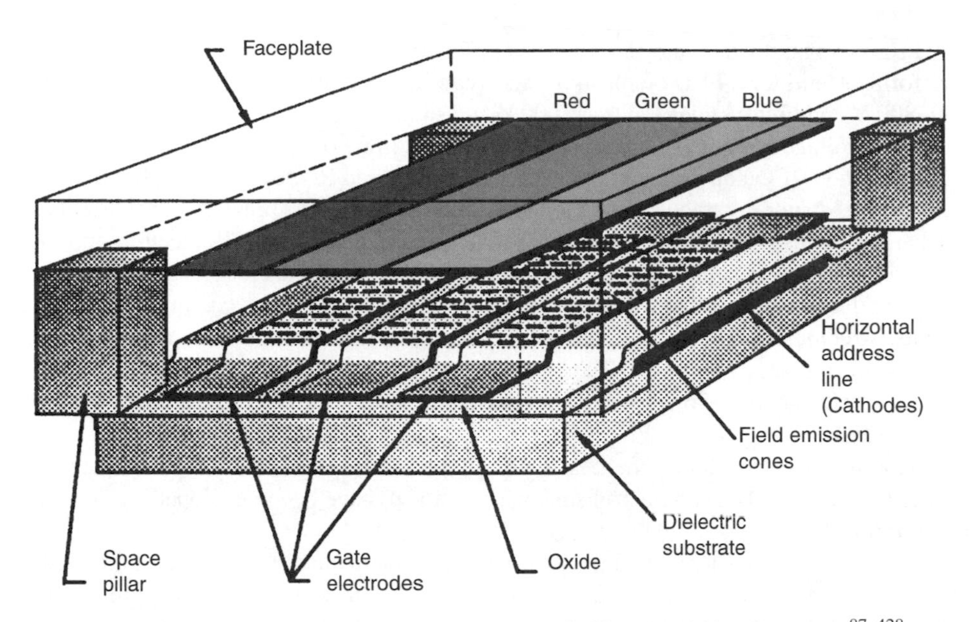

Figure 8.6. The principle of a flat-panel display's color pixel based on field-emitter arrays.[87, 428]

Tip cathodes used in FEDs should keep their shape, withstand the pressure differential on the casing walls, have conductivity of 10^8–10^{12} Ohm and be resistant to oxidation by residual oxygen.

The size of an individual pixel in latest FEDs is usually 0.3 mm. At the present time efforts are undertaken to reduce it down to 0.1 mm, which will increase the display resolution. Each pixel comprises more than 1500 tips distributed over the matrix surface at a density of 10^6–10^7 cm^{-2}, making the fluorescence practically uniform within the pixel. The current density from one pixel in this case exceeds 50 μA.[482]

Above the tips, strips of the control electrode (three per each pixel, with gates) are situated. A pixel is activated by applying a potential corresponding to the threshold value for FE to the multi-emitter matrix and the required bias potential to the control electrode, which sets the signal magnitude. The signal can be modulated by varying the pixel turn-on time. A pixel can also be controlled by individual lines or rows. The electrons emitted from every pixel are accelerated in the vacuum gap between anode and cathode (screen) and impinge upon a luminescent screen. If the separation between the anode (screen) and cathode is much larger than the size of pixel, the images of pixels can overlap reducing the display resolution. To prevent this, in some designs a system of conducting parallel electrodes comprising the bases of every separate pixel is introduced in the vacuum gap.

It should be noted that the state-of-the-art lithographic technology of FEAs does not allow fabrication of FE cathode of large operating surface area. It is reported that the work is underway on display panels of an operating area up to 15 cm^2.[481] According to reports by LELI Company, panels of an area up to 1 m^2 are being developed.

A relatively inexpensive alternative technology of multi-emitter matrices for display panels uses dielectric matrices with randomly distributed ultra-fine particles of semiconductors or conductors, which represent point emission source.[484]

FEDs can be divided into several types differing in the operating voltage and the size and form of emitters. FEDs of low-voltage type are operating at voltages in the range 100–800 V and high-voltage types at 1000 V and higher.

High-voltage FEDs have longer service life than low-voltage displays and give better color rendition. If the operating voltage has to be increased to raise the phosphor emission efficiency, the anode–cathode separation should be increased to prevent an occurrence of discharge. Intrinsic divergence of the electron beam produced in field electron emission increases the spot size on the screen necessitating focusing of the electron beam.

Therefore, in high-voltage FEDs, which have higher brightness than low-voltage ones, the precise color separation poses problems related to switching between separate phosphor strips within a pixel.

An advantage of the low-voltage FEDs is mainly their lower cost compared with high-voltage FEDs.

Motorola has developed the best high-voltage FEDs and the Futaba company the best low-voltage FEDs, and microdisplays 3 cm in size have been developed by the MDT Corporation.[481]

LETI (France) has fabricated the first monochrome display.[457–459] One of the models announced by LETI[460] is a 6-in. black and white display, in which the screen luminance of 200 Cd/cm^2 has been achieved. The accelerating voltage between the cathode and anode is 250 V and the voltage across the gap between the gates and the multiple-tip FE cathode is 80 V. The matrix of the FE cathode contained 256×256 pixels, each of an area 0.12 mm. Power consumption is less than 1 W.

Companies such as LETI, Motorola, and Futaba successfully introduce this technology into production of commercial full-color displays. At present the research by these companies aims at increasing the area of the displays (up to 1 m^2) and the efficiency of phosphors.[481]

A low-voltage FED developed by Futaba (anode voltage of 800 V) has a 7-in. panel with the attainable screen luminance of up to 300 Cd/cm^2.[481]

Pix Tech Company has already produced a pilot batch of monochrome panels with a screen diagonal of 7 and 12.1 in. Candenset Technologies Corporation and Sony plan for 2001 production of a full-color FED with a screen diagonal of 14 in.

At the IVMC (1999, 2000) conference it was reported that a number of firms began production of the FE displays. In particular, Motorola will invest about a billion dollars in production of FEDs.

8.4.2. Comparative Characteristics of the Display Types: Advantages of Field Emission Displays

For displaying and processing graphic information color and monochrome CRT displays and liquid-crystal displays (LCD) are now predominantly used. In CRT displays, electron beams in vacuum pass through a shadow mask producing visible pixels at the spots, where under the electron impact a layer of phosphor starts to emit light of different colors. The most promising area of the CRT technology is in HDTV television, which provides a resolution of more then150 lines/mm and the pixel size less than 0.1 mm and makes possible fabrication of large-format monitors with a screen diagonal of more than 20 in.

Standard CRT designs meeting requirements with respect to information content, resolution, and cost in some cases cannot be used because they are bulky (the monitor depth and weight grow fast with the screen size), breakable, consume much power, and also degrade the equipment characteristics. For example, CRTs are impossible to build into portable computers, they have poor efficiency when used in technological and cosmic installations.

Close competitors of CRTs are LCD. In an LCD, a thin layer of sheet material transmits or blocks, where appropriate, the backlight (passing color filters). The sheet material is patterned into small cells, each cell corresponding to a pixel. The advantages of LCDs are their small weight and depth, and extremely low power consumption. But these displays are costly, rather slow, have poor contrast, require backlight, and noted for considerable dependence of their performance on the ambient temperature. Particularly interesting is the work on development of liquid-crystal full-color displays carried out by some Japanese and American companies such as Sony, Pix Tech, Nokia, Motorola, and Tektronix.

The main problem with producing large-area LCD displays is making defect-free matrices. The current state of the LC-indicators (production of matrices, LC-materials, packages, activating integrated microcircuits) reached such a high level that the wide use of active matrix addressing became possible. Because of successful development of indicators with active matrix addressing for miniature pocket television sets, fabrication of large screens for computers, automated work sites, and domestic electronic information systems became feasible.

As for production of large displays, the most suitable for them appears to be the technology of plasma panel displays, in which the phosphor is activated by UV radiation from the plasma of electric discharge.

In the last five years, as noted above, flat-panel displays based on FEAs (multi-emitter matrices) have found increasingly wider application in electronic devices.

The FED's principle of operation makes them considerably more compact than CRTs. So, for the same screen area of, say, 12.5 cm, the FED has a thickness of 1.5 cm and a weight of 285 g compared with CRT's 27 cm and 1450 g, respectively.

Advantages of the matrix addressing scheme for FED control compared with that of CRT are abandonment of the cathode heating, low-efficiency deflection system, and the shadow mask for confining the electron beam to a specified point on the phosphor. FEDs are also practically instantaneous, afford large viewing angles, have ultra-short frame switching time, low power consumption, high brightness and uniformity of screen luminance required for displaying and processing of images, stable operation in a wide temperature range, and compactness.

Image luminance and contrast provided by FEDs are practically independent of the viewing angle within up to 160°. Highly efficient FED designs (3–40 lm/W) have a brightness of up to 3500 lm that can be changed in increments as small as 0.05 lm. High speed corresponding to a response time of 20 ns ensures good dynamic definition of the image. Use of phosphorus as a fluorescent substance provides high quality of color separation, wide spectrum of color shades and 256 hues of gray (8-bit gray scale). In particular, FEDs can reproduce scores of the human body color hues, making them suitable for applications in television for medicine. FEDs demonstrate stable operating characteristics in a wide temperature range (from −40 to +85°C). Because of small distance between the FE tips and the anode (0.2 mm) the size of the spot on the screen (or the pixel) is only about 0.33 mm.

Table 8.1. Comparative data on parameters of various types of displays for 1999

Parameter	CRT	LCD	FED
Brightness, Cd/m^2	$>500^a$	>50	>300
Contrast	$>100:1$	$>12:1$	$100:1$
Viewing angle	$>100^{\circ b}$	$>80^\circ$	$>160^\circ$
Pixel size, mm	0.3^c	0.3	0.32
Image stability	Fair	Excellent	Fair
Response time	50 μs	>200 ms	20 ns
Power consumption, W	>200	1^d	1^d
Maximum screen size (diagonal), in.	38	30	12.1

Notes:
[a] Can be operated under direct sun illumination.
[b] The value corresponds to HDCRT flat-panel display.
[c] For EG Electronics 78FT II 795FT CRT-monitors, the pixel size is 0.24 mm;, for HDCRT, it is less than 0.15 mm.
[d] LCDs with backlight.

Drawbacks of FEDs include possible considerable fluctuations of the emission current (30 percent on the average) even with absolutely stable applied voltage. Occasional current surges can bring about destruction of the electrodes. Such occurrences can be prevented by perfecting the FED design.

The potential of FEDs make them especially attractive for creating system for reproduction and processing of optical information in medical apparatus, measuring devices, video system for the military, and so on.

Table 8.1 gives comparative data on parameters of various types of displays for 1999. FED characteristics have been taken from advertising information of the Pix Tech Company posted in the Internet.

8.5. OTHER APPLICATIONS OF FIELD EMISSION

Recent years have seen an explosive growth of VME as scores of its applications emerged in totally unexpected areas and continue to expand. Traditional application area of the FEAs is radio industry devices, such as vacuum integrated circuits for fast processing of signals; miniature microwave tubes for generation of ultrahigh (higher than 1 GHz), and high (below 0.5 GHz) frequencies; flat display panels for high-resolution television; and high capacity fast memory cells for operation in a parallel mode at rates of 10 Gbit/s.[498] VME devices find application in neighboring areas of high-current electronics, lithography, surface diagnostics, electronic devices resistant to high temperatures, and radiation; and in generation of high-power electron beams.

Mention should be made of the use of FE cathodes as sources of relativistic electrons for cosmic research, as well as, in particle accelerators and nuclear reactors; in multiple-beam electron sources for lithography; high-intensity electron sources in devices for surface states diagnostics; and in free-electron lasers.[499]

Above (Section 8.4) FEDs have been considered and the use of field emitters as high-intensity point sources of electrons (see Chapter 7).

Below we briefly present the major ideas, designs and techniques relating to latest achievements in VME and some other applications.

8.5.1. Miniature Field Emission Triodes and Amplifiers Based on Field Emission Arrays (FEAs)

First designs of the current-controlled electronic devices with needle-like FE cathodes have been disclosed by Dyke.[27, 89] However, these devices did not find wide application because needle cathodes could not operate for a long time under forevacuum conditions and their internal resistance was considerable.

The possibility of producing miniature vertical and horizontal microtriodes (vacuum transistors), as mentioned above (Section 8.1) was first predicted by Shoulders in 1961.[102]

First vacuum FE triode was reported by Gray in 1986.[500] Originally, it was a *planar (lateral)* vacuum FE triode (transistor), in which a multi-emitter tip cathode was structurally combined with a control electrode in the form of metal teeth. In the planar microtriode electrons move straight upward and their flow converges at the anode-sink at the same surface (Fig. 8.7).

In *three-dimensional* microtriodes[501] of size less than 3 μm, the most efficient collection of electrons at the anode is achieved on their trajectory upward from the tip (Fig. 8.8).

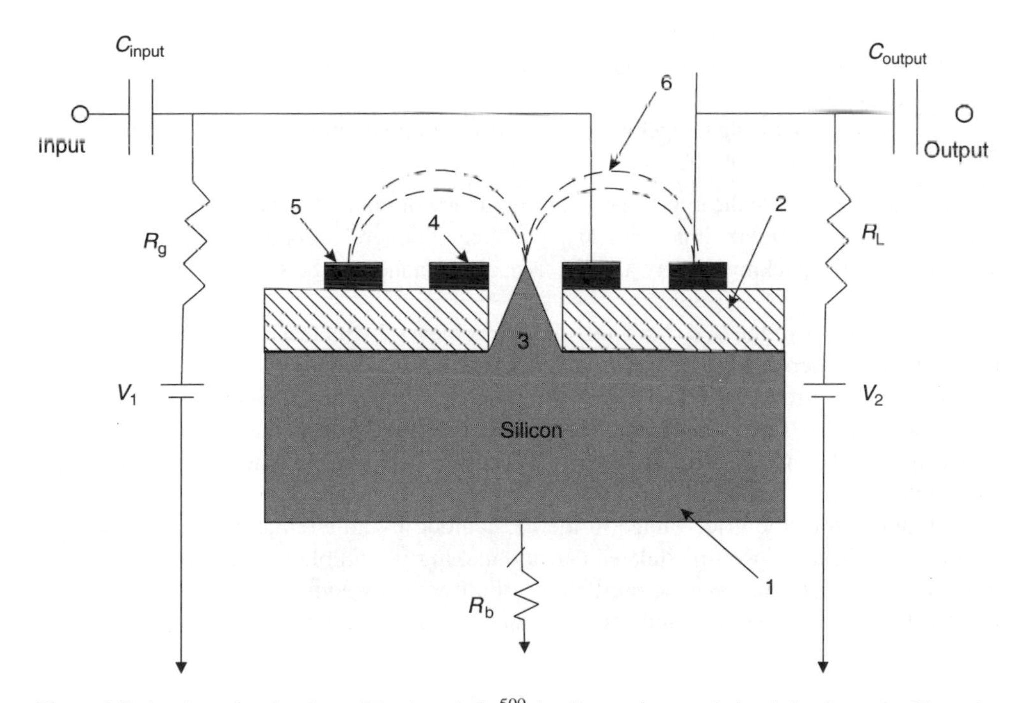

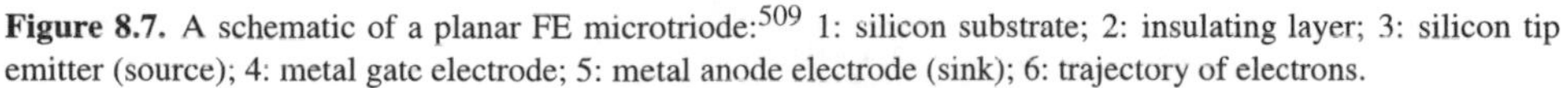

Figure 8.7. A schematic of a planar FE microtriode:[509] 1: silicon substrate; 2: insulating layer; 3: silicon tip emitter (source); 4: metal gate electrode; 5: metal anode electrode (sink); 6: trajectory of electrons.

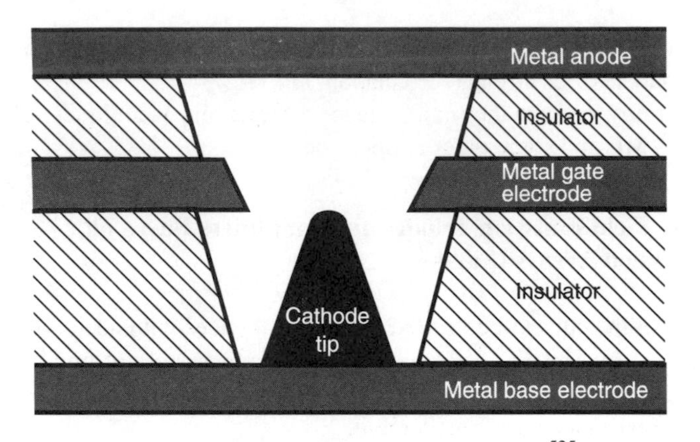

Figure 8.8. A schematic of a vertical FE triode.[525]

The technology of planar microtriodes is less complicated compared with that of three-dimensional devices; however, planar devices have a principle limitation on the use of metal tip or blade emitters in contact with a layer of dielectric because the electrons can tunnel from the Fermi level of the metal into the conduction band of dielectric at lower fields than necessary for tunneling into vacuum.*

In a microtriode, by varying the potential of the control grid electrode (Fig. 8.9), the anode current can be modulated.

Vacuum microtriodes have a number of advantages over solid-state devices:

1. higher power (more than 10 W);
2. resistance to ionizing radiation up to the level capable of destroying the anode material;
3. stability in a wide temperature range (from liquid helium temperature to temperatures in excess of 1000 K).

In order to increase the upper operating frequency of microtriodes from 1.3 GHz to the centimeter range (30 GHz) it is necessary to reduce the interelectrode capacitances as well as to increase the packing density and the transconductance of the static characteristics of the triodes.

Steeper slopes of the I–V characteristics are achieved by increasing the density of tips, use of blade emitters, and materials with low work function values.

The interelectrode capacitances can be lowered by reducing the cell size of the grid electrode down to 1 μm or smaller, reducing the permittivity of the layer of dielectric between the gate and the base, reducing the distance between the gate and the base, and using ultrasharp FE cathodes.

Modulation of the grid voltage in the FE cathode by an alternating signal produces an electronic beam of well-modulated density, making it possible to excite in the electrodynamic system electromagnetic oscillations of a frequency equivalent to an input signal. On this basis have been proposed and fabricated various modifications of amplifiers with

* We do not consider here the technologies of microtriodes based on achievements of the solid-state electronics, which have been discussed in detail elsewhere.[214, 445, 500, 502]

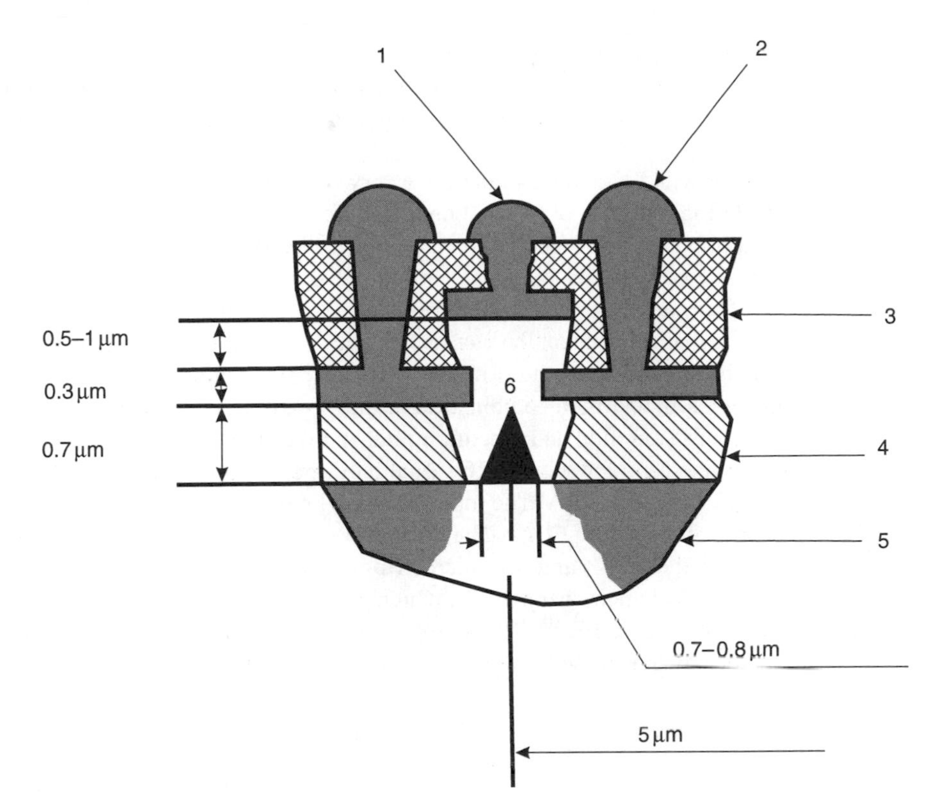

Figure 8.9. Structure of a three-dimensional silicon FE microtriode:[509] 1: anode; 2: grid; 3: intercomponent insulation; 4. insulation layer of silicon oxide SiO_2; 5: silicon substrate; 6: silicon tip emitter.

modulation of the emission current by the signal.[504] In the most thoroughly studied devices of this type, femitron, clistrode, the amplification coefficient is 10–12 dB with an efficiency of up to 70 percent.

Wide amplification band, high efficiency, and compactness are the features of the amplifiers with FE cathodes, which led to their wide use in cosmos-based, intersatellite and mobile communication systems, cosmic radars, local information networks, generators, detectors for radioastronomy, and so on.

8.5.2. Microwave Devices for Millimeter Wave Amplification

Such, properties of the FE cathodes, as no need of heating, lack of inertia (the time of flight of electrons being less than 10^{13} s for a cathode–anode distance of 0.5 μm), high average current density (more than 10^3 A/cm^2 from an array surface[87]), and nonlinearity of the current-voltage characteristics make them especially suitable for use in microwave devices. Narrow energy distribution curve of the FE electrons (0.22 eV at room temperature and 0.14 eV at liquid nitrogen temperature, see[381]) lowers the intrinsic noise of microwave devices. Due to sharpness of the electron beams produced by arrays of emitter structures their dimensions can be reduced, which also reduces the potential at the collector and raises device efficiency. Highly intense and narrow electron beams obtainable from the FEAs make possible size and weight reduction of microwave tubes. The introduction of electron collector serves to increase efficiency by returning the emitted electrons back into the device.

Microwave devices with FE cathodes can have a very wide range of operating frequencies (from a few GHz to thousands of GHz), linear frequency characteristic, insignificant variations of the output power and other parameters in the operating range, fast frequency tuning, and low intrinsic noise levels.

Microwave devices are superminiature triodes of a size much less than the wavelength.[505, 506] The main element of the most electronic microwave devices for frequencies up to 1000 GHz is a high-current thin-film FEA (Fig. 8.10) of submicron dimensions. Such a cathode possesses an optimum combination of the basic device parameters: current density, total current, pulse duration and repetition rate, and the stability of operation.

The transmission line for a low-power RF signal is formed by a FEA and a grid with a pitch equal to the spacing of the FEA. The input RF signal modulates the electron beams produced by every row of the matrix FEAs. The microwave device is designed (Fig. 8.10a) as an FEA (3) with tip cathodes (4) and two microstrip lines having a common metal strip with holes (1) and cylindrical microchannels (2), which plays the role of a grid and to which a potential is applied causing the FE.[506] To the upper metal strip (5) a potential is applied capable of ensuring the passage of the electron beam through the microchannels. The entire

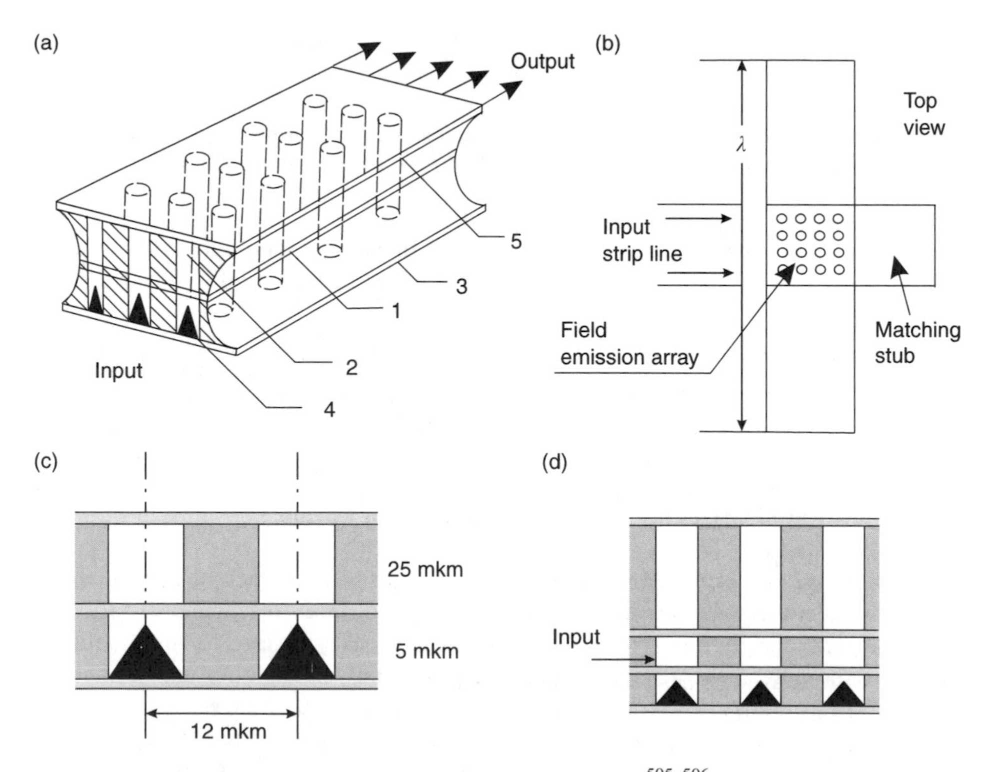

Figure 8.10. A schematic of microwave amplifier with the Spindt cathode:[505, 506] (a) construction of the device; (b) resonant amplifier; (c) distributed amplifier; (d) schematic of tetrode amplifier.

system is evacuated. The height of the strip lines is from a few to tens of microns. In this way, a system of microtriodes is obtained, which have grid and anode circuits coupled by traveling electromagnetic waves propagating along the strip lines.

Two patterns of strip lines have been suggested: perpendicular (Fig. 8.10b) and parallel (Fig. 8.10c, distributed amplifier).

For the resonant amplifier the following parameters have been obtained: frequency −60 GHz; width of the operating frequency band 1 percent ; power more than 20 W; gain factor 10–20 dB; and efficiency >50 percent.[506, 507]

In a distributed amplifier the frequency band can be much wider but because of higher wave resistance the efficiency is lower.[507]

By adjusting the area of FEAs, required cathode currents can be obtained from vacuum electronic microwave devices. The cathode is usually made as a matrix of conical tips on a silicon substrate. The cathode material is usually W or Mo;[87] besides, carbides of Zr or Tl, as well as, heavily doped silicon can be used.[500] Advances in modern microelectronics technology made possible fabrication of FE cathodes with practically inertialess control of the emission current at frequencies up to 1000 GHz.

Because of high nonlinearity of their I-V characteristics, with FE cathodes more effective modulation of the emission is possible than with hot cathodes, which is made use of in diodes, clystrodes, and other devices.[508] Due to this circumstance, in a diode with FE apart from high FE current density an efficient generation at microwave frequencies can be reached.

In a clystrode the integrated electron beam excites a microwave signal and is accelerated by the voltage applied between the cathode and the grid, and then the kinetic energy of the beam is transformed into the energy of harmonic microwave oscillations, when it passes the resonator.

Structures prepared by methods of the VME can be used for creating analogs of many traditional types of electrical vacuum microwave transit-time devices, for example, traveling wave tubes (TWT), backward-wave tubes (BWT), clystrons, distributed amplifiers and so on.[498] These devices are used as heterodynes in radios, driving oscillators in amplifying circuits, or self-contained generators in some types of radio equipment.

In a TWT a low-power RF signal is fed to the input and passes through a transmission line formed by the FEA and the cathode grid. The cathode, isolating and anode grids have a pitch equal to that of the FEA and are made of wires of different diameters; this and corresponding potentials on the grids enables electron beams to be formed from isolated tip cathodes. Dimensions of the TWT as well as the electrode potentials are matched to the input signal frequency in such a way that the input signal passing through the transmission line formed by the anode and anode grid is amplified by the modulated electron beam from every row of the tips.

The principle of operation of the BWT is based on interaction of the electron beam with a backward harmonic of the RF field wave. Prototypes of BWTs generate wide-band oscillations in a range of 137–311.5 GHz as the control voltage is swept through a range of 0.5–5 kV with an average peak signal power of 100 μW[509]

Wide-band amplifiers with distributed amplification on thin-film FE cathodes (Mo tip cathodes for a supply voltage of 100 V) with modified control electrode and heat-resistant anode in the form of transmission line can be used for fabrication of triode and tetrode amplifiers with distributed interaction for a output power of more than 10 W in a frequency range of 0–20 GHz. Such amplifiers are more compact than TWTs. Thin-film FE cathodes for a current density of up 100 A/cm^2 at accelerating voltages not exceeding 3 kV and operating currents up to 20 mA represent components of a new class of ultrasmall TWTs of the submillimeter range.

8.5.3. Nanolithography

High brightness (10^9 A/cm^2 · sr) of the electron beam as well as the high density of energy released on the target make the FE cathodes suitable for precision fabrication of the elements of submicron structures by electron lithography, microengraving, and so forth. The divergence, angle of the electron beam can be reduced using apertures cutting out the most intensely emitting spot on the tip surface or using adsorption of foreign atoms onto the FE cathode surface (e.g., Zr on W), which reduce the work function for electrons. Tips with narrow solid angles can be formed using thermal-field treatment of the cathode (see Chapter 7).

By now a large number of works have been published, in which a possibility of producing nanostructures without using electron beam lithography is demonstrated. The most thoroughly developed is the area of "resist-free" lithography in which under a potential difference of a few volts between the probe and the substrate and a specific ambient near the probe incidence the surface is modified because of a local chemical reaction (e.g., oxidation). This method was successfully applied for forming nanostructures on a silicon surface passivated with hydrogen,[510] as well as on thin films of some metals, such as titan.[511]

Of special interest for FEAs fabrication is the technique of multiple-beam resist-free electron lithography capable of producing tip emitters of smaller size and greater density. As a source of electrons in the multiple-beam lithography matrix-controlled field emitter arrays are used.[509] The smallest hole in a film of electronic resist obtainable by this technique is 0.5 μm in diameter. The hole size is limited by aberration effects in simple electronic lens of the multiple-beam lithography units and electron scattering processes in the substrate.

8.5.4. Vacuum Magnetic Sensors

Magnetic field sensors (e.g., Vacuum Magnetic Sensors (VME)) are devices based on the use of ultrasmall tip field emitters displaying high sensitivity, low noise level, and considerable thermal stability. VMS detects the deviation of an electron beam incident on the anode caused by the Lorentz force thus giving a measure of the magnetic field induction. The first demonstration of VMSs and their potential for practical application was done by Sugiyama.[512] A VMS designs described in[513] comprised a conical silicon emitter tip positioned centrally in an orifice of a pulling gate electrode, four lens for focusing the electron beam, and a two-section anode. The tip was made by reactive ion etch and subsequent thermal oxidation of a silicon substrate with deposited layer of SiO_2 1 μm in diameter as a masking coat. The multilayer emitter structure was formed by subsequent deposition on the substrate of a 0.5 μm thick layer of dielectric (SiO_2) and a 0.2 μm thick film of niobium. The unit dimensions were 2 × 6 cm for an emitter–anode distance of 2 mm. The emission threshold voltage applied to the gate was 60 V; at 80 V on the gate and 200 V applied to anode the emission current was 0.6 μA. The tip was positioned exactly at the center of the gap between two anode sections, ensuring, in the absence of magnetic field,

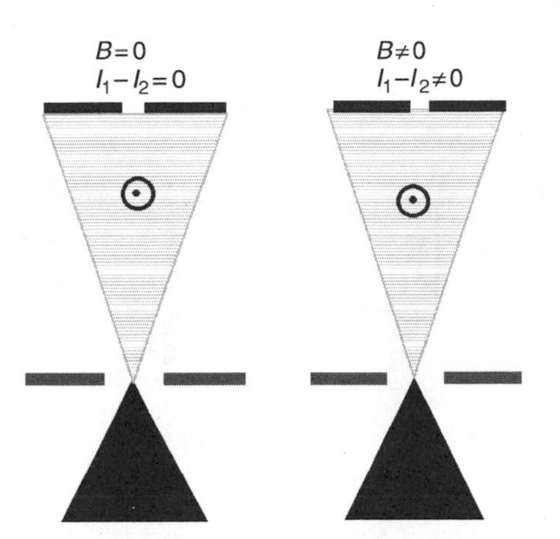

Figure 8.11. Principle of the magnetic sensor.[105]

equal current flows in circuits closing the anode sections. Application of the magnetic field in the plane of the multilayer emitter created a difference of currents in the circuits of the two-section anode proportional to the field induction (Fig. 8.11). The sensor was able to register magnetic field inductions of less than 10 mTl.

8.5.5. External pressure sensors

The main element of the sensor is the FE cathode positioned at a distance of about 1 μm from an elastic membrane, the collector. The pressure sensors may have diode or triode configurations.[526–529] In both cases the external pressure deflects (bends) the collector electrode causing variations of the current in the circuit corresponding to the magnitude of pressure (Fig. 8.12).

8.5.6. Mass-spectrometers with field emission cathodes

Mass-spectrometers with electronic ionization are widely used as sensitive and fast versatile detectors of molecular beams. Cryopumping is used to bring the residual gas pressure down to 10^{-10} Pa.

Use of hot cathodes in mass-spectrometers introduces an interfering effect of additional fragmentation of the molecules in the sample analyzed as a result of heating and illumination by the cathode. FE considerably simplifies the interpretation of mass-spectra because no heating or illumination is produced by FE cathodes. In the FE ionizer molecules are ionized as a result of losing or acquring electrons due to the applied field. Molecules ionized in an area of high field (10^2–10^3 MV/cm) experience much less fragmentation than in ionization by electronic impact. In Ortoff and Swanson,[514] a durable water-etch resistant iridium FE cathode was used as ionizer. Of special significance for the development of FE cathodes for use in ionizers are graphite fibers 5–8 μm in diameter or graphite filaments 1 mm in diameter. Such FE cathodes are non-sensitive to ion bombardment and spark discharges reducing the service life of metal FE cathodes.[515]

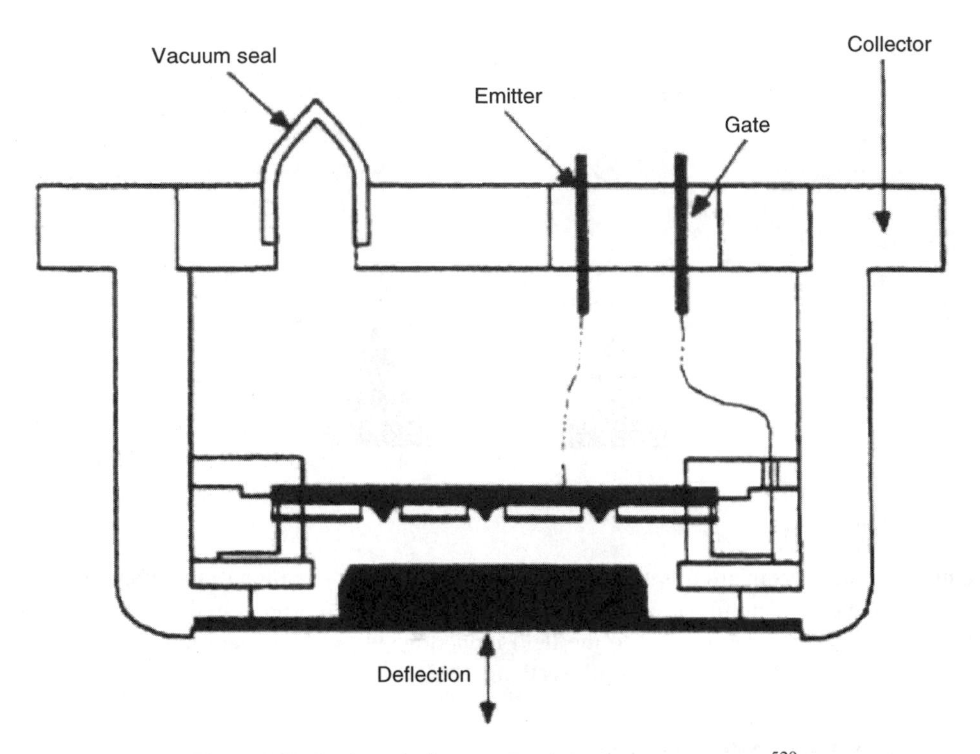

Figure 8.12. A schematic diagram of a triode-type pressure sensor.[529]

FEAs considerably increase the ionization volume and thus the sensitivity of the method.

For mass-spectrometry of nonvolatile compounds a field desorption ionizer has been proposed.[515] In field desorption a large electric field gradient is used for inducing electron tunneling from the sample surface, on which the substance studied is adsorbed. When a sample molecule looses an electron, it leaves the positively charged surface going over to the gas phase. The ions thus created enter the mass-spectrometer, are separated according to their mass-to-charge ratio, and detected. Field desorption has a number of advantages over other ways of producing ions, especially of polar and nonvolatile substances, namely the molecules are not vaporized by heating but are available in the adsorbed form right in the ionization zone. The field desorption ionization is less rough than electron impact ionization.

8.5.7. Use of Field Emission in Gas Lasers

Electron beams of energy of hundreds of keV and current density 10^2–10^4 A/cm^2 are used for electron excitation of gas lasers. The cross-section shape and dimensions of the electron beam should coincide with the working gas excitation chamber surface, on whose area the laser power is dependent in direct proportion. The electron beam has rectangular cross-section with uniform current density. The working gas excitation efficiency depends on uniformity of the input beam with constant current density over the excitation area of the gas laser. Usually, a blade FEC and a planar thin anode transparent for emitted electrons are used.

8.6. ARRAYS MADE OF CARBON NANOCLUSTERS

One of the major problems of the modern VME is fabrication of high-efficiency, low-voltage FEEs. Such field emitters should meet the following basic requirements: large working area, efficient emission at low voltages (from a few to tens of volts) and fields (1–10 V/μm), high density of emission centers, and a controllable and reproducible fabrication procedure. The traditional method of fabricating field emitters is based on the use of multi-needle FE cathodes and precision technological processes based on electron lithography techniques ac.

As a cathode material, metals and semiconductors are usually used which, unfortunately, have a rather high work function (4–5 eV). An alternative development path in VME is a relatively inexpensive technology of planar FE cathodes based on materials with low electron affinity or having high surface microroughness to ensure high local amplification factor β for the electric field. These materials include films and coats of various allotropic forms of carbon such as nanoclusters.[446, 461, 462]

From analysis of the data accumulated so far, two main approaches to the development of highly efficient field emitters based on carbon nanoclusters can be outlined.[461, 462] The first approach relies on planar FE cathodes with coats of diamond-like films or thin carbon films having high density of ordered emission centers.[153] In the other approach use is made of a FE multi-needle cathode with a coat of carbon nanoclusters of a thickness from hundreds of nanometers to tens of micrometers. The coat lowers the electronic work function and the emission threshold field strength, and also reduces fluctuations of the emission current, thus improving the emission characteristics.[463, 464]

Carbon nanoclusters are self-organizing structures and open a principally new way of fabricating FE cathodes with unique characteristics, such as *very low average magnitude of the threshold emission fields*. In addition, carbon nanoclusters ensure high emission stability, uniformity of the emission current density over the cathode surface and so on. Considerable theoretical and practical interest in the ways of obtaining efficient and stable field electron emission from carbon nanoclusters is reflected in the fact that in the past two years a considerable number of publications on this subject appeared. However, the advances in this field have not yet been presented in a systematic manner. A review of the most recent achievements was presented in Sinitsyn et al.[446]

Carbon materials of nanocluster structure having unique properties for use as FEEs can be divided into the following two basic groups: (1) diamond-like films and (2) coats based on fullerenes and carbon nanotubes.

Major preparation methods for these materials are electric arc vaporization, magnetron and laser sputtering of graphite, as well as ion-plasma deposition with the use of plasma injected with a gas containing hydrocarbons. Electric arc vaporization of graphite should be recognized as one of the most promising methods of producing carbon nanoclusters because of its ability to yield various allotropic forms of carbon from diamond to graphite, including intermediate forms: nanotubes and atomic clusters. It can also yield a variety of surface morphologies from nanoscale nonuniformities (50–200 nm) to carbon fibers, and coats of large area can be obtained.

Investigations of the emission properties of various carbon nanoclusters[461, 462, 466, 467] confirmed that FE from these materials takes place in electric fields more than two orders of magnitude lower than in metals and semiconductors. The current–voltage characteristics investigated in a wide range of currents (4–5 orders of magnitude) give a positive evidence that the mechanism of electron emission from these materials is linked to FE (Fig. 8.13).

No clear interpretation exists at the moment of the extremely low threshold field for the electron emission from carbon nanoclusters. A number of suggestions have been made to explain this phenomenon.

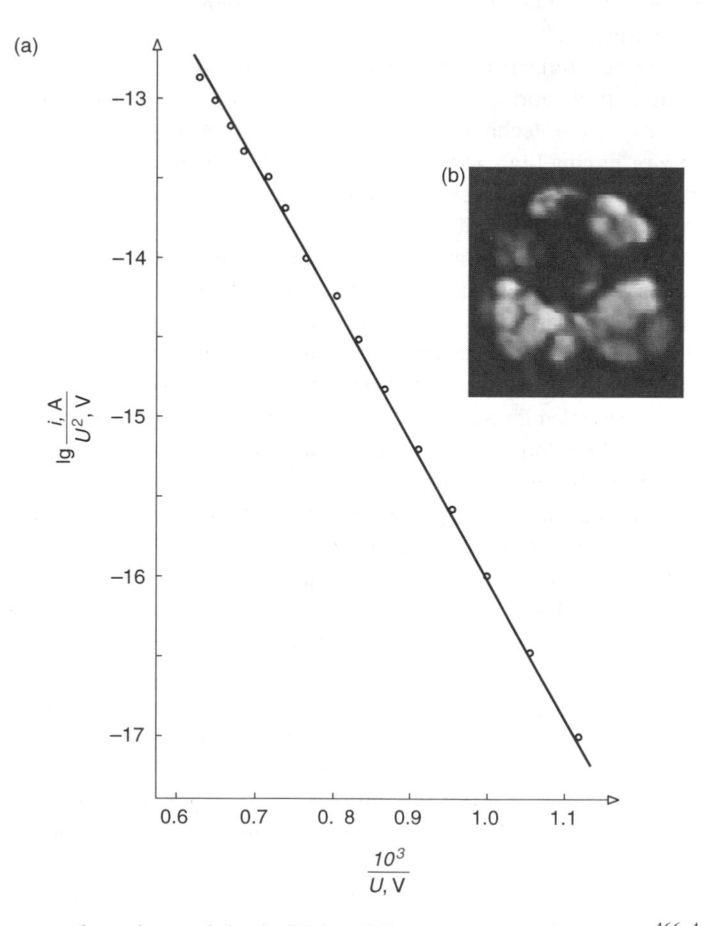

Figure 8.13. Current–voltage characteristics for FE from fullerene structure and nanotubes:[466, 467] (a) Current–voltage characteristics; (b) FE image of the cathode made from cathode deposit.

However, studies of the surface geometry of diamond-like films and films of amorphous carbon with the use of atomic force and tunneling microscopes [462, 468, 516] did not reveal any microstructures with a field enhancement factor β exceeding 5–10, so that the experimentally observed lowering of the threshold field by more than two orders of magnitude could not be accounted for.

2. To explain the FE mechanism in diamond-like films, a fundamental assumption has been made, now shared by many researchers, that the electron affinity for the emitter surface could be negative.[468]

However, this assumption raises some doubts because lowering of the threshold field for initiating the emission has been observed not only for diamond-like carbon but for a number of other carbon materials, such as films and coats of nanotubes and fullerenes as well as for some wide bandgap semiconductors.[463, 466, 467, 470] Besides, this interpretation is unable to explain the lack of enhanced emissive ability in macro-crystals of diamond and lowering of the emission threshold with transition from microcrystalline to nanocrystalline structure in diamond films.[517]

3. In studies,[470, 518, 519] a new approach has been proposed to explain the low values of electric fields, at which the FE from carbon nanoclusters is observed. This approach is based on a hypothesis that in the near-surface region (0.2–1 nm), a negative charge is accumulated in deep trap states within the bandgap and on defects as well as induced by polarization of adsorbed dipole-type molecules of impurities. This charge causes drastic increase of the electric field close to the surface of emitting centers. Calculations carried out for diamond-like films[470] have shown that at a concentration of the charges trapped by these centers of about 10^{16} cm^{-3}, the local electric field strength near the surface of nanoclusters can exceed 10^7 V/cm even without an externally applied voltage.

In the framework of this model, the experimentally observed relationships receive proper explanation. Becomes understandable the dependence of emission characteristics on film thickness, the effect of hysteresis in current–voltage characteristics of FEEs[462, 518] (with increasing current charging of the deep trap states takes place and the field strength in localized sites is changing). Ultimately, small values of the threshold voltage for emission become explainable: the external voltage is required not so much for creating a field at the surface as for transporting the electrons in the direction normal to the film.

4. Gulyaev and Sinitsyn[446] suggested that the key role is played by structural defects causing local lowering of the electronic work function.

Topologically, the defects can be produced by missing carbon atoms in the structure's network, introduced atoms of other elements, mutual interaction of the nanostructure components, or specific geometry of the nanoclusters formed.

Quantum-chemical calculations of model structures based on isolated or interacting fullerenes and tubelenes have shown that they have lower ionization potential and hence lower electron work function than graphite.[446]

In Zakharchenko et al.,[520] an analysis of the electronic spectra of single- and multiple-wall carbon nanotubes with open and capped ends was carried out. The inside of the tubes was doped with atoms of the elements of groups I–IV. It has been shown that the cylindrical part of the tubes has metallic conductivity with the Fermi level at about 5.5 eV; doping of the nanotubes with metal atoms (tin, lead) causes widening of the conduction band and lowering of the barrier at the interface with vacuum down to 3 eV. Near the dome cap at the top of a nanotube a forbidden band exists, which is characteristic of fullerene clusters.

A conclusion has been made that some walls in open-ended nanotubes are polarized by an external electric field and the mobility has been estimated of the electrons tunneling through the potential barrier at the surface of the walls arising due to a positive charge induced at their surface. It was shown that the positive charge induced at the end of a nanotube can significantly reduce the width of the potential barrier in the cylindrical part of the nanotubes and thus considerably enhance the FE current. The induced charge can be significantly increased due to ionization of atoms of different species, for example, oxygen, chemically bonded to a carbon atom.

Measurements of the energy distribution function of the electrons emitted from carbon nanotubes[521] have shown that the electrons are emitted not from the wide quasi-continuous band (as in metals) but from certain narrow energy levels having a half width of 0.3 eV corresponding to states localized at the ends of nanotubes, possibly, related to changes in the overlap of electronic π-orbitals in the area of the spherical caps of nanotubes.

In Chen et al.[522] a dependence of the emission current on orientation of nanotubes in an external field has been established. It is shown that film coats of nanotubes aligned parallel to the substrate have lower threshold field strength compared with coats of nanotubes oriented at a normal. This result is an evidence that the dominant emission from the walls is due to stacking faults of individual graphite layers, which can be observed in a scanning tunneling microscope.

Experiments with doping of films of carbon nanoclusters with atoms of tin, lead, and others[523] showed that doping of a carbon nanotube with, for example, tin atoms may cause severe deformation of its structure and, as a result, lowering of the ionization potential of the nanotube by 2 eV and corresponding reduction of the work function for electrons.

In studies by Rakhimov and co-workers,[461] FE images of diamond-like films were obtained and dependence of the emission current on the electric field strength investigated. It has been found that the emission current originated at isolated areas on the sample surface. High density of emission centers ($10^6\,\mathrm{cm}^{-2}$) and high surface uniformity of the emission current were observed. It was demonstrated that the centers of efficient emission consisted of a large number of spots (of a size 50–100 nm) located in the intercrystalline boundary areas. An emission mechanism has been proposed based on field enhancement at conducting intercrystalline "ribs" and supported by data of scanning electron microscopy indicating that the current from intercrystalline areas (of higher conductivity) is much larger than from apexes of nanocrystalline grains. It is interesting that in the limiting case of a mainly amorphous diamond-like film, high emissive properties were not observed. Rakhimov's data for the diamond-like films also confirm the hypothesis of the role of defect sites, exemplified in this case by the boundaries of nanocrystalline grains.

One of the most important tasks of our investigations was a quantitative study of the relationship between emission characteristics and kinetics of the FE process, on the one hand, and the topology of emitting surfaces based on fullerenes and other carbon nanoclusters, on the other.[462, 466, 467, 532, 533]

High-resolution atomic force microscopy was used in Baskin et al., (1999), Dyuzhev et al. (1998, 1999)[462, 466, 467] to determine dimensions of carbon clusters and the curvature radii of micro-irregularities at the surface of diamond-like films and fullerene coats.

Shown in Fig. 8.14 is a computer reconstruction of the surface element of a diamond-like film obtained by means of scanning atomic force microscopy. Characteristic heights

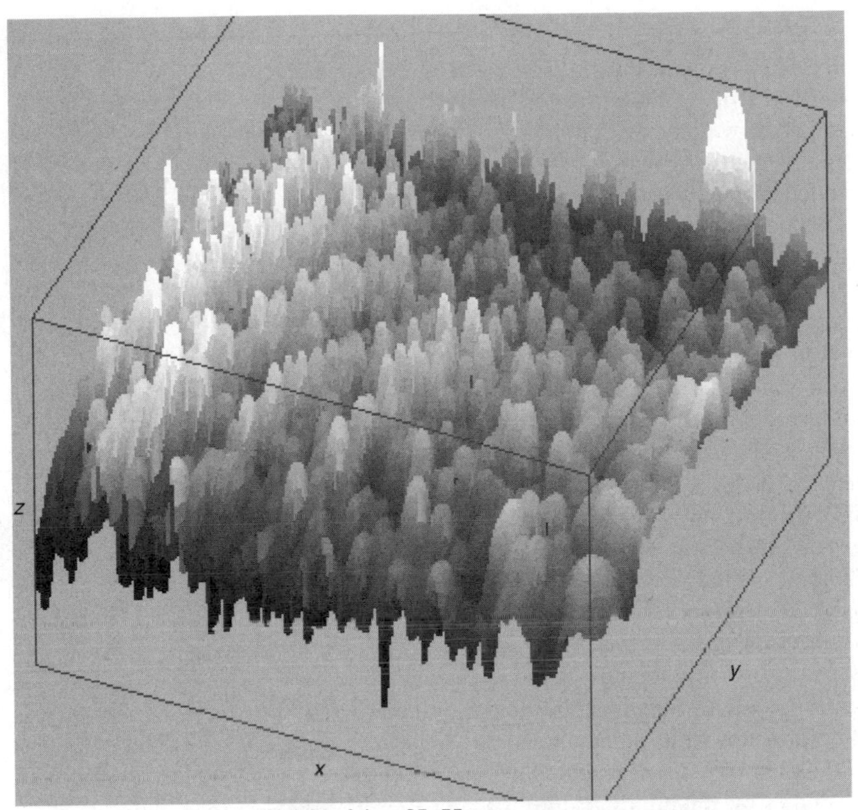

Figure 8.14. Atomic force microscopy image of diamond-like film.[453, 462]

of the pyramidal nanocrystals are 30–50 nm and the apex radii of the nanoclusters in some cases are as small as 5 nm. The density of micro-protrusions of small dimension in some spots at the surface amounts to 10^9 cm^{-2}.

Presented in Fig. 8.15 is a micrograph of adsorbed close-packed spatially ordered aggregates of C_{60} molecules of a size about 20 Å obtained by vacuum evaporation onto a Mo substrate.[453]

The obtained FE images of diamond-like and fullerene films reveal the presence of separate small-area centers with a high current density, up to 10^5 A/cm^2.

Most specimens had current–voltage characteristics obeying the Fowler-Nordheim (FN) law.

Carbon nanotubes is a potentially promising material for vacuum nanoelectronics. Such features of the nanotubes as their small dimensions (with curvature radii from 1 to 10 nm), electrical conductivity (100 Ω^{-1} cm^{-1}), mechanical strength, and chemical stability make nanotubes a candidate material for future microelectronic elements.[465]

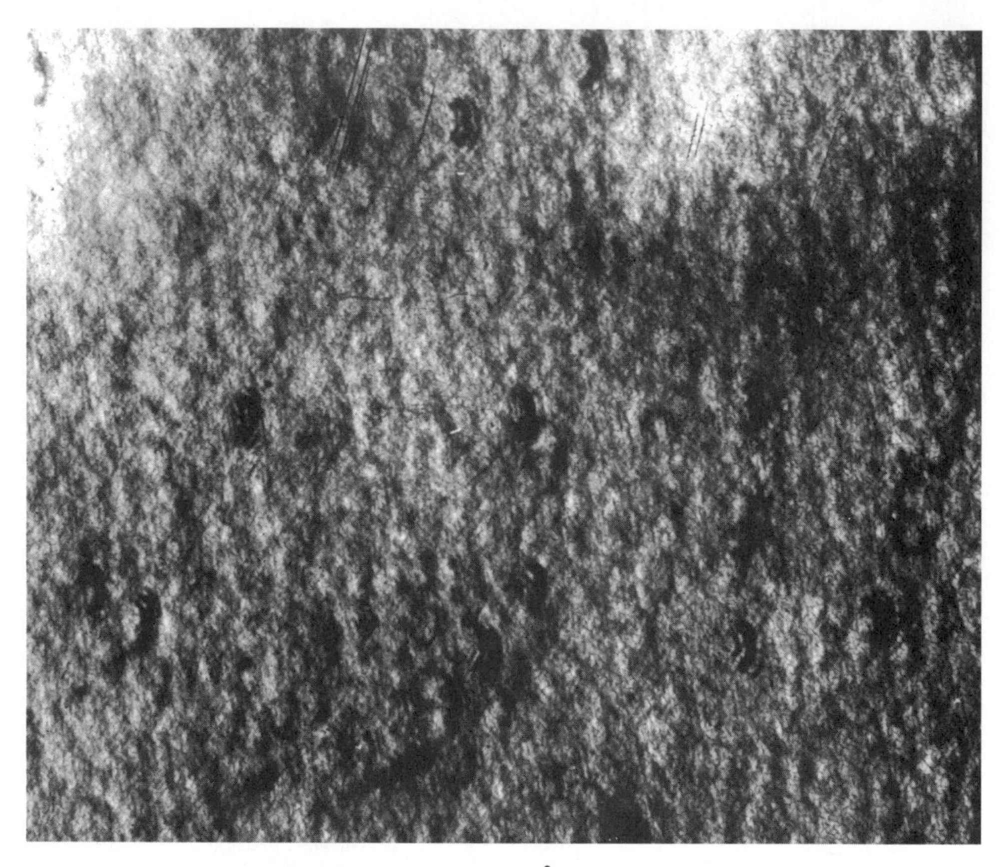

Figure 8.15. Transmission electron microscopy image of surface with fullerene structures.[453, 467]

Gulyaev et al.[446] have developed a technique for producing films and coats consisting mainly of single- and multiple-wall carbon nanotubes. The FE characteristics of carbon nanotubes are close to those of diamond-like films. According to Sinitsyn et al.,[446] electronic work function for films of single-wall nanotubes found from the slope of current–voltage characteristic in the FN coordinates is less than 0.5 eV. The FE current density averaged over the cathode surface reached $3\ \mathrm{A/cm^2}$.

A special variety of the carbon materials is the so-called "cathode deposit" forming during arc discharge vaporization of graphite directly on the cathode.[467, 532–534] It might prove a very suitable material for the fabrication of FEAs.

Detailed electron microscopy studies including scanning electron microscopy, transmission electron microscopy, and high-resolution tunneling microscopy have shown that the cathode deposit contains closely packed nanotubes, which in the growth process aligned along a preferred direction related to the direction of electric field (Fig. 8.16).

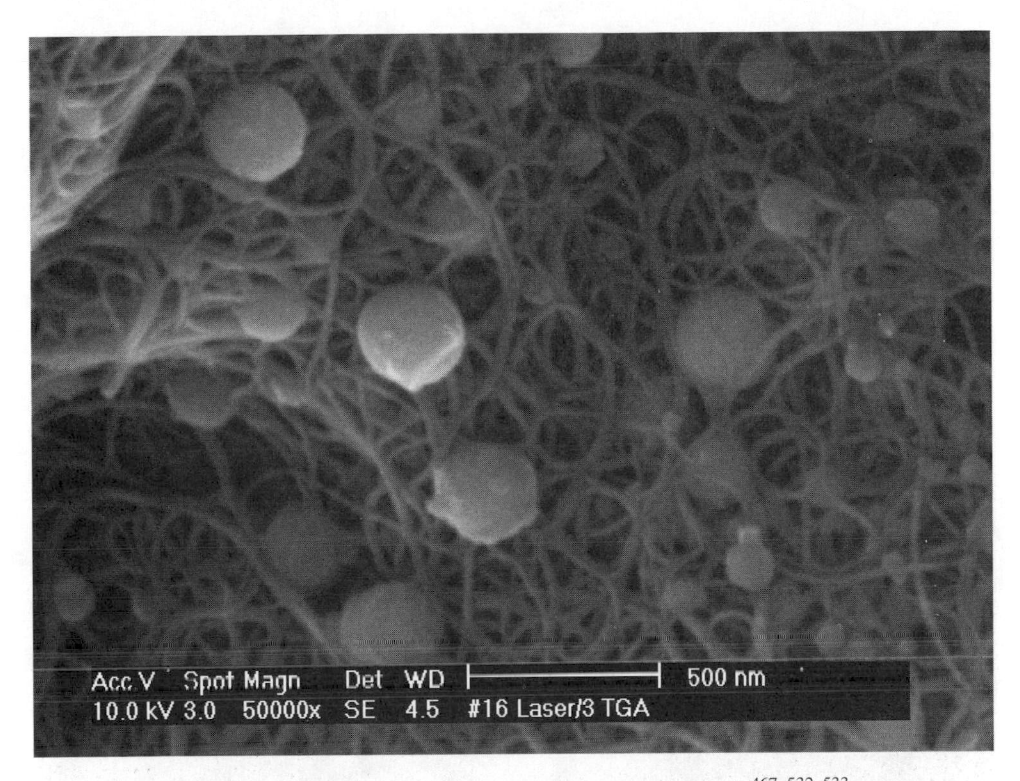

Figure 8.16. Scanning electron microscopy nanotubes micrograph.[467, 532, 533]

Distribution of the carbon nanotubes in the cathode deposit has been studied and a method developed for fabricating FE cathode arrays from the deposit material.

Direct observations of the emitting surface of the deposit in a FE microscope have revealed high density and uniformity of emission centers and low operating voltages. This opens the prospect of developing a technology of fabricating large-area FE cathodes taking advantage of self-organization of carbon clusters directly in the growth process. In this case, there is no need, in a special high-precision technology, for creating a regular pattern of nanotubes on a substrate of different material. FE images of a multi-emitter cathode made of the deposit are shown in Fig. 8.17.[530]

The studies of carbon nanoclusters are now in their initial stage, with the number of publication growing exponentially with time. The fact that a number of new devices for VME based on carbon materials have already been demonstrated can be considered a considerable technological achievement.

One of these entirely new types of devices are ecologically safe flat display panels with high brightness and resolution.[449]

FE cathodes based on diamond-like films and carbon nanotubes made possible fabrication of flat monolithic display panels characterized with high operating stability and low emission current fluctuations (not exceeding 10 percent).

Figure 8.17. FE image for nanotubes.[530, 532, 533]

A display with an emitter based on diamond-like carbon film described in Kim et al.[471] possesses a resolution of 60 lines per in. at an average emission current of 500 mA and operating voltage of 800 V. The diagonal of this monochrome panel is 2 in. and the screen brightness is $100\,cd/m^2$.

A monochrome display based on carbon nanotubes reported in Choi et al.[472] has an effective screen area of $130\,cm^2$ with resolution of 30 lines per in.; its brightness can reach $450\,cd/m^2$ at an operating voltage of 800 V and average current of 1.5 mA.

Fabrication with the use of nanotubes of a full-color display with a diagonal of 9 in. was reported in Choi et al.[524] The FE cathode is made by rubbing in the prepared matrix of a paste based on ground nanotubes. The pixel size is $200 \times 20\,\mu m$. At an electric field strength of $3\,V/\mu m$, the image brightness is 800, 200, and $150\,cd/m^2$ for green, red, and blue lights, respectively. The high brightness is achieved at a grid-cathode voltage of 100–300 V and anode voltage of 500–1000 V. The authors mention good reproducibility of the fabrication process.

8.7. RESUME

1. VME as a principally new field of electronics has taken shape.
2. Outstanding results have been achieved in developing FEAs for planar low-voltage FEDs, amplifiers, microwave devices, and so on.
3. It has been shown that there are no limitations for TV applications of multi-emitter FEAs at frequencies up to 330 MHz.
4. No principle limitations exist for the operation of high-frequency devices up to tens of GHz, which will have been fabricated in the near future.
5. Point FE cathodes have already found application in optoelectronic devices, in microscopy, spectroscopy, and holography.
6. VME have opened a possibility of integrating in one unit a point electron source and controlling electrodes for electron microscopy, mass-spectroscopy, Auger spectroscopy, and other applications (Baptist, Gray).

In 1996, Henry Gray had excellently portrayed the future of VME. His prophecies are coming true almost word for word now. It seems to me appropriate, as a homage to his memory, to quote his own words by way of a resume.

"Leading industrial manufacturers all over the world have already started commercial production of some types of vacuum microelectronic devices. Investments into this field are expected to amount to 10 billion dollars during the coming five to ten years. For instance, the Motorola company alone invested about 900 million dollars into the production of field emission displays in 1999."

REFERENCES

1. R. W. Wood, A new form of cathode discharge and the production of X-rays, together with some notes of diffraction, *Phys. Rev.* **5**(1), 1–10 (1897).
2. W. Schottky, Überkalte und warme elektronenentladungen, *Z. Physik* **14**, 63–106 (1923).
3. R. A. Millikan and C. F. Eyring, Laws governing the pulling of electrons out of metals by intense electrical fields, *Phys. Rev.* **27**(1), 51–67 (1926).
4. B. S. Gossling, The emission of electrons under the influence of intense electric fields, *Phil. Mag.* **1**(3), 609–635 (1926).
5. R. A. Millikan and C. C. Lauritsen, Dependence of electron emission from metals upon field strengths and temperatures, *Phys. Rev.* **33**(4), 598–604 (1929).
6. R. H. Fowler and L. W. Nordheim, Electron emission in intense electric fields, *Proc. Royal Soc. (London)* **A119**(A781), 173–181 (1928).
7. L. W. Nordheim, The effect of the image force on the emission and reflection of electrons by metals, *Proc. Royal Soc. (London)* **A121**(A788), 626–639 (1928).
8. J. E. Henderson and R. I. Badgley, The work required to remove a field electron, *Phys. Rev.* **38**(3), 590 (1931).
9. J. E. Henderson, R. H. Dahlstorm, and F. R. Abott, The energy distribution of electrons in field emission, *Phys. Rev.* **41**(3), 261 (1932).
10. J. E. Henderson and R. H. Dahlstorm, The energy distribution in field emission, *Phys. Rev.* **55**(5), 473–481 (1939).
11. G. M. Fleming and J. E. Henderson, A search for temperature changes accompanying field emission at high temperatures, *Phys. Rev.* **54**(3), 241 (1938).
12. G. M. Fleming and J. E. Henderson, The energy losses attending field current and thermionic emission of electrons from metals, *Phys. Rev.* **58**(10), 887–894 (1940).
13. E. W. Müller, Zur Geschwindigkeitsverteilung der Elektronen bei der Feldemission, *Z. Physik* **120**(5/6), 261–269 (1943).
14. E. W. Müller, Feldemission, *Ergebn. d. exakt. Naturwiss* **27**, 290–360 (1953).
15. R. D. Young, Theoretical total-energy distribution of field emitted electrons, *Phys. Rev.* **113**(1), 110–114 (1959).
16. R. D. Young and E. W. Muller, Experimental measurement of the total energy distribution of field emitted electrons, *Phys. Rev.* **113**(1), 115–120 (1959).
17. A. G. J. van Oostrum, Validity of the Fowler–Nordheim model for field electron emission, *Philips Res. Rep. Suppl.* **11**, 1–102 (1966).
18. L. W. Swanson and L. C. Crouser, Total energy distribution of field-emitted electrons and single plane work functions for tungsten, *Phys. Rev.* **163**(3), 622–641 (1967).
19. L. W. Swanson and A. E. Bell, Recent advances in field electron microscopy of metals, *Adv. Elect. Electron Phys.* **32**, 193–309 (1973).
20. E. W. Müller, Versuche zur Theorie der Elektronenemission unter der Einwirkung hoher Feldstarken, *Z. Techn. Physik* **37**(22/23), 838–842 (1936).

21. E. W. Müller, Elektronenmikroskopische Beobachtungen von Feldkathoden, *Z. Techn. Physik* **106**(9/10), 541–550 (1937).
22. G. Ehrlich, An atomic view of adsorption, *Brit. J. Appl. Phys.* **15**(4), 349–364 (1964).
23. R. Gomer, *Field Emission and Field Ionization.* (Harvard University Press, Cambridge, Massachusetts, 1961), 195 p. 2 ed (AIP, New York, 1993).
24. I. L. Sokol'skaya, Field-emission microscopy application for surface diffusion and self-diffusion investigations, in: *Surface Diffusion and Flowing*, edited by Ya. E. Geguzin (Nauka, Moscow, 1969), pp. 108–148.
25. M. I. Elinson and G. F. Vasil'ev, *Field Emission.* (Fizmatgiz, Moscow, 1958), p. 272.
26. G. N. Shuppe and N. S. Zakirov, *Electron Emission from Metal Crystals.* (Izdat. SAGU, Tashkent, 1959).
27. W. P. Dyke and W. W. Dolan, Field emission, in: *Advances in Electronics and Electron Physics*, Vol. 8, edited by L. Marton (Academic Press, New York, 1956), pp. 89–185.
28. G. N. Fursey, V. N. Shrednik, and G. A. Mesyats, High voltage cold cathodes, in: *Cold Cathodes*, edited by M. I. Elinson (Sovetskoe Radio, Moscow, 1974), pp. 165–304.
29. A. Modinos, *Field, Thermionic, and Secondary Electron Emission Spectroscopy* (Plenum, New York, 1984), 375 p.
30. R. Haefer, Experimentelle Untersuchungen zur Prüfung der wellenmechanischen Theorie der Feldelectronenemission, *Z. Phys.* **116**(9/10), 604–609 (1940).
31. W. P. Dyke and J. K. Trolan, Field emission: large current densites, space charge and vacuum arc, *Phys. Rev.* **89**(4), 799–808 (1953).
32. W. P. Dyke, J. K. Trolan, E. E. Martin, and J. P. Barbour, The field emission initiated vacuum arc. I. Experiments on arc initiation, *Phys. Rev.* **91**(5), 1043–1054 (1953).
33. W. W. Dolan, W. P. Dyke, and J. K. Trolan, The field emission initiated vacuum arc. II. The resistively heated emitter, *Phys. Rev.* **91**(5), 1054–1057 (1953).
34. J. P. Barbour, W. W. Dolan, J. K. Trolan, E. E. Martin, and W. P. Dyke, Space-charge effects in field emission, *Phys. Rev.* **92**(1), 45–51 (1953).
35. E. E. Martin, J. K. Trolan, and W. P. Dyke, Stable, high density field emission cold cathode, *J. Appl. Phys.* **31**(5), 782–789 (1960).
36. W. P. Dyke, F. M. Charbonnier, R. W. Strayer, R. L. Floyd, J. P. Barbour, and J. K. Trolan, Electrical stability and life of the heated field emission cathode, *J. Appl. Phys.* **31**(5), 790–805 (1960).
37. T. J. Lewis, Theoretical interpretation of field emission experiments, *Phys. Rev.* **101**(6), 1694–1698 (1956).
38. P. H. Cutler and D. Nagy. The use of a new potential barrier model in the Fowler–Nordheim theory of field emission, *Surf. Sci.* **3**(1), 71–94 (1965).
39. W. P. Dyke, Advances in field emission, *Sci. Amer.* **210**(1), 108–118 (1964).
40. M. I. Elinson, V. A. Gor'kov, and G. F. Vasil'ev, Field emission from rhenium, *Radiotekhnica i Elektronika* **3**(3), 307–312 (1958).
41. M. I. Elinson, Influence of gas adsorption at the surface of emitter upon field-electron emission, *Radiotekhnica i Elektronika* **3**(3), 438–439 (1958).
42. M. I. Elinson and G. F. Vasil'ev, Investigation of the field emission from lanthanum hexaboride, *Radiotekhnika i Electronika* **3**(7), 945–953 (1958).
43. M. I. Elinson and G. A. Kudinzewa, Field emission cathodes of high-melting metal compounds, *Radio Eng. Electron. Phys.* **7**, 1417–1423 (1962).
44. M. I. Elinson and G. A. Kudintseva, Field-electron cathodes of metallic high-melting compounds, *Radiotekhnika i Electronika* **7**(9), 1511–1518 (1962).
45. I. I. Gofman, O. D. Protopopov, and G. N. Shuppe, Pulse technique in the study of field emission from tungsten emitter, *Izvestiya Akad. Nauk UzSSR, Fizika- Matematika* **6**, 72–77 (1959).
46. I. I. Gofman and G. N. Shuppe, Pulse technique in the study of field emission from tungsten single crystels, *Radiotekhnika i Elekronika* **4**, 1215–1216 (1959).
47. G. N. Fursey, Field emission from tungsten monocrystal preceding the vacuum arc development, *Radiotekhnika i Elekronika* **6**(2), 298–302 (1961).
48. I. L. Sokol'skaya and G. N. Fursey, Investigation of the phenomena preceding the tungsten emitter destruction by pulses of high-density field-electron emission current. *Radiotekhnika i Elekronika* **7**(9), 1474–1483 (1962).
49. G. N. Fursey and G. K. Kartsev, Stability of field emission and migration processes preceding development of a vacuum arc, *Zh. Tech. Fiz.* **40**(2), 310–319 (1970).

50. I. L. Sokol'skaya and G. N. Fursey, The influence of different coatings upon the character of phenomena preceding the distruction of tungsten emitter by pulses of high density field-electron emission current, *Radiotekhnika i Elekronika* **7**(9), 1484–1494 (1962).
51. G. N. Fursey, V. E. Ptitsyn, and D. N. Krotevich, Spontaneous migration of the surface atoms at maximum current densities of the field-electron emission initiating vacuum breakdown, *Proc. 11th ISDEIV, Berlin, GDR.* (September 24–28, 1984), **1**, 69–71.
52. D. N. Krotevich, V. E. Ptitsyn, and G. N. Fursey, Spontaneous build-up of a field-emission cathode at maximum current densities, *Zh. Tech. Fiz.* **55**(3), 625–627 (1985).
53. G. N. Fursey, G. K. Kartsev, G. A. Mesyats, D. I. Proskurovskii, and V. P. Rotshtein, Field emission initiated vacuum arc at extremely high field at high current densities. *Proc. 9th Int. Conf. Phenomena Ionized Gases, Bucharest*, (1969), p. 88.
54. G. K. Kartsev, G. A. Mesyats, D. I. Proskurovskii, V. P. Rotshtein, and G. N. Fursey, Investigation of the time characteristics of the transition of field emission to vacuum arc, *Dokl. Akad. Nauk USSR* **192**(2), 309–312 (1970).
55. G. N. Fursey and S. A. Shakirova, Localization of field emission in small solid angles, *Zh. Tech. Fiz.* **36**(6), 1125–1131 (1966).
56. V. G. Pavlov, A. A. Rabinovich, and V. N. Shrednik, High local current densities of field emission on steady state process, *Zh. Tech. Fiz.* **45**(10), 2126–2134 (1975).
57. G. N. Fursey, L. M. Baskin, D. V. Glazanov, A. O. Yevgen'ev, A. V. Kotcheryzhenkov, and S. A. Polezaev, The specific features of field emission from submicron cathode surface areas at high current densities. *7th Int. Conf. Vacuum Microelectronics, Portland, Oregon* (July 30–August 3, 1995). Technical Digest (Electron Device Society), pp. 504–508, *J. Vac. Sci. Technol. B* **16**(1), 232–238 (1998).
58. M. I. Elinson, V. A. Gor'kov, and G. F. Vasil'ev, Study of a method to reduce bombardment of field-electron cathode by ions of residual gases, *Radiotekhnika i Electronika* **2**(2), 204–218 (1957).
59. V. A. Gor'kov, M. I. Elinson, and V. B. Sandomirskii, On the role of space charge in initiating of high-density field-electron emission current, *Radiotekhnika i Electronika* **7**(9), 1495–1500 (1962).
60. L. M. Baskin, L. L. Anan'ev, D. A. Borisov, A. A. Kantonistov, and G. N. Fursey, Ion bombardment suppression effect of a field emission cathode, *Radiotekhnika i Electronika* **28**(12), 2462–2464 (1983).
61. G. N. Fursey, Field emission in a microwave field, *J. Vac. Sci. Technol. B.* **13**(2), 558–565 (1995).
62. F. M. Charbonnier, J. P. Barbour, J. L. Brewster, W. P. Dyke, and F. J. Grundhauser, Intense nanosecond electron beams, *Trans. IEEE* **NS 14**(3), 789–793 (1967).
63. W. P. Dyke and F. M. Charbonnier, Advanced electron tube techniques, *Proceeding of the VI National Conference* (1963).
64. M. I. Elinson, V. A. Gor'kov, A. A. Yasnopol'skaya, G. A. Kudintseva, Study of pulse field emission at high current densities, *Radiotekhnika i Elektronika* **5**(8), 1318–1326 (1960).
65. G. N. Fursey, Power electron sources using field emission, in: *Cold Cathodes*, edited by M. I. Elinson (Sovetskoe Radio, Moscow, 1974), pp. 241–269.
66. G. N. Fursey and G. A. Mesyats, Explosive field emission at the initial stages of vacuum discharges, in: *Cold Cathodes*, edited by M. I. Elinson (Sovetskoe Radio, Moscow, 1974), pp. 269–303.
67. G. N. Fursey and P. N. Vorontsov-Vel'yamnov, Qualitative model of vacuum arc initiation, *Zh. Tech. Fiz.* **37**(10), 1870–1888 (1967).
68. S. P. Bugaev, A. M. Iskol'ski, G. A. Mesyats, and D. I. Proskurovskii, Electron-optical observation of initiation and development of pulse breakdown of short vacuum gap, *Zh. Tech. Fiz.* **37**(12), 2206–2208 (1967).
69. E. A. Litvinov, G. A. Mesyats, and D. I. Proskurovskii, Field emission and explosive electron emission processes in vacuum discharges, *Uspekhi Fiz. Nauk* **139**(2), 265–302 (1983).
70. G. N. Fursey, Field emission and vacuum breakdown, *IEEE Trans. Electron Insulation* **EI-20**, 659–670 (1985).
71. A. V. Crewe, D. N. Eggenberger, J. Wall, and L. M. Welter, Electron gun using a field emission source, *Rev. Sci. Instrum.* **39**(4), 576–583 (1968).
72. A. V. Crewe, J. Wall, and L. M. Welter, High-resolution scanning transmission electron microscope, *J. Appl. Phys.* **39**(13), 5861–5868 (1968).
73. A. V. Crewe and J. Wall, High-resolution capabilities in scanning microscopy. *Proc. 7th Int. Congr. Electron Microsc., Grenoble, France, 1970.* Paris, Franscaise de Microscope Electroniques, 1970, pp. 1–2.
74. A. V. Crewe, J. Wall, and J. Langmore, Visibility of single atoms, *Science* **168**(3937), 1338–1340 (1970).

75. P. Morin and F. Simondet, An electrostatic-field emission microscope for spectroscopic Auger-electron analysis, *J. Physique Colloq.* **45**, 307–308 (1984).
76. B. Reihl and J. K. Gimzewski, Field-emission scanning Auger microscope (FESAM), *Surf. Sci.* **189**, 36–43 (1987).
77. I. P. Zhizhin, G. N. Fursey, S. A. Shakirova, and A. V. Shishatskiy, For a problem on the getting of electron tips by field emission and magnetic optics, 24th All-Union Conference on Field emission electronics. Tashkent (1970), section 4, p. 34. *Izvestiya Akad. Nauk USSR* **35**(2), 302–306 (1971) (in Russian).
78. N. Samoto, N. Tamura, R. Shimizu, S. Namba, H. Hashimoto, and K. Gamo, A stable high-brightness electron-gun with Zr/W-tip for nanometer lithography. 1. Emission properties in Schottky-emission and thermal field-emission regions, *Jpn J. Appl. Phys. Part 1-Regular Papers and Short Notes* **24**(6), 766–771 (1985).
79. R. W. Devenish, D. J. Eaglesham, C. J. Humphreys, and D. M. Maher, Nanolithography using field-emission and conventional thermionic electron sources, *Ultramicroscopy* **28**, 324–329 (1989).
80. H. W. Fink, Point source for ions and electrons, *Physica Scripta* **38**, 260–263 (1988).
81. A. Tonomura, Applications of electron holography using a field-emission electron-microscope, *J. Electron Microsc.* **33**, 101–115 (1984).
82. H. W. Fink, I. Stoecker, and H. Schmid, Holography with low-energy electrons, *Phys. Rev. Lett.* **65**(10), 1204–1206 (1990).
83. H.-W. Fink, H. Schmid, H. J. Kreuzer, and A. Wierzbicki, Atomic resolution in lensless low-energy electron holography, *Phys. Rev. Lett.* **67**(12), 1543–1546 (1991).
84. J. J. Sáenz, N. García, H. De Raerdt, and V. T. Binh, Electron emission from small microtips, *Colloque de Physique, Colloque C8* **50**(11), C8-73–C8-78 (1989).
85. T. Kawasaki, M. Tomita, A. Tonomura, T. Miyada, T. Matsuda, and J. Endo, Development of 350 kv holography electron-microscope. 2. Magnetic-field superimposed field-emission electron-gun, *J. Electron Microsc.* **38**, 295 (1989).
86. R. Morin, Point source physics: application to electron microscopy and holography, *Microsc. Microanal. Microstruct.* **5**, 1–10 (1994).
87. I. Brodie and C. A. Spindt, Vacuum microelectronics, in: *Advances in Electronics and Electron Physics*, Vol. 83 (Academic Press, New York, 1992), pp. 1–106.
88. H. H. Busta, Vacuum microelectronics—1992. *J. Micromech. Microeng.* **2**(2), 43–74 (1992).
89. F. M. Charbonnier, J. P. Barbour, L. F. Garrett, and W. P. Dyke, Basic and applied studies of field emission at microwave frequencies, *Proc. IEEE* **51**(7), 991–1004 (1963).
90. M. I. Elinson and A. V. Gor'kov, The device for generation of short-term electronic packages, *Inventor's Certificate #137545. Bulletin of Inventions* (1961), No 8.
91. I. Brodie, Application of field emission cathodes to microwave power tubes. *Tri-Service Microwave Power Tube Conf.*, Monterey, California, USA, (1976).
92. I. Brodie and C. A. Spindt, The application of thin film field emission cathodes to electronic tubes, *Appl. Surf. Sci.* **2**(2), 149–163 (1979).
93. S. T. Smith and H. F. Gray, Distributed amplifier possibilities using field emitter arrays, *1st Int. Conf. Vacuum Microelectronics* (1988), Williamsburg, Virginia, USA.
94. D. Palmer, D. Temple, C. Ball, H. F. Gray, and G. McGuire, Low capacitance high-transconductance, silicon FEAs for microwave amplifier applications, *7th Int. Conf. Vacuum Microelectronics* (July 1994). Technical Digest (Societe Francaise du Vide, Paris), p. 297.
95. M. Garven, M. A. Kodis, J. L. Shaw, K. T. Nguyen, and H. F. Gray, FEA emission characterisation and beam transport for linear microwave power amplifiers, *7th Int. Conf. Vacuum Microelectronics, Portland, Oregon* (July 30–August 3, 1995). Technical Digest (Electron Device Society), pp. 231–234.
96. Yu. V. Gulyaev, A. Gritsenko, V. Efimov, Y. Zakharchenko, and N. Sinitsyn, Functional possibility of vacuum circuits based on distributed microwave systems, *3rd Int. Conf. Vacuum Microelectronics* (1990), Monterey, California, USA.
97. Yu. F. Zakharchenko, N. I. Sinitsyn, and Yu. V. Gulyaev, Distributed generator with extented interaction based on field emitter arrays, *9th Int. Conf. Vacuum Microelectronics*, *St. Petersburg, Russia* (July 7–12, 1996). Technical Digest (Nevsky Courier, St. Petersburg), pp. 610–613. *J. Vac. Sci. Technol. B* **15**(2), 533–534 (1997).
98. P. N. Laily, Y. Goren, and E. A. Nettesheim, A broadband distributed amplifier based on thin-film field-emission cathodes, *1st Int. Conf. Vacuum Micrielectronics* (1988), Williamsburg, Virginia, USA.

99. W. A. Anderson, Frequency limits of electronic tubes with field emission cathodes. *Vacuum Microelectronics 89 (Inst. Phys. Conf. Ser. 99)* edited by R. Turner (Institute of Physics, Bristol, 1989), pp. 217–221.
100. H. G. Kosmahl, A wide-bandwidth high-gain small-size distributed amplifier with field-emission triodes (FETRODEs) for the 10 to 300 GHz frequency range, *IEEE Trans. Electron Devices* **ED-36**(11), 2728–2737 (1989).
101. V. I. Makhov, Low voltage pulsed magnetron with heating free excitation, *9th Int. Conf. Vacuum Microelectronics, St. Petersburg, Russia* (July 7–12, 1996). Technical Digest (Nevsky Courier, St. Petersburg), pp. 449–452.
102. K. R. Shoulders, Microelectronics using electron beam activated machining techniques, in: *Advances in Computing*, edited by F. L. Alt (Academic, New York, 1961), pp. 135–293.
103. C. A. Spindt, C. E. Holland, I. Brodie, J. B. Mooney, and E. R. Westerberg, Field emitter arrays applied to vacuum fluorescent display, *IEEE Trans. Electron Devices* **ED-36**(1), 225–228 (1990).
104. H. G. Gray, Vacuum microelectronics 1996: Where we are and where we going, *9th Int. Conf. Vacuum Microelectronics, St. Petersburg, Russia* (July 7–12, 1996). Technical Digest (Nevsky Courier, St. Petersburg), pp. 1–3.
105. R. Baptist, Trends and developments of vacuum microelectronics in Europe. *9th Int. Vacuum Microelectronics Conf. (July, 7–12 1966)*, St. Petersburg, Russia. *Le Vide: Science Tecnique et Applications* **52**(282) 499–516 (1996).
106. R. E. Burgess, H. Kroemer, and J. M. Houston, Corrected values of Fowler–Nordheim field emission function $\vartheta(y)$ and S(y), *Phys. Rev.*, **90**, 515 (1953).
107. V. I. Makhov, Field emission cathode technology and its application, *4th Int. Vacuum Microelectronics Conf.* Technical Digest (Nagahama, Japan, 1991), pp. 40–43.
108. L. D. Karpov, A. P. Genelev, V. A. Drach, V. S. Zasenkov, Y. V. Mirgorodski, and A. N. Tikhonski, Patterning of vertical thin film emitters in field emission arrays and their emission characteristics, *9th Int. Conf. Vacuum Microelectronics, St. Petersburg, Russia* (July 7–12, 1996). Technical Digest (Nevsky Cur'er, St. Petersburg), pp. 501–504.
109. L. D. Karpov, A. P. Genelev, Y. V. Mirgorodski, and A. N. Tikhonski, Patterning and electrical testing of field emission arrays with novel emitter geometries, *9th Int. Conf. Vacuum Microelectronics, St. Petersburg, Russia* (July 7–12, 1996). Technical Digest (Nevsky Courier, St. Petersburg), pp. 542–546.
110. A. K. Ganguly, P. M. Philips, and H. F. Gray, Linear theory of a field-emitter-array distributed amplifier, *J. Appl. Phys.* **67**(11), 7098–7110 (1990).
111. C. A. Spindt, C. E. Holland, P. R. Schwoebel, and I. Brodie, Maximizing field emitter-array transconductance for microwave applications, *8th Int. Conf. Vacuum Microelectronics, Portland, Oregon* (July 30–August 3, 1995). Technical Digest (Electron Device Society), p. 137.
112. B. Gorfinkel and J. M. Kim, Co-development of field emission displays at scientific research institute "Volga" (Russia) and Samsung Advanced Institute of Technology (Korea), *9th Int. Conf. Vacuum Microelectronics, St. Petersburg, Russia* (July 7–12, 1996). Technical Digest (Nevsky Courier, St. Petersburg), pp. 552–556. *J. Vac. Sci. Technol. B*, **15**(2), 524–527 (1997).
113. C.-C. Wang, J.-C. Wu, C.-M. Huang, Data line driver design for a 10″480 × 640 × 3 color FED, *9th Int. Conf. Vacuum Microelectronics, St. Petersburg, Russia* (July 7–12, 1996). Technical Digest (Nevsky Courier, St. Petersburg), pp. 557–561.
114. P. Vaudine and R. Meyer, 'Microtips' fluorescent display, *Proc. IEDM 91 (Washington DC)* (IEEE, New York, 1991) pp. 197–200.
115. C. Curtin, The field emission display: a new flat panel technology, *Invited Paper, Conf. Record of 1991 Int. Display Res. Conf. (IDRC 91, San Diego, C4)* pp. 12–15.
116. A. Ghis, R. Meyer, P. Rambaud, F. Levy, and T. Leroux, Sealed vacuum devices: fluorescent microtip displays, *IEEE Trans. Electron Devices* **ED-38**(10), 2320–2322 (1991).
117. A. Ting, C. M. Tang, T. Swyden, D. McCatrhy, and M. Peckerar, Field-effect controlled vacuum field-emission cathodes. *4th Int. Vacuum Microelectronics Conf.* Technical Digest (Nagahama, Japan, 1991), pp. 200–201.
118. L. Zhang, A. Q. Gui, and W. N. Carr, Lateral vacuum microelectronic logic gate design, *J. Micromech. Microeng.* **1**(2), 126–134 (1991).
119. I. I. Gofman, Study of electrostatic emission of electrons from tungsten in wide range of current densities, *Fizika Tv. Tela* **4**(8), 2005–2015 (1962).

120. A. E. Bell and L. W. Swanson, Total energy distributions of field-emitted electrons at high current density, *Phys. Rev. B* **19**, 3353–3364 (1979).
121. G. Lea and R. Gomer, Evidence of electron–electron scattering from field emission, *Phys. Rev. Lett.* **25**, 804–806 (1970).
122. J. W. Gadzuk and E. W. Plummer, Hot-hole-electron cascades in field emission from metals, *Phys. Rev. Lett.* **26**, 92–95 (1971).
123. V. T. Binh, S. T. Purcell, N. Garcia, and J. Doglioni, Field-emission electron spectroscopy of single-atom tips, *Phys. Rev. Lett.* **69**(17), 2527–2530 (1992).
124. M. Yu. Sumetskii and G. V. Dubrovskii, Tunneling through axially-symmetrical barrier, *Yadernaya Fizika* **29**, 1406–1413 (1979).
125. M. Yu. Sumetskii, Tunneling probability through multidimensional potential barriers, *TMF* **45**, 64–75 (1980).
126. M. Yu. Sumetskii, About dimensional effects in the tunneling process, *Zh. Tekhn. Fiz.* **54**(11), 2227–2232 (1981).
127. B. B. Rodnevich, Energy spectrum of field emission, *9th Int. Conf. Vacuum Microelectronics, St. Petersburg, Russia* (July 7–12, 1996). Technical Digest (Nevsky Courier, St. Petersburg), pp. 77–80.
128. J. W. Gadzuk, Many-body tunneling-theory approach to field emission of electrons from solids, *Surf. Sci.* **15**, 466–482 (1969).
129. J. W. Gadzuk, Band-structure effects in the field-induced tunneling of electrons from metals, *Phys. Rev.* **182**(2), 416–426 (1969).
130. F. I. Itskovich, On the theory of field-electron emission from metals, *Zh. Eksper. i Teor. Fiz.* **50**(5), 1425–1437 (1966).
131. F. I. Itskovich, On the theory of field-electron emission from metals, *Zh. Eksper. i Teor. Fiz.* **52**(6), 1720–1735 (1967).
132. F. V. Bunkin and M. V. Fedorov, Cold emission of electrons from surface of metals in strong field of radiation, *Zh. Eksper. i Tekh. Fiz.* **48**, 1341–1345 (1965).
133. L. V. Keldysh, Ionization in the field of a strong electromagnetic wave, *J. Exp. Theo. Phys.* **47**, 1945–1957 (1964).
134. M. Buttiker and R. Landauer, Traversal time for tunneling, *Phys. Rev. Lett.* **49**, 1739–1742 (1982).
135. D. G. Sokolovski and L. M. Baskin, Traversal time in quantum scattering, *Phys. Rev. A* **36**, 4604–4611 (1987).
136. N. Muller, W. Eckstein, W. Heiland, and W. Zinn, Electron spin polarization in field emission from EuS-coated tungsten tips, *Phys. Rev. Lett.* **29**(25), 1651–1654 (1972).
137. N. Muller, Electron spin polarization in field emission from nickel: effects of surface adsorption, *Phys. Lett. Ser. A* **54**, 415–416 (1975).
138. M. Landolt and M. Compagna, Spin polarization of field-emitted electrons and magnetism at the (100) surface of Ni, *Phys. Rev. Lett.* **38**, 663–666 (1977).
139. M. Landolt, M. Compagna, J. N. Chazalviel, Y. Yafet, and B. Wilkens, New tunneling experiments from strong ferromagnets and predictions of the band theory: Resolution of a controversy, *J. Vac. Sci. Technol.* **14**, 468–470 (1977).
140. M. Landolt and M. Compagna, Demagnetization of the Ni(100) surface by hydrogen adsorption, *Phys. Rev. Lett.* **39**(9), 568–570 (1977).
141. G. Chrobok, M. Hofmann, G. Regenfus, and R. Sizmann, Spin polarization on field-emitted electrons from Fe, Co, Ni and rare-earth metals, *Phys. Rev. Ser. B* **15**(1), 429–440 (1977).
142. G. Baum, E. Kisker, A. H. Mahan, W. Raith, and B. Reihl, Field emission of monoenergetic spin-polarized electrons, *Appl. Phys.* **14**, 149–153 (1977).
143. M. Landolt and M. Compagna, Electron spin polarization in field emission, *Surf. Sci.* **70**, 197–210 (1978).
144. M. Landolt and Y. Yafet, Spin polarization of electrons field emitted from single-crystal iron surfaces, *Phys. Rev. Lett.* **40**, 1401–1403 (1978).
145. A. I. Baz', Ya. B. Zeldovich, and A. M. Perelomov, Scattering, reactions and decay in nonrelativistic quantum mechanics. (Wiley, New York, 1969), p. 544.
146. E. Kisker, G. Baum, A. H. Mahan, W. Raith, and B. Reihl, Electron field emission from ferromagnetic europium sulfide on tungsten, *Phys. Rev. Ser. B* **18**(5), 2256–2275 (1978).
147. G. N. Fursey, A. V. Kocheryzhenkov, and V. I. Maslov, The quantity of elementary acts and statistics of the field emission, *Surf. Sci.* **246**(1–3), 365–372 (1991).

148. G. N. Fursey, A. V. Kocheryzhenkov, V. I. Maslov, and A. P. Smirnov, Multi-particle tunneling during field emission from YBa(2)Cu(3)O(7-d), *Pis'ma v Zhurnal Tekhnicheskoi Fiziki* **14**(20), 1853–1856 (1988).
149. R. Fischer and H. Neumann, Feldemission aus Halbleitern, *Fortschr. Phys.* **14**, 603–692 (1966).
150. G. N. Fursey and O. I. L'vov, News in the investigation of field-electron emission from semiconductors (review), in: *R. Fischer and Kh. Noyman, Field-electron emission from semiconductors*, translated by G. N. Fursey and B. I. Saprykin, edited by I. L. Sokol'skaya (Nauka, Moscow 1971), pp. 137–215.
151. L. M. Baskin, O. I. L'vov, and G. N. Fursey, General features of field emission from semiconductors, *Phys. Stat. Sol. (b)* **47**, 49–62 (1971).
152. G. N. Fursey, Early field emission studies of semiconductors. *Appl. Surf. Sci.* **94/96**, 44–59 (1996).
153. A. T. Rakhimov, B. V. Seleznev, N. V. Suetin, A. V. Kandidov, V. A. Tugarev, and I. A. Leont'ev, Examination of electron field emission efficiency and homogeneity from CVD diamond films, *Proc. Appl. Diamond Conf. 1995, Applications of Diamond Films and Related Materials: 3rd Int. Conf., National Institute of Standards and Technology, Gaithersburg, MD* (August 21–24, 1995). Edited by Albert Feldman, National Institute of Standards and Technology Gaithersburg, MD, USA, pp. 11–14.
154. Yu. V. Gulyaev, L. A. Chernozatonskii, Z. Ya. Kosakovskaya, A. L. Musatov, N. I. Sinitsin, and G. V. Torgashov, Carbon nanotubes structures a new material of vacuum microelectronics, *9th Int. Conf. Vacuum Microelectronics, St. Petersburg, Russia* (July 7–12, 1996). Technical Digest (Nevsky Cur'er, St. Petersburg), pp. 5–9.
155. Yu. V. Gulyaev, N. I. Sinitsyn, G. V. Torgashov, Sh. T. Mevlyut, A. I. Zhbanov, Yu. V. Zakharchenko, Z. Ya. Kosakovskaya, L. A. Chernozatonskii, O. E. Glukhova, and I. G. Torgashov, Work function estimate for electrons emitted from nanotube carbon cluster films. *9th Int. Conf. Vacuum Microelectronics, St. Petersburg, Russia* (July 7–12, 1996). Technical Digest (Nevsky Cur'er, St. Petersburg), pp. 206–210. *J. Vac. Sci. Technol. B* **15**(2), 422–424 (1997).
156. Yu. V. Gulyaev, N. I. Sinitsyn, G. V. Torgashov, Yu. A. Grigoriev, V. I. Shesterkin, A. G. Veselov, Yu. V. Shvetsov, and V. C. Semyonov, Emission of low-voltage multi-tip carbon matrices coated by carbon clusters, *9th Int. Conf. Vacuum Microelectronics, St. Petersburg, Russia* (July 7–12, 1996). Technical Digest (Nevsky Cur'er, St. Petersburg), pp. 519–521.
157. F. Y. Chuang, C. Y. Sun, H. F. Cheng, W. C. Wang, C. M. Huang, and I. N. Lin, Electron emission characteristics of pulsed laser deposited diamond-like films, *9th Int. Conf. Vacuum Microelectronics, St. Petersburg, Russia* (July 7–12, 1996). Technical Digest (Nevsky Cur'er, St. Petersburg), pp. 334–338.
158. H. Kim, Y. W. Choi, J. O. Choi, H. S. Jeong, and S. Ahn, I–V characteristics of volcano type Si FEAs with diamond-like carbon coating, *9th Int. Conf. Vacuum Microelectronics, St. Petersburg, Russia* (July 7–12, 1996). Technical Digest (Nevsky Courier, St. Petersburg), pp. 344–348.
159. L. W. Nordheim, Die Theorie der Elektronenemission der Metalle, *Z. Physik* **30**(7), 177–196 (1929).
160. N. Fröman and P. O. Fröman, *JWKB Approximation. Contribution to the Theory*. (New-Holland Publishing Company, Amsterdam, 1965).
161. D. Bohm, *Quantum Theory*. (Prentice-Hall, New York, 1952).
162. W. W. Dolan, Current density tables for field emission theory, *Phys. Rev.* **91**(3), 510–511 (1953).
163. E. Guth and C. J. Mullin, Electron emission of metals in electric fields, *Phys. Rev.* **61**(5–6), 339–348 (1942).
164. E. L. Murphy and R. H. Good, Thermionic emission, field emission and the transition region, *Phys. Rev.* **102**(6), 1464–1473 (1956).
165. S. G. Christov, General theory of electron emission from metals, *Phys. Stat. Sol.* **17**(1), 11–26 (1966).
166. I. S. Andreev, Investigation of electron emission from metal in the range of transition from cold emission to thermionic field emission, *Zh. Tekh. Fiz.* **22**(9), 1428–1441 (1952).
167. W. W. Dolan and W. P. Dyke, Temperature and field emission of electrons from metals, *Phys. Rev.* **95**(2), 327–332 (1954).
168. M. I. Elinson, F. F. Dobryakova, V. F. Krapivin, and Z. A. Malina, Yasnopol'skaya, On the theory of field- and thermionic-field emission of metals and semiconductors, *Radiotekhnika i Electronika* **6**(8), 1342–1353 (1961).
169. L. D. Landau and E. M. Lifschitz, *Quantum Mechanics*. (Nauka, Moscow, 1982).
170. R. Z. Bakhtizin and Yu. M. Yumaguzin, Study of the high-energy tail of the tungsten field-electron distribution, *Poverkhnost': Fiz., Khim., Mekhan.* **7**, 51 (1987).
171. R. Z. Bakhtizin and Yu. M. Yumaguzin, High energy tails in the total energy-distribution of field emitted electrons from tungsten, *Phys. Stat. Sol. B* **147**, 241 (1988).

172. G. Binning, H. Rohrer, Ch. Gerber, and E. Weibel, Surface studies by scanning tunneling microscopy, *Phys. Rev. Lett.* **49**(1), 57–61 (1982).
173. G. N. Fursey and D. V. Glazanov, Field emission properties of ultrasmall Zr spots on W, *J. Vac. Sci. Technol. B* **13**(3), 1044–1049 (1995).
174. G. N. Fursey, L. M. Baskin, D. V. Glazanov, A. O. Yevgen'ev, A. V. Kotcheryzhenkov, and S. A. Polezhaev, The specific features of field emission from submicron cathode surface areas at high current densities, *J. Vac. Sci. Technol. B*, **16**(1), 232–237 (1998).
175. P. H. Cutler, Jun He, and N. M. Miskovsky, Theory of electron emission in high fields from atomically sharp emitters: validity of the Fowler–Nordheim equation, *J. Vac. Sci. Technol. B* **11**(2), 387–391 (1993).
176. Jun He, P. H. Cutler, N. M. Miskovsky, T. E. Feuchtwang, T. E. Sullivan, and M. Chung, Derivation of the image interaction for non-planar pointed emitter geometries:application to filed emission I–V characteristics, *Surf. Sci.* **246**, 348–364 (1991).
177. A. A. Porotnikov and B. B. Rodnevich, On the structure of the potential barrier at a metal boundary, *Appl. Mech. Tech. Phys.* **5**, 80–86 (1978).
178. V. N. Shrednik, Field emission microscopy from metal-film covers. Ph. D. Dissertation, Ioffe Institute of Physics and Technology, USSR Academy of Sciences, Leningrad (1965).
179. V. N. Shrednik, Borders of applicability of the Fowler-Nordheim theory and corrections to the theory, in: *Cold Cathodes*, edited by M. I. Elinson (Sovetskoe Radio, Moscow, 1974), pp. 169–177.
180. I. M. Lifshits, M. Ya. Azbel', and M. N. Kaganov, *Electronic Theory of Metals*. (Nauka, Moscow, 1971).
181. D. N. Krotevich, Study of surface microgeometry and structure in the pre-explosion phase of electron emission. Ph. D. Dissertation, Prof. M. A. Bonch-Bruevich Electrotechnical Institute of Telecommunication, Leningrad (1985).
182. G. N. Fursey, L. M. Baskin, and D. V. Glazanov, Study of general emissive parameters of sub-nanometer field emission emitter. *42nd Int. Field Emission Symp., Madison, Wisconsin, USA*, (August 7–11, 1995).
183. B. A. Politzer and P. H. Cutler, Model calculation of field emission from 3d bands in nickel, *Surf. Sci.* **22**(2), 277–289 (1970).
184. B. A. Politzer and P. H. Cutler, Theory of electron tunneling from d bands in solid. Application to field emission, *Mater. Res. Bull.* **5**(8), 703–720 (1970).
185. A. M. Brodskiy and Yu. A. Gurevich, *Theory of Field Electron Emission from Metals*. (Nauka, Moscow, 1973).
186. L. M. Baskin. *To the theory of field emission from semiconductors*. Ph. D. Dissertation, Institute of automatic control systems and radioelectronic, Tomsk (1975).
187. I. O. Kulik and I. K. Yanson. *Jozephson Effect in Superconducting Tunnel Structures*. (Nauka, Moscow, 1970), p. 144.
188. F. R. Schrieffer, *Theory of Superconductivity*. (Benjamin, New York, 1964).
189. A. A. Abrikosov, L. P. Gor'kov, and I. E. Dzyaloshinski, *Methods of Quantum Field Theory in Statistical Physics*. (Fizmatgiz, Moscow, 1962).
190. J. W. Gadzuk and E. W. Plummer. Field emission energy distribution (FEED), *Rev. Modern Phys.* **45**(3), 487–548 (1973).
191. R. Young, Search for field emission of "paired" or correlated electrons. *Proc. 14th Field Emission Symp.*, NBS, Maryland and Georgetown University (1967), pp. 7–10.
192. T. Engel and R. Gomer, Search for field-emission of electron pairs. *Proc. 14th Field Emission Symp.*, NBS, Maryland and Georgetown University (1967), p. 6.
193. J. Bardeen, The image and van der Waals forces at a metallic surface, *Phys. Rev.* **58**(8), 727–735 (1940).
194. G. N. Fursey and I. D. Tolkacheva, High field emission current and the effects preceding vacuum breakdown for Ta and Mo emitters, *Radiotechnika I Electronika* **8**(7), 1210–1221 (1963).
195. G. N. Fursey, Impulse field emission from rhenium, *Zh. Tekh. Fiz.* **34**(7), 1312–1316 (1964).
196. T. E. Stern, B. S. Gosling, and R. H. Fowler, Further studies in the emission of electrons from cold metals, *Royal Soc. Proc.* **A124**, 699–722 (1929).
197. N. B. Isenberg, Role of a volume charge in spherical electron projectors. *Zh. Tekh. Fiz.* **24**(11), 2079–2082 (1954).
198. N. B. Isenberg, Influence of a volume charge on a form of $lnI = f(I/V)$ characteristic of field cathodes, *Radiotekhnika I Electronika* **9**(12), 2147–2155 (1964).

199. A. S. Kompaneets, Influence of a volume charge of field emission, *Radiotekhnika I Electronika* **5**(8), 1315–1317 (1960).
200. A. S. Kompaneets, Influence of a volume charge on field emission, *Doklady AN USSR* **128**(6), 1160–1162 (1959)
201. R. N. Poplavskii, Potential distribution in a spherical capacitor in case of saturation current, *Zh. Tekh. Fiz.* **20**(2), 149–159. (1950).
202. V. A. Godyak, L. V. Dubovoi, and G. R. Zabolotskaya, Calculation of field-emission current of relativistic electrons confined by a space charge. *Zh. Eksp. Tekh. Fiz.* **57**(11), 1795–1798 (1969).
203. L. M. Baskin, V. A. Godyak, O. I. Lvov, G. N. Fursey, and L. N. Shirochin, Influence of space charge of relativity electrons on field emission. *Zh. Tekh. Fiz.* **57**(11), 1795–1798 (1972).
204. G. N. Fursey, A. A. Antonov, and B. F. Gulin, Study of tungsten field emission in nanosecond range of pulse durations, *Vestnik Leningrad. University*, **10**, 71–74 (1971).
205. J. Bardeen, Theory of the work function, *Phys. Rev.* **49**(9), 653–663 (1936).
206. H. J. Juretchke, Exchange potential in the surface region of a free electron metal, *Phys. Rev.* **93**(5), 1140–1144 (1953).
207. T. L. Loucks, P. H. Cutler, The effect of correlation of the surface potential of a free electron metal, *J. Phys. Chem. Solids* **25**, 105–113 (1964).
208. T. J. Lewis, Some factors influencing field emission and the Fowler–Northeim law, *Proc. Phys. Soc.* **B68**, Part II, No. 431B, 938–943 (1955).
209. P. H. Cutler and J. J. Gibbons, Model for the surface potential barrier and the periodic deviation on the Schottky effect, *Phys. Rev.* **111**(2), 394–402 (1958).
210. G. G. Bertold, A. Kuppermann, and T. E. Phipps, Application of numerical methods to the theory of the periodic deviations in the Schottky effect, *Phys. Rev.* **128**(2), 524–531 (1962).
211. G. F. Vasil'ev, Influence of potential barrier shape at an emitter–vacuum boundary and electric field distribution along emitter surface upon the current–voltage characteristics of field emission, *Radiotekhnika i Elektronika* **5**(11), 1857–1861 (1960).
212. F. Seitz, *Modern Theory of Solids*. (McGraw-Hill, New York, 1940), p. 163.
213. S. C. Miller and R. H. Good, A WKB-type approximation to the Schrödinger equation, *Phys. Rev.* **91**(1), 174–179 (1953).
214. C. A. Spindt, A thin film field emission cathode, *J. Appl. Phys.* **39**, 3504–3505 (1968)
215. M. Yu. Sumetskii, Transparency of non-symmetric tunnel micro-contact and semi-classical theory of scanning tunneling microscope. *Zh. Exper. i Teor. Fiz.* **94**(3), 7–22, (1988).
216. A. A. Lucas, H. Morawitz, G. R. Henry, J. P. Vigneron, Ph. Lambin, P. H. Cutler, and T. E. Feuchtwang, Tunneling through localized barriers with application to scanning tunneling microscopy: New scattering theoretic approach and results, *J. Vac. Sci. Technol. A* **6**(2), 296–299 (1988).
217. Z. H. Huang, T. E. Feuchtwang, P. H. Cutler, and E. Kazes, Wentzel–Kramers–Brillouin method in multidimensional tunneling, *Phys. Rev.* **A41**(1), 32–41 (1990).
218. W. P. Dyke, J. K. Trolan, W. W. Dolan, and G. Barnes, The field emitter: fabrication, electron microscopy, and electric field calculations. *J. Appl. Phys.* **24**, 570–576 (1953).
219. G. N. Fursey and D. V. Glazanov, Deviation from the Fowler–Nordheim theory and peculiarities of field electron emission from small-scale objects (Review). *J. Vac. Sci. Technol. B* **16**(2), 910–915 (1998).
220. P. R. Schwoebel and I. Brodie, Surface-science aspects of vacuum microelectronics, *J. Vac. Sci. Technol. (B)* **13**(4), 1391–1410 (1995).
221. H.-W. Fink, H. Schmid, and H. J. Kreuzer, In-line holography, using low-energy electrons and photons: applications for manipulation on a nanometer scale, *J. Vac. Sci. Technol. (B)* **13**(6), 2428–2431 (1995).
222. G. N. Fursey, L. M. Baskin, D. V. Glazanov, A. O. Yevgen'yev, A. V. Kotcheryzhenkov, and S. A. Polezhaev, Specific features of field emission from submicron cathode surface areas at high current densities, *J. Vac. Sci. Technol. (B)* **16**(1), 232–237 (1998).
223. G. N. Fursey, Study of electron emittion in extremely high electric field and in the condition of transition to the vacuum arc, Ph. D. dissertation, Institute Of Semiconductors (Russian Academy Of Science), Novosibirsk (1973), pp. 1–393.
224. M. S. Aksyonov, L. M. Baskin, V. M. Zhukov, N. F. Fedorov, and G. N. Fursey, Increase of the region of localization of the Nottingham effect during field-electron emission under low temperature conditions, *Izv. Akad. Nauk SSSR: Ser. Fiz* **43**(3), 543–546 (1979).

225. L. W. Swanson, L. C. Crouser, and F. V. Charbonnier, Energy exchanges attending field electron emission, *Phys. Rev.* **151**, 327–340 (1966).
226. P. H. Levine, Thermoelectric phenomena associated with electron-field emission, *J. Appl. Phys.* **33**, 582–587 (1962).
227. W. B. Nottingham, Remarks on energy losses attending thermoionic emission of electrons from metals, *Phys. Rev.* **58**, 906–907 (1940).
228. S. I. Anisimov, B. L. Kapeliovich, and T. L. Perel'man, Emission of electrons from the surface of metals induced by ultrashort laser pulses, *Zh. Eksper. i Teor. Fiz.* **66**(2), 776–781 (1974).
229. N. F. Mott and E. H. Jones, *The Theory of the Properties of Metals and Alloys*, (Oxford university press, London, 1936), pp. 305–314.
230. R. S. Stratton, Energy distribution of field emitted electron, *Phys. Rev.* **135**, A, 794–805 (1964).
231. J. Houston, Private communication, See also [225].
232. M. Drechsler and G. Jaehing, Heating and cooling of thin wires by field emission, *Z. Naturforsch*, **18a**, 1367–1369 (1963).
233. I. Engle and P. H. Cutler, The effect of different surface barrier models on the Nottingham energy exchange process, *Surf. Sci.* **8**, 288 (1967).
234. W. B. Nottingham, Remarks on energy losses attending thermionic emission of electrons from metals, *Phys. Rev.* **59**, 907 (1941).
235. E. A. Litvinov, G. A. Mesiats, and A. F. Shubin, Calculation of thermo-field emission preceding the explosion of micro-emitters caused by electron field-emission pulses, *Izv. Vysshikh Uchebnikh Zavedenii SSSR: Ser. Fiz.* **4**, 147–151 (1970).
236. V. A. Nevrovskii and V. I. Rakhovskii, Thermal instability of microprotrusion on a cathode during vacuum breakdown, *Zh. Tekh. Fiz.* **50**(10), 2127–2135 (1980).
237. J. Mitterauer, P. Till, and M. Haider, The initiation of cathode induced vacuum breakdown by dynamic field emission from microprotrusions. *Proc. 7th Int. Symp. Discharges Electrical Insulation Vacuum, 1976, Proc. ISDEIV, Novosibirsk* (1976), pp. 83–87.
238. D. V. Glazanov, L. M. Baskin, and G. M. Fursey, Kinetic of the pulsed heating of field-emission cathode point with real geometry by a high-density emision current, *Zh. Tekh. Fiz.* **59**(5), 60–68 (1989).
239. V. A. Gor'kov, M. I. Elinson, and G. D. Yakovleva, Theoretical and experimental investigations of pre-arc phenomena during field-electron emission, *Radiotekhnika i Elektronika* **7**(9), 1501–1510 (1962).
240. F. M. Charbonnier, R. W. Strayer, L. W. Swanson, and E. E. Martin, Nottingham effect in field and T-F emission: heating and cooling domains, and inversion temperature, *Phys. Rev. Lett.* **13**, 397–401 (1964).
241. L. D. Landau and E. M. Lifshitz, *Electrodynamics of Continua*. (Nauka, Moscow, 1982).
242. Ibid., p. 620.
243. Tables of Physical Constants, edited by I. K. Kikoin, Moscow, Atomizdat, 1976.
244. G. N. Fursey, D. V. Glazanov, and S. A. Polezhaev, Field emission from nanometer protrubances at high current density, *IEEE Trans. Dielect. Electric. Insul.* **2**(2), 281–287 (1995).
245. D. V. Glazanov, L. M. Baskin, and G. N. Fursey, Numerical modeling of the development of thermal instability of the filed emission initiating vacuum breakdown. *XI Int. Symp. Discharges Electrical Insulation Vacuum, Berlin*, German Democratic Republic, September 24–28, 1984, *Pro. ISDEIV*, (1984), pp. 65–68.
246. L. M. Baskin, D. V. Glazanov, and G. N. Fursey, Influence of thermoelastic stresses on the destruction processes of field emission cathode points and the transition to explosive emission, *Zh. Tekh. Fiz.* **59**(5), 130–133 (1989).
247. D. N. Krotevich, V. E. Ptitsyn, and G. N. Fursey, Observation of the fine structure of rearranged microcrystal surface by pulsed filed-emission microscopy, *Fiz. Tverd. Tela* **28**(12), 3722–3724 (1986).
248. M. I. Kaganov, I. M. Lifshitch, and L. V. Tantarov, Relaxation between electrons and the crystalline lattice, *Zh. Eksper. i Teor. Fiz.* **31(2)**, 232–237 (1956).
249. G. N. Fursey, L. M. Baskin, D. V. Glazanov, A. O. Evgen'ev, A. V. Kotcheryzhenkov, and S. A. Polezhaev, The specific features of field emission from submicron cathode surface areas at high current densities, *J. Vac. Sci. Technol. B* **16**(1) 232–238 (1998).
250. G. N. Fursey, B. F. Gulin, and A. A. Antonov, Investigation of the field-electron emission from tungsten in the nanosecond range of pulse durations, *Vestnik Leningradskogho Universiteta: Seria 4. Physics & Chemistry* **2**, 71–74 (1971).
251. M. I. Elinson and G. F. Vasil'ev, Investigation of the filed emission from lanthanum hexaboride, *Radio Eng. Electron.* **3**(7), 123–124 (1954).

252. M. I. Elinson and G. A. Kudnitseva, Field-electron cathode of metallic high-melting compounds, *Radioteknika i Electronika* **7**(9), 1511–1518 (1962).
253. G. N. Fursey and S. A. Shakirova, Field emission localization within small solid angles, *Proc. Electronics Technique Conf. Electron-beam and photo-electron Devices* **1**, 99–102 (1969).
254. L. M. Baskin, R. Z. Bachtizin, V. G. Valeev, and G. N. Fursey, *Advances in the Field Emission Investigations*, Supplement to the Russian Translation of A. Modinos, *Field, Thermonic, and Secondary Electron Emission Spectroscopy*, edited by G. N. Fursey (Nauka, Moscow, USSR, 1985), pp. 232–302 (in Russian).
255. G. N. Fursey, L. L. Anan'ev, and D. A. Borisov, Field emission in microwave field. *Proc. IV All-Union Seminar on Microwave Relativistic Electronics, Moscow, USSR* (1985), pp. 1–20 (in Russian).
256. H. F. Gray, The electronics of the 21st century, nanoelectronics, *Vacuum Microelectronics* **8**, Electro/89, 1–6 (1989).
257. S. T. Smith and H. F. Gray, Distributed amplifier possibilities using field-emitter arrays, *Proc. 1st Int. Conf. Vacuum Microelectronics, Williamsburg, Virginia, USA* (1988), pp. 7–9.
258. R. L. Parker, RF vacuum microelectronics, *7th Int. Conf. Vacuum Microelectronics, Grenoble, France* (1994).
259. Yu. V. Gulyaev, I. S. Nefedov, N. I. Sinitsyn, G. T. Torgashov, Yu. F. Zakharchenko, and A. I. Zhbanov, *Distributed microwave amplifier on field emitter arrays with a nonhomogeneous energy collector*, Revue "le vide, les Couches Minces"-Supplemeny au No. **271**, 84–87 (1994).
260. V. Ambageokar, U. Eckern, and G. Schoen, Quantum dynamics of tunneling between superconductors, *Phys. Rev. Lett.* **48**(25), 1745–1748 (1982).
261. P. F. Voss and R. A. Webb, Macroscopic quantum tunneling in 1 μm Nb Josephson junctions, *Phys. Rev. Lett.* **47**(4), 265–268 (1981).
262. A. M. Gabovich, V. M. Rosenbaum, and A. L. Loitenko, Dynamic image forces in three-layer systems and field emission, *Surf. Sci.* **186**(3), 523–549 (1987).
263. E. P. Wigner, Lower limit for the energy derivative of the scattering phase shift, *Phys. Rev.* **98**(1), 145–147 (1955).
264. T. E. Hartman, Tunneling of a wave packet, *Appl. Phys.* **33**(12), 3427–3433 (1962).
265. A. Goldberg, H. M. Schey, and J. L. Schwartz, Computer-generated motion pictures of one-dimensional quantum-mechanical transmission and reflection phenomena, *Amer. J. Phys.* **35**(3), 177–193 (1967).
266. P. Schnupp, The tunneling time of an electron and the image force, *Thin Solid Films* **2**(3), 177–183 (1968).
267. A. I. Baz, Life-time of intermediate states, *J. Nucl. Phys. (USSR)* **4**(2), 252–260 (1966) (in Russian).
268. A. I. Baz, A quantum mechanical calculation of collision time, *J. Nucl. Phys. (USSR)* **5**(1), 229–235 (1967) (in Russian).
269. V. F. Rybachenko, Time of particle penetrating through barrier, *J. Nucl. Phys. (USSR)* **5**(4), 895–901 (1967) (in Russian).
270. M. Buttiker, Larmor precision and the traversal time for tunneling, *Phys. Rev. B* **27**(10), 6178–6188 (1983).
271. R. Feinman and A. Gibbs, *Quantum Mechanics and Path Integrals* (McGraw-Hill, New York, 1969).
272. D. G. Sokolovski, Quantum time of flight and adiabatic limit in scattering problems, *Report of the Ministry High Schools: Physics. (USSR)* **3**, 52–58 (1988) (in Russian).
273. L. M. Baskin and D. G. Sokolovski, Traversal time in the process of tunneling, *Report of the Ministry of High Schools: Physics. (USSR)* **3**, 26–29 (1987) (in Russian).
274. D. G. Sokolovski and L. M. Baskin, Time of flight at tunneling, *J. Tech. Phys. (USSR)* **55**(9), 1838–1840 (1985) (in Russian).
275. Linfield College, McMinnville, Ore., *Engineering Report* (1953).
276. L. L. Anan'ev, M. M. Bogatskii, D. A. Borisov, A. A. Kantonistov, and G. N. Fursey, Method of field emission investigation in microwave electric field, *Instrum. Tech. Exp. (USSR)* **5**, 165–168 (1983) (in Russian) [*Instrum. Exp. Tech. (USA)* **26**, 1181 (1983)].
277. L. M. Baskin, A. A. Kantonistov, G. N. Fursey, and L. A. Shirochin, Peculiarities in explosive emission of liquid metals in microwave field, *Rep. USSR Academy of Sciences (USSR)* **296**(6), (1987) 1352 (in Russian).
278. A. A. Kantonistov, I. N. Radchenko, G. N. Fursey, and L. A. Shirochin, Field emission of liquid metals in alternating fields, *Colloque de Physique*, Colloque C8, supplement au Tome 50, **11**, C8-203–207 (1989).
279. D. V. Glazanov. *Thesis of kandidate dissert.*, Tomsk, 1985 (in Russian).
280. L. L. Anan'ev, D. A. Borisov, A. A. Kantonistov, and G. N. Fursey, Study of field emission at high current densities in microwave field, *Proc. IV All-Union Symp. High-Current Electronics*, Tomsk, pp. 26–29 (1982) (in Russian).

281. M. I. Elinson and V. A. Gorkov, Some peculiarities of field emission cathode operation in microwave field, *Radiotech. Electron. (USSR)* **6**(2), 336–339 (1961) (in Russian).
282. R. M. Voronkov, V. F. Gass, V. A. Danilichev, and I. A. Smirnov, The testing of injector with field emission cathode in accelerator, *Pribory i Tekhnica Experimenta (USSR)* **4**, 18–20 (1974) (in Russian).
283. L. D. Landau and E. M. Lifschitz, *Mechanics* (Nauka, Moscow, 1965).
284. L. M. Baskin, E. A. Sazonov, L. M. Chernych, and G. N. Fursey, Influence of transient effects on parameters of microwave diode with non-heated cathode, *Radiotech. Electron. (USSR)* **33**(10), 2208–2210 (1988) (in Russian).
285. E. V. Muller and T. T. Tsong, *Field-ion microscopy* (American Elsevier Publishing Company, Inc., New York, 1969).
286. L. Tonks, A theory of liquid surface rupture by uniform electric field, *Phys. Rev.* **48**, 562–568 (1935).
287. Ya. Frenkel, On a theory of liquid surface instability in electric field in vacuum, *Zh. Tekh. Fiz.* **6**, 347–351 (1935).
288. K. Hata, R. Ohya, S. Hishigaki, H. Tamura, and T. Noda, Stable field emission of electrons from liquid metal, *Jpn J. Appl. Phys.* **26**(6), L896–L898 (1987).
289. L. M. Baskin, G. N. Fursey, A. A. Kantonistov, B. N. Movchan, I. N. Radchenko, and L. A. Shirochin, Microstructure formation at the liquid surface in microwave fields, *Proc. 14th Int. Symp. Discharges Electrical Insulations Vacuum, Santa-Fe, CA*, (1990), p. 112.
290. N. Fursey, L. A. Shirochin, and L. M. Baskin, Field emission processes from liquid-metal surface, *J. Vac. Sci. Technol. B* **15**(2), 410–421 (1997).
291. J. R. Arthur, Surface structure and surface migration by field emission microscopy, *J. Phys. Chem. Solids* **25**, 583–591 (1964).
292. G. Busch and T. Fischer, Feldemission aus Silizium, *Phys. Kondens. Materie* **1**, 367–393 (1963).
293. J. R. Arthur, Gallium arsenide surface structure and reaction kinetics: field emission microscoipy and surface migration, *J. Appl. Phys.* **37**(6), 3057–3064 (1966).
294. F. G. Allen, Field emission from Silicon and Germanium; Filed decorbtion and surface migration, *Chem. Solids* **19**, 87–99 (1961).
295. V. G. Ivanov, On preparing germanium emitters and obtaining a field emission picture of pure germanium, *Radi. Electr.* **10**(3), 576–578 (1965).
296. J. R. Arthur, Photosensitive field emission from p-type Germanium, *J. Appl. Phys.* **36**, 3221–3227 (1965).
297. V. G. Ivanov, G. N. Fursey, and I. L. Sokol'skaya, Study of field emission from germanium, *Fiz. Tverd. Tela* **9(4)**, 1144–1148 (1967) (in Russian).
298. L. Ernst, Über die Oberflächenreinigung von Germaniumspitzen durch Felddesorption im Feldemissions-mikroskop, *Phys. Status Solidi* **7**, K61–K64 (1964).
299. L. Ernst, Study of the field desorption of oxidized germanium tips in a field emission microscope, *Phys. Status Solidi* **14**, K107–K109 (1966).
300. L. Ernst, Feldemissions-Mikroskopie von Germanium, *Phys. Status Solidi* **19**, 89 (1967).
301. L. A. D'Assaro, Field emission from silicon, *J. Appl. Phys.* **29**, 33–34 (1958).
302. R. L. Perry, Experimental determination of the current density electric field relationship of silicon field emitters, *J. Appl. Phys.* **33**, 1875–1883 (1962).
303. G. N. Fursey and V. G. Ivanov, Field-electron emission from p-type Si cleaned by field desorption, *Fiz. Tverd. Tela* **9**(6), 1812–1818 (1967) (in Russian).
304. G. N. Fursey, V. G. Egorov, S. P. Manokhin, and M. R. El'Nimr, Relaxation effects during field emission from silicon, *Fiz. Tverd. Tela* **15**(5), 1360–1363 (1973) (in Russian).
305. S. G. Truxillo, J. C. Blair, N. G. Einspruch, and R. Stratton. High-field electron emission from indium arsenide, *J. Chem. Phys.* **44**, 1724 (1966).
306. W. R. Savage, High-field electron emission from indium antimonide, *Solid State Commun.* **1**, 144–147 (1963).
307. J. Marien and J. Loosveld, Field emission studies on clean symmetrical surfaces of cadmium sulfide, *Phys. Status Solidi (a)* **48**, 213 (1971).
308. A. J. Melmed and R. J. Stern, Field-ion microscopy of silicon, *Surf. Sci.* **49**, 645–656 (1975).
309. A. J. Melmed and E. I. Givargizov, unpublished research, NBS, 1979.
310. E. I. Givargizov, *Krist. Tech.* **10** (1975) 473 (in Russian); in: *Current Topics in Materials Science*, edited by E. Kaldis (North-Holland, Amsterdam, 1978), p. 79.

311. U. Apker and E. A. Taft, Field emission from photoconductors, *Phys. Rev.* **88**, 1037–1038 (1952).
312. I. L. Sokol'skaya and G. P. Shcherbakov, Study of strong field on field-electron emitters of CdS crystal, *Fiz. Tverd. Tela* **31**(1), 167–175 (1961).
313. G. P. Shcherbakov and I. L. Sokol'skaya, Experimental investigation of energy distribution of filed-emitted electrons from CdS single crystals, *Fiz. Tverd. Tela* **4**(12), 3526–3536 (1962).
314. O. V. Golovanova and A. I. Klimin, Sb_2S_3 field-electron emission, *Radi. Electr.* **10**, 491–497 (1965) (in Russian).
315. P. G. Borzjak, A. F. Yatsenko, and L. S. Miroshnichenko, Photo-field-emission from high-resistance silicon and germanium, *Phys. Status Solidi* **14**, 403–411 (1966).
316. A. F. Yatsenko, Photo-field emission from high-resistance p-type semiconductors, *Elektr. Tekhnika*, Ser. 4: *Elektr.-luchevye i Fotoelektr. Pribory* (in Russian) **2**, 169–175 (1968).
317. Z. P. But, A. E. Kravzov, and A. F. Yatsenko, Field electron emission from semiinsulator galium arsenide p-type, *XIII Conf. Emission Electronics, Moscow* (1968), p. 42.
318. H. Neumann, Field emission from In_2S_3, *Z. Naturforsch.* **22a**(7), 1012–1019 (1967).
319. H. Neumann, Field emission from silver-doped cadmium antimonide, *Z. Naturforsch.* **23a**(8), 1240 (1968).
320. L. M. Baskin, O. I. L'vov, and G. N. Fursey, General features of field emission from semiconductors, *Phys. Status Solidi (b)* **47**, 49–62 (1971).
321. N. V. Egorov, G. N. Fursey, and M. K. El'Nimr, Preservation of semiconductor field emission cathode properties, *Zh. Tech. Fiz.* **44**(5), 1117–1118 (1974) (in Russian).
322. R. R. Stead and O. W. Richardson, USA Patent No. 2.871.111.
323. N. D. Morgulis, On Shottky effect in compound semiconductor cathodes, *Zh. Eksp. i Teor. Fiz.* **16**, 959–954 (1946) (in Russian).
324. R. Stratton, Fieldemission von Halbleitern, *Proc. Phys. Soc.* **B68**, 746–756 (1955).
325. R. Stratton, Theory of field emission from semiconductors, *Phys. Rev.* **1251**, 67–82 (1962).
326. M. S. Kagan, N. M. Lifshits, A. L. Musatov, and A. A. Sharonov, Field-electron emission from high-resistivity germanium, *Fiz. Tverd. Tela.* **6**, 723–727 (1964) (in Russian).
327. I. L. Sokol'skaya and G. P. Scherbakov, Nonlinearity of the current-voltage characteristics of field emitter from CdS single crystals, *Fiz. Tverd. Tela* **4**(1), 44–51 (1962) (in Russian).
328. G. N. Fursey and N. V. Egorov, Field emission from p-type Si, *Phys. Status Solidi* **32**, 23–29 (1969).
329. G. N. Fursey and R. S. Bakhtizin, Nonlinear current-voltage characteristics of p-type Ge, *Fiz. Tverd. Tela* **11**, 3672–3674 (1969) (in Russian).
330. M. I. Elinson, A. G. Zdan, V. F. Krapivin, Zh. B. Lipkovskii, V. N. Lutskii, and V. B. Sandomirskii, The theory of "non contact" emission of hot electrons from semiconductors, *Radiotekhnika i Elektronika* **10**, 1288 (1965).
331. N. D. Morgulis, On the problem of filed-electron emission of complex semiconductor cathodes, *Zh. Tech. Fiz.* **17**(9), 983–986 (1947).
332. S. G. Christov, Phys. United theory of thermionic and field emission from semiconductors, *Phys. Status Solidi* **21**, 159–173 (1967).
333. P. G. Borzyak, A. F. Yatsenko, and L. S. Miroshnichenko, Photo-field—emission from high-resistance silicon and germanium, *Phys. Status Solidi* **14**, 403–411 (1966).
334. G. N. Fursey, I. L. Sokolskaya, and V. G. Ivanov, Field emission from p-type germanium, *Phys. Status Solidi* **22**, 39–46 (1967).
335. G. N. Fursey, M. I. Kaplan, and O. I. Lvov, On the theory of field-electron emission from p-type semiconductors, *Vestnik Leningrad Univ.* **16**, 167–170 (1968).
336. R. Seiwatz and M. Green, Space-charge calculations for semiconductors, *J. Appl. Phys.* **29**, 1034–1340 (1958).
337. E. M. Conwell, High field transport in semiconductors (Academic Press, NY 1967), W. Shockley, *Bell Syst. Tech. J.* **30**, 990 (1951).
338. A. F. Yatsenko, On a model of photo-field-emission from p-type semiconductors, *Phys. Status Solidi (a)* **1**(2), 333–348 (1970).
339. Z. P. But, L. S. Miroschenko, and A. F. Yatsenko, Temperature dependence of field emission from p-type silicon with various bulk doping, *Ukr. fiz. Zh.* **17**, 949–955 (1972).
340. R. Fischer, Field emission from Tamm surface states in the one-dimensional model, *Phys. Status Solidi* **13**, K5–K8 (1966).

341. L. M. Baskin, O. L. Lvov, and G. N. Fursey, On the theory of filed emission from p-type semiconductors, *Phys. Status Solidi (a)* **42**, 757–767 (1977).
342. G. N. Fursey and L. M. Baskin, Features of field emission from semiconductors, *Mikroelectronika* **26**(2), 117–122 (1997) (in Russian).
343. K. Yokoo, Functional field emission for high frequency wave application, *Proc. 12th Int. Microelectronics Conf.*, Darmstadt, Germany, (July 6–9, 1999), 206–207.
344. K. Yokoo, New approach for high frequency electronics based on field emission array, *Proc. 2nd Int. Workshop Vacuum Microelectronics*, Wroclaw, Poland, (July 11–13, 1999), pp. 38–40.
345. V. G. Ivanov, V. S. Danilova, and O. L. Volosova, Oxygen adsorption on atomically clean surface of gallium arsenide, *Vestnik Novgorod University*. Ser. Estestv. i Tekhn. Nauki No. 3, 8–12 (1996).
346. V. G. Ivanov, Phenomenon of intrinsic breakdown in the field emission p-type semiconductors, *IX Int. Vacuum Microelectronics Conf., St. Petersburg, Russia*, pp. 148 (1996).
347. V. G. Ivanov, T. G. Rozova, G. N. Fursey, and T. P. Smirnova, Field emission from germanium during oxygen adsorption, *Fiz. Tverd. Tela* **16**, 495–501 (1974).
348. T. G. Rozova, V. G. Ivanov, and G. N. Fursey, Influence of a strong electric field on the adsorption of oxygen on germanium, *Fiz. Tverd. Tela* **17**(1), 64–66 (1975).
349. V. G. Ivanov, T. P. Smirnova, and G. N. Fursey, Threshold effect of oxygen electric adsorption on the individual faces of single crystal germanium in a field-emission microscope, *Poverkhnost*, **12**, 128–130 (1986) (in Russian).
350. G. N. Fursey and N. V. Egorov, Stable semiconductor field-emission cathode, *Zh. Tech. Fiz.* **42**(5), 1090–1092 (1972).
351. L. M. Baskin, N. M. Egorov, V. E. Pritsin, and G. N. Fursey, Effect of deep capture centers on the emissivity of wide-gap semiconductor field-emission cathodes, *Pis'ma Zh. Tech. Fiz.* **5**, 1345–1348 (1979).
352. L. M. Baskin and G. N. Fursey, Decisive role of the deep states in a vacuum breakdown initiated in presence of dielectrical inseptions, *Proc. 13th ISDEIV*, Paris, Part 1, pp. 31–33 (1988).
353. L. Ernst, Adsorption of oxygen on atomically clean germanium surfaces examined in the field emission microscope, *Surf. Sci.* **6**, 487–491 (1967).
354. V. G. Ivanov and Zabotin, A study of adsorption of oxygen on atomically clean germanium surfaces in field emission microscope, *XIV All-Union Conf. Field Emission Electronics*. Tashkent, pp. 70–71 (1970).
355. R. Z. Bakhtizin and V. I. Stepanov, Influence of the adsorption of SiO molecules on field-electron emission from p-type Ge. *Fiz. Tverd. Tela* **14**(1), 294–296 (1972).
356. I. L. Sokol'skaya, V. G. Ivanov, and G. N. Fursey, Study of barium adsorption on germanium by field-emission microscopy, *Phys. Status Solidi* **21**, 789–795 (1967).
357. R. Z. Bakhtizin and I. L. Sokol'skaya, Titanium adsorption on the p-type Ge single crystals, *Fiz. Tverd. Tela*, **12**(10), 2815–2820 (1970).
358. I. L. Sokol'skaya, N. V. Mileshkina, and R. Z. Bakhtizin, Gold adsorption on clean germanium field emitters, *Phys. Status Solidi (a)* **14**(2), 417–422 (1972).
359. N. V. Mileshkina and R. Z. Bakhtizin, Oxygen adsorbtion on the atomically clean surface of germanium field emitter, *Surf. Sci.* **29**, 644–652 (1972).
360. V. G. Ivanov, Experimental study of field emission from semiconductors. Doctor's dissertation. St. Petersburg, State University, pp. 1–316 (1999).
361. J. T. Trujillo, A. G. Chakovskoi, and C. E. Hunt, Effect of vacuum conditions on low frequency noise in silicon field emission devices. *Proc. IXth Int. Vacuum Microelectronics Conference, St. Petersburg*, pp. 133–137 (1996).
362. N. V. Egorov, G. N. Fursey, A., and V. Kotcheryzhenkov. Kinetics effects in field emission from high-resistive n-type Si. *Fiz. Tverd. Tela* **15**(3), 892–894 (1973).
363. P. G. Borzjak, V. F. Bibik, A. E. Kravzov, B. V. Stetsenko, A. F. Yatsenko, Kinetics of field electron emission from hieghomic semiconductors, Report to the XIII All-Union Conference on Field Emission Electronics, Moscow, p. 41, (1968).
364. I. L. Sokol'skaya and A. I. Klimin, Field-electron emission from cadmium sulphide and selenide. I Influence from an electric field and temperature, *Vestnik Leningradskogo Universiteta:* Serya Fiz. **4**(1), 34–41 (1961).
365. B. V. Stetsenko, A. F. Yatsenko, and L. S. Miroshnichenko, Effect of surface states on transient processes in silicon field cathodes, *Phys. Status Solidi (a)* **1**(2), 349–355 (1970).
366. G. N. Fursey and P. G. Shlyakhtenko, Kinetics of p-type Ge filed emission, *Fiz. Tverd. Tela* **12**(2), 645–647 (1970).

367. G. N. Fursey, N. V. Egorov, and S. P. Manokhin, Transient effects associated with field emission of electrons from silicon, *Fiz. Tverd. Tela* **14**(6), 1686–1690 (1972).
368. S. A. Pshenichnyuk, Yu. M. Yumaguzin, and R. Z. Bakhtizin. Energy distribution of electrons emitted from diamond like film in a strong electric field, *Pis'ma Zh. Tekh. Fiz.* **25**(15), 46–52 (1999).
369. R. Z. Bakhtizin, Yu. M. Yumaguzin, and S. A. Pshenichnyuk, Field emission properties of diamond thin films. *9th Int. Vacuum Microelectronics Conf.*, St. Petersburg, Russia, (July 7–12, 1996), Technical Digest, p. 658.
370. C. Gazier, Multiple-electron events from field emission, *Phys. Lett. A* **35**(4), 243–244 (1971).
371. M. Hermann, Untersuchung verschiedener Elektronenemission-sprozesse auf Mehrfachemission, *Z. Phys.* **184**, 352–354 (1965).
372. G. N. Fursey, M. M. Mokhasne, N. V. Egorov, V. S. Ponomaryov, and V. N. Shchemelev, Statistics of filed-electron emission, *Fiz. Tverd. Tela* **18**(2), 631–632 (1976).
373. M. M. Mokhasne, V. S. Ponomaryov, N. V. Egorov, G. N. Fursey, and V. N. Schemelev, Procedure for investigation of the statistics of field-electron emission, *Pribory i Tekhnika Eksperimenta* **19**(2), 150–152 (1976).
374. N. P. Afanas'eva, N. V. Egorov, A. V. Kocheryzhenkov, and G. N. Fursey, A method of investigation of the elementary acts number during the filed-electron emission, *Pribory i Tekhnika Experimenta* **25**(5), 141–142 (1982).
375. G. N. Fursey, N. V. Egorov, and A. V. Kocheryzhenkov, Field emission statistics for various faces of a tungsten single crystal, *Pis'ma Zh. Tekh. Fiz.* **7**(13), 798–801 (1981).
376. G. N. Fursey, A. V. Kocheryzhenkov, V. I. Maslov, A. L. Shmaev, and L. N. Borisov, Apparatus for study of multi particle acts of field emission at low temperatures, *Prib. Tekh. Eksper.* **33**(1), 149–151 (1990).
377. G. N. Fursey, V. I. Maslov, and A. V. Kocheryzhenkov, Search for correlated electrons in field emission from superconducting niobium, *34th Field Emission Symp., Osaka* (July 13–18, 1987), (Abstract) p. 67.
378. A. V. Kocheryzhenkov, V. I. Maslov, and G. N. Fursey, Statistics of field-electron emission from tungsten and niobium at helium temperatures, *Fiz. Tverd. Tela* **29**(8), 2471–2472 (1987).
379. G. N. Fursey, A. V. Kocheryzhenkov, V. I. Maslov, and A. L. Shmaev, *Royal Microscopical Soc. Proc., 36th Int. Field Emission Symp., Oxford* (1989), Vol. 24, art 4, p. 34.
380. V. I. Maslov, G. N. Fursey, and A. V. Kocheryzhenkov, Investigations of the quantity of elementary acts of field emission with time resolution of 5μs–100 ps, *COLLOQUE DE PHYSIQUE*, Colloque C8, supplement au $n°$11, Tome 50 (novembre 1989), C8-113–C8-116.
381. L. W. Swanson and L. C. Crouser, Angular confinement of field electron and ion emission, *J. Appl. Phys.* **40**(12), 4741–4749 (1969).
382. L. W. Swanson, Comparative study of the zirconiated and built-up W thermal-field cathode, *J. Vac. Sci. Technol.* **12**(6), 1228–1233 (1975)
383. A. Tonomura, Electron holography: a new view of the microscopic, *Phys. Today* 22–29 (April 1990).
384. A. Tonomura, Electron-holographic interference microscopic, *Adv. Phys.* **41**(1), 59–103 (1992).
385. I. L. Sokolskaya, Surface migration of tungsten atoms in electric field, *Zh. Techn. Fiz.* (USSR) **26**(6), 1177–1184 (1956) (in Russian).
386. N. A. Gorbatyi and G. N. Shuppe, Dependence of absorption binding on the metallic crystal surface, *Zh. Tech. Fiz.* **24**(8), 1364–1375 (1955) (in Russian).
387. I. D. Ventova and G. N. Fursey, Surface self-diffusion in critical build-up, *Zh. Tech. Fiz.* (USSR) **43**(11), 2432–2440 (1973) (in Russian).
388. I. D. Ventova, G. N. Fursey, and S. A. Polezhaev, Formation of microscopic protuberances at the vertex of a sharp metal tip in a strong electric field. Critical Build-up. II., *Zh. Tech. Phys.* **47**(4), 849–856 (1977) (in Russian).
389. V. N. Shrednik, V. G. Pavlov, A. A. Rabinovich, and B. M. Shaikhin, Action of strong electric field and heating on metallic points, *Izv. Akad. Nauk SSSR*: Ser. Fiz. (USSR) **38**(2), 296–301 (1974) (in Russian).
390. V. G. Pavlov, A. A. Rabinovich, and V. N. Shrednik, Field errosion of Mo, Ta, Re, *Fiz. Tverd. Tela* (USSR) **18**(2), 631–632 (1976) (in Russian).
391. V. M. Zhukov and S. A. Polezhaev, Surface evolution of microcrystal on the top of the emitter during the thermo-field treatment, *Zh. Tech. Fiz.* (USSR) **57**(6), 1133–1136 (1987) (in Russian).
392. M. Drechsler, Kristallstufen von 1 bis 1000 Å, *L. Electrochemie*, **61**(48), (1957).
393. I. A. Becker, The use of the field emission electron microscope in adsorption studies of W on W and Ba on W, *Bell. Syst. Tech. J.* **30**, 907–932 (1951).

394. G. N. Fursey, Critical features of field emission processes, *IV Int. Vacuum Microelectronics Conf., Nagahama, Japan* (1991), pp. 1–2.
395. G. N. Fursey, B. N. Movchan, and V. A. Shvarkunov, Direct electron microscope observations of the formation of microscopic protuberances on the close-packed faces of single-crystal tungsten in a strong electric field, *Pisma v Zhurn. Tech. Fiz.* (USSR) **16**(20), 42–46 (1990) (in Russian).
396. V. E. Ptitsyn and G. N. Fursey, Effect of field emission on single-crystal point shapes, *Izv. Akad. Nauk SSSR:* Ser. Fiz. (USSR) **52**(8), 1513–1517 (1988) (in Russian).
397. Yu. A. Vlasov, O. L. Golubev, and V. N. Shrednik, Metal point shape change due to interaction between electrostatic "capillary" forces, *Izv. Akad. Nauk SSSR:* Ser. Fiz. (USSR) **52**(8), 1538–1543 (1988) (in Russian).
398. V. G. Butenko, Yu. A. Vlasov, O. L. Golubev, and V. N. Shrednik, Point sources of electrons and ions using microprotrusion on the top of a tip, *Surf. Sci.* **266**, 165–169 (1991).
399. I. D. Ventova and G. N. Fursey, Build-up of the tip of a microcrystal into polyhedron I, *Zh. Tech. Fiz.* (USSR) **47**(4), 844–848, (1977) (in Russian).
400. V. N. Shrednik, V. G. Pavlov, A. A. Rabinovich, and B. M. Shaikhin, Growth of tips in the directions normal to close-packed faces by heating in the presence of an electric field, *Phys. Status Solidi A*, **23**, 373–380 (1974).
401. M. Drechsler, Kristallstufen von 1 bis 1000 A. Herstellung der studen in feldemissiosmikroskopen durch electrishe felder. Nessung der stufenhofen. Eine feldbidungsenergie, *Z. Electrochem.* **61**, 48–55 (1957).
402. V. M. Zhukov and S. A. Polezhaev, Surface evolution of point field-emission cathodes during pulse regime of work, *Rad. i Electr.* (USSR) **33**(8), 1741–1747 (1988).
403. J. A. Becker, The use of the field emission electron microscope in adsorption studies of W on W and Ba on W, *Bell. Syst. Tech. J.* **30**, 907–932 (1951).
404. V. G. Pavlov, B. M. Shaikhin, and V. N. Shrednik, Study of the tips shape evolution in the process of self-diffusion in strong electric field, *I All-Union Conf. Field-Ion Microscopy, Kharkov* (1976), pp. 77–80 (in Russian).
405. D. M. Pautov and I. L. Sokolskaya, Reduction of divergence of an electron beam emitted by a field-electron cathode, *Zh. Tech. Fiz.* (USSR) **41**(9), 1999–2001 (1971) (in Russian).
406. G. N. Fursey, Destruction of a field emitter initiating a vacuum breakdown, *Proc. 3th Int. Symp. Discharges Electr. Insul. Vacuum, Paris* (1968), pp. 113–118.
407. I. P. Zhizhin, G. N. Fursey, S. A. Shakirova, and A. V. Shishatskii, On the problem of obtaining small electron probes by means of a field-electron emitters and magnetic optics, *Izv. Akad. Nauk. SSSR Ser. Fiz.* **35**(2), 302–306 (1971) (in Russian) [*Bull. Acad. Sci. USSR—Phys. Ser.*, **35**, 277–281, 1971].
408. G. N. Fursey, R. Z. Bakhtizin, V. E. Ptitsyn, Yu. M. Yumaguzin, and V. A. Shvarkunov, Emission characteristics of field-emission cathodes with emission localized into a small solid angles, *Izv. Akad. Nauk SSSR: Ser. Phys.* (USSR) **52**(7), 1250–1253 (1988) (in Russian).
409. L. W. Swanson and N. A. Martin, Field electron cathode stability studies: Zirconium/Tungsten thermal-field cathode, *J. Appl. Phys.* **46**(5), 2029–2050 (1974).
410. D. Tuggle, L. W. Swanson, and J. Orloff, Application of a thermal field emission source for high resolution, high current e-beam microprobes, *J. Vac. Sci. Technol.* **16**(6), 1699–1703 (1979).
411. V. N. Shrednik and G. A. Odisharia, Anisotropy of heat of migration heats and adsorbent–adsorbent binding energies in a metallic-film systems Zr-W, Hf-W, Hf-Mo, Zr-Nb, Zr-Ta, *Izv. Akad. Nauk SSSR: Ser. Fiz.* (USSR) **33**(3), 536–543, 1969 (in Russian).
412. H. Hosoki, T. Yamamoto, H. Todoroko, H. Kawase, and C. Hirai, Field emission cathode and method of fabricating same, United States Patent No. 4379250 (Publ. April 5, 1983), Hitachi Ltd Tokyo, Japan.
413. V. N. Shrednik, Investigation of zirconium atomic layers on tungsten crystal faces by means of electron and ion projectors, *Fiz. Tverd. Tela* (USSR), **3**(6), 1750–1761 (1961) (in Russian).
414. T. Inoue and M. Nakada, A novel zirconiating technique for high brightness Zr/W emitter, *Surf. Sci.* **246**, 87–93 (1991).
415. J. E. Wolfe, Electron-beam cathode having a uniform emission pattern, US Patent, 13.04.1982, 4324999 (Burroughs Corp. 30.04.1980, No. 145043).
416. G. N. Fursey, Field emission localization in narrow solid angles, *6th Int. Vacuum Microelectrnics Conf., Newport, USA* (1993). Technical Digest pp. 30–31.
417. G. N. Fursey, D. A. Polezhaev, and V. A. Shvarkunov, Development of thermo-field cathode for Auger-spectrometer (unpublished) (1990), pp. 1–42.

418. L. W. Swanson, US Patent 3817592 (1974).
419. T. Inoue and M. Nakada, A novel zirconiating technique for high brightness Zr/W emitter, *Surf. Sci.* **246**, 87–93 (1991).
420. J. Orloff, Thermal field emission for low voltage scanning electron microscopy, *J. Microsc.* **140**(3), 303–311 (1985).
421. G. N. Fursey, Produce of a small emission segment during Zr layer of W forming in electric field (to be published.)
422. D. Gabor, A new microscopic principle, *Nature* (London), **161**, 777–778 (1948).
423. Y. Nakaizumi, K. Kanda, T. Watanabe, M. Yamada, and T. Ohno, Development of Hitachi Model S-806 scanning electron microscope for wafer inspection, *Hitachi Instrument News*, No. 17, 20–25 (1985).
424. E. Zeitler and M. Ohtsuki, Young's experiment with electrons, *Ultramicroscopy* **2**, 147–148 (1977).
425. C. Py and R. Baptist, Low-energy electrons interferences with Spindt-type microtips, Revue "Le Vide, Les Couches Minces", Supplement au N 271, pp. 17–20 (1994).
426. V. T. Binh and N. Garcia, On the electron and metallic ion emission from nanotips fabricated by field-surface-melting technique: experiments on W and Au tips, *Ultramicroscopy* **42–44**, 80–90 (1992).
427. L. A. Chernozatonskii, Z. Ya. Kosakovskaya, Yu. V. Gulyaev, N. I. Sinitsyn, G. V. Torgashov, and Yu. F. Zakharchenko. Influence of external factor on electron field emission from thin-film nanofilament carbon structures, *J. Vac. Sci. Technol. B* **14**(3), 2080–2082 (1996).
428. I. Brodie and P. R. Schwoebel, Vacuum microelectronic devices, *Proc. IEEE* **82**(7), 1006–1034 (1994).
429. J. P. Calame, H. F. Gray, and J. L. Shaw, Analysis and design of microwave amplifiers employing field-emitter arrays, *J. Appl. Phys.* **73**(3), 1485–1504 (1992).
430. G. N. Fursey, Field Emission in Vacuum Microelectronics. The Proceedings for the 23rd and 24th ISTC Japan Workshop on Advanced Nanotechnologies in Russia/CIS, November, 2002, Tokyo and 8 November, Kyoto Japan pp. 63–106 (*Appl. Surf. Sci.* **215**, 113–134, 2003).
431. M. E. Crost, K. Shoulders, and M. E. Zinn, Thin electron tube with electron emitters at the intersection of crossed conductors, US Patent 3500102 (1970).
432. R. N. Thomas, R. A. Wickstrom, D. K. Schroder, and H. C. Naihanson, Fabrication and some applications of large-area silicon field emission arrays, *Solid-State Electron.* **17**, 155–163 (1974).
433. D. O. Smith, J. S. Judge, M. Trongello, and P. R. Thomton, Micro-structure field emission electron source, US Patent 3970887 (1976).
434. C. A. Spindt, I. Brodie, L. Humphrey, and E. R. Westerberg, Physical properties of thin field emission cathodes with molybdenum cones, *J. Appl. Phys.* **47**, 5248–5263 (1976).
435. A. M. E. Hoeberechts, Field emission device, US Patent 4995133 (1978).
436. C. A. Spindt and W. Aberth, The volcano field ionization source. *Proc. 27th Int. Field Emission Symp.*, edited by Y. Yashiro and N. Igata, (Dept. of Metallurgy and Materials Science, University of Tokyo, Tokyo, Japan) (1977).
437. C. A. Spindt, C. E. Holland, and R. D. Stowell, Field emission cathode array development for high current density applications, *Appl. Surf. Sci.* **16**, 268–276 (1983).
438. G. Bennig, H. Rohrer, C. Gerber, and E. Weibel, 7 × 7 reconstruction on Si(111) resolved in real space, *Phys. Rev. Lett.* **50**, 120 (1983).
439. P. M. Lally, C. D. Mack, and C. A. Spindt, Experiments with a field emission cathode in TWT. *Presented at the Microwave Power Conf.* Naval Postgraduate School, Monterey. CA, (1986).
440. H. F. Gray, G. J. Campisi, and R. F. Greene, A vacuum field effect transistor using silicon field emission arrays. *IEDM Dig. Tech. Papers* (Washington, DC), pp. 766–799, (1986).
441. C. E. Holland, C. A. Spindt, I. Brodie, J. Mooney, and E. R. Westerberg, Matrix addressed cathodoluminiscent display, *Int. Display Conf. London, UK* (1987).
442. V. I. Makhov, Ballistic filed emission devices, *2nd Int. Conf. Vacuum Microelectronics, Bath, UK IOP Conf. Ser. 99* (1989).
443. K. Betsui, Fabrication and operation of silicon micro-field-emitter arrays, *Presented at the Japanese Physical Society Meet* (1990).
444. R. B. Marcus, K. Chin, D. Liu, W. J. Orvis, D. R. Diavlo, C. E. Hunt, and J. Trujillo, Formation of silicon tips with <12 nm radius, *Appl. Phys. Lett.* **56**, 236–238 (1990).
445. C. A. Spindt, C. E. Holland, A. Rosengreen, and I. Brodie, Field-emitter arrays for vacuum microelectronics, *IEEE Trans. Electron Devices* **38**(10), 2355 (1991).
446. N. I. Sinitsyn, Yu. V. Gulyaev, N. D. Devjatkov, et al., Carbon nanoclusters structures as a one of materials of emission electronics in the future, *Radiotekhnika* (in Russian), No. 2, 9 (2000).

447. C. A. Spindt, Microfabricated field-emission and field-ionization sources, *Surf. Sci.* **266**, 145, (1992).
448. C. A. Spindt. C. E. Holland, A. Rosengreen, and I. Brodie, Progess in field emitter array development for high-frequency operation, *IEDM Dig. Tech. Papers, Washington, DC*, (1993), p. 749.
449. Yu. V. Gulyaev, I. S. Nefedov, N. I. Sinitsyn, et al., Distributed microwave amplifer on field emitter arrays with a nonhomogeneous energy collector, *J. Vac. Sci. Tech. B* **13**(2), (1995).
450. C. A. Spindt, FEA's state of art and applications, 2nd *Int. Workshop on Vacuum Microelectronics, Wroclaw, Poland*, (July 11–13, 1999).
451. G. A. Dyuzhev, G. N. Fursey, A. V. Kotcheryzhenkov, D. V. Novikov, and V. M. Oichenko, Carbon clusters produced by electric arc evaporation of graphite and their field emission characteristics *11th Int. Vacuum Microelectronics Conference Proc. IVMC'98, The Grove Park Inn Asheville, NC, USA* July 19–24, 1998. pp. 198–199.
452. G. A. Dyuzhev, G. N. Fursey, A. V. Kotcheryzhenkov, D. V. Novikov, V. M. Oichenko, and V. S. Boikov, Structural and emission characteristics of field emitters based on carbon nanoclasters. *12th Int. Vacuum Microelectronics Conf., Darmstadt, Germany. Proc. IVMC'99* (July 6–9 1999), pp. 316–318; G. N. Fursey, Explosive emission processes. *2 the Int. Workshop on Vacuum Microelectronics, Wroclaw, Poland, Proc. IWVM'99* (July 11–13, 1999), p. 44.
453. G. N. Fursey, D. V. Novikov, and V. M. Oichenko, Alternative field emission cathodes for multi-emitter systems based on diamond-like films and tubelenes. *13th Int. Vacuum Microelectronics Conf.*, Guangzhou, China. *Proc. IVMC'2000*. (August 14–17, 2000).
454. V. I. Makhov, N. A. Duzhev, A. I. Kozlov, B. I. Gorfinkel, and E. N. Petrov, Flat-panel fluorescent display with silicon matrixed cold cathode, *1st Int. Conf. Vacuum Microelectronics, Williamsburg, VA* (1988).
455. M. E. Crost, K. Shoulders, and M. E. Zinn, Thin electron tube with electron emitters at the intersection of crossed conductors, US Patent 3500102, (1970).
456. J. P. Biberian, A serial addressing method for field emission tip flat panel TV, *1st Int. Conf. Vacuum Microelectronics, Williamsburg, VA* (1988).
457. R. Meyer, A. Ghis, P. Rambaud, and F. Muller, Microchip fluorescent display, *Proc. Japan Display*, (1985), p. 513.
458. R. Meyer, A. Ghis, P. Rambaud, and F. Muller, Development of a matrix array of cathode emitters on a glass substrate for flat panel display applications, *1st Int. Conf. Vacuum Microelectronics, Williamsburg, VA* (1988).
459. R. Meyer, Recent development on microtips display at LETI, *4th Int. Conf. Vacuum Microelectronics, Nagahama, Japan* (1991).
460. R. Meyer, Private communication (1993).
461. A. T. Rakhimov, V. A. Samorodov, E. S. Soldatov, et al., Study of emission and structure characteristic correlation of diamond films by scanning tunneling microscopy, *Poverkhnost* **7**, 47, (1999) (in Russian).
462. L. M. Baskin, G. N. Fursey, V. I. Ivanov-Omskiy, et al., The microgeometry and field emission properties of diamond-like films, *Int. Conf. Phys. Dielectrics, St. Petersburg* (1997) (in Russian).
463. N. I. Sinitsyn, Yu. V. Gulyaev, G. V. Torgashov, et al., Thin films consisting of carbon nanotubes as a new material for emission electronics, *J Appl. Surf. Sci.* **111**, 145, (1997).
464. G. G. Sominski and T. A. Tumareva, Field electron emitter with fullerene coverages. *12th Int. Vacuum Microelectronics Conf., Darmstadt, Germany*. Proc. *IVMC'99*, (July 6–9, 1999), p. 304.
465. G. A. I. Amaratunga and S. R. Silva, Nitrogen containing hydrogenated amorphous carbon for thin film field emission cathodes, *Appl. Phys. Lett.* **68**(18), 2529–2531 (1996).
466. G. A. Dyuzhev, G. N. Fursey, A. V. Kotcheryzhenkov, et al., Carbon clusters produced by electric arc evaporation of graphite and their field emission characteristics. *11th Int. Vacuum Microelectronics Conf. Proc. IVMC'98 The Grove Park Inn Asheville, NC, USA* (July 19–24, 1998), pp. 198–199.
467. G. A. Dyuzhev, G. N. Fursey, A. V. Kotcheryzhenkov, et al., Structural and emission characteristics of field emitters based on carbon nanoclusters. *12th Int. Vacuum Microelectronics Conf. Proc. IVMC'99, Darmstadt, Germany* (July 6–9, 1999), pp. 316–318.
468. J. Robertson M. W. Mechanism of electron field emission from diamond and diamond-like carbon, *Proc. 11 IVMC* (1998), p.162.
469. O. M. Kuttel, O. Groning, Ch. Emenegger, E. Mailland, and L. Schlapbach, Electron field emission from nanotube and other carbon containing films, *Proc. 11 IVMC* (1998), p. 194.
470. G. N. Fursey and L. M. Baskin, Particularly of field emission from semiconductors, *Mikroelektronika* **26**(2), 117–122 (1997) (in Russian).

471. H. Kim, J. W. Huh, J. M. Kim, et al., Development of a diamond-like carbon based field emission display. *11th Int. Vacuum Microelectronics Conf. Proc. IVMC'98, The Grove Park Inn Asheville, NC, USA* (July 19–24, 1998), pp. 192–193.
472. W. B. Choi, D. S. Chung, S. S. Hong, et al., Carbon nanotube based field emission displays. *12th Int. Vacuum Microelectronics Conf., Darmstadt, Germany Proceeding IVMC'99* (July 6–9, 1999), pp. 310–311.
473. W. Stocker, H. W. Fink, and R. Morin, Low-energy electron and ion projection microscopy, *Ultramicroscopy* **31**, 379–384 (1989).
474. R. Morin, A. Gargani, and F. A Bel simple UHV electron projection microscopy, *Microsc., Microanal., Microstruct.* **1**, 289–297 (1990).
475. R. Morin and A. Gargani, Ultra-low-energy-electron projection holograms, *Phys. Rev. B* **48**, 6643–6645 (1993).
476. E. S. Snow and P. M. Campbell, Fabrication of Si nanostructures with an atomic force microscope, *Appl. Phys. Lett.* **64**(15), 1932 (1994).
477. Kazuhiko Matsumoto, Masami Ishii and Kazuhito Segawa, *J. Vac. Sci. Technol. B* **14**(2), 1331 (1996).
478. K. Tjaden, Chemical mechanical polishing in FED manufacturing, *Chemical Mechanical Polishing Workshop*, ICMCTF (1996).
479. E. I. Givargizov, Field emitters based on diamond-covered silicon tips, *Microelectronika* **26**(2), 102–106 (1997) (in Russian).
480. E. I. Givargizov and A. T. Rachimov, Field emission cathodes on nanocrystal and nanodiamond films (physics, technology, application), *Uspekhi Fiz. Nauk* **170**(9), 996–999 (2000) (in Russian).
481. J. Lee and J. Nam, The technologies of FED devices, 2nd *Int. Workshop Vacuum Microelectronics, Wroclaw, Poland* (July 11–13, 1999), pp. 26–31.
482. S. Ahn, N. Lee, H. Jeonge, et al., Development of 3.5-inch full-color field emission display panels in the low and high voltage packaging, *12th IVMC, Darmstadt* (1999), pp. 28–29.
483. R. Meyer and R. Baptist, Field emission display: the ultimate thin film flat CRT technology, *Electronic Display 94*, Germany, (1994).
484. A. P. Burden, H. E. Bishop, M. Brieriey, et al., Field emitting inks for consumer-priced broad-area flat-panel display, *12th IVMC, Darmstadt*, (1999), pp. 62–63.
485. G. A. Mesyats and G. N. Fursey, Explosive electron emission of initial stages of vacuum discharges, *Kholodnie Katodi* (Radio, Moskva, 1974), pp. 269–292 (in Russian).
486. G. N. Fursey, Field emission and vacuum breakdown, *IEEE Trans. Electrical Insulations* **EI-20**(4), 659–670 (1985).
487. G. A. Mesyats, Ectons in a vacuum discharge: breakdown, the spark, and the arc, – M.: Nauka, (2000), p. 424.
488. G. N. Fursey and L. A. Shirochin, Explosive emission processes in high electric fields, *Uzbek J. Phys.* **2**(1), 53–62 (2000).
489. M. Benjamin and R. O. Jenkins, The distribution of autoelectronic emission from single crystal metal points, *Proc. Royal Soc.(A)* **176**, 262 (1940).
490. F. J. Blatt, *Physics of Electronic Conduction in Solids* (McGraw-Hill Book Company, 1968).
491. E. W. Muller, Weitere Beobachtungen mit dem Feldelectronenmikroskop, *Z. Phys.* **108**, 668 (1938).
492. E. W. Muller, Das Feldionenmikroskop, *Z. Phys.* **131**, 136–142 (1951/52).
493. M. J. Troyon, Energy spread of different electron-beams. 2. Field-emission electron-beams, *J. Microscopie Spectroscopie Electroniques* **13**(1), 49–64 (1988).
494. B. F. Coll, Present status and future outlook of microtip and CNT based field emission displays, *Int. Topical Meet. Field Electron Emission from Carbon Materials, Moscow* July 2–4, (2001).
495. A. P. Barsukov, Displays. casting engineering, *Film and TV Engineering* No. 11, 53–62 (2000) (in Russian).
496. A. Pur, The future of imaging techniques, *PC Magazine*, 76 (March 25, 1997).
497. D. J. Rose, On the magnification and resolution of the field electron microscope, *J. Appl. Phys* **27**(3), 215–226 (1956).
498. K. Yokoo and H. Ishizuka, RF application using field emitter arrays, *9th Int. VMC, St. Petersburg* (1996), pp. 490–500.
499. B. F. Bondarenko, Yu. L. Ruhbakov, E. G. Sheshin, and A. A. Shuyka, Field emission cathodes and devices using them, *Review of Electronic Engineering*, **4**, 4(814), 2–58 (1981) (in Russian).
500. H. F. Gray, All-silicon gridded FEAs, *Presented at 1980 Tri-Services Cathode Workshop, Greffis AFB, Rome*, No. 4 (April 1980).

501. I. K. Yokoo, Vacuum microelectronics, *J. Vac. Soc. Japan* **32**(2), 63–67 (1989).
502. J. K. Cochran, A. T. Chapman, R. K. Feeney, and D. N. Hill, Review of field emitter array cathodes, *Int. Electron Devices Meet.* (1980), pp. 462–466.
503. P. V. Philips, R. E. Neidert, C. Hor, C. A. Spindt, *Technical Digest of IVMC-93* (1993), p. 155.
504. A. B. Gurinovich, A. A. Kuraev and A. K. Sinitsyn, The studies of TVT with cathode modulation optimized on power efficiency, *Radiotechnika and Electronika* **45**(2), 1493 –1498 (2000) (in Russian).
505. I. Brodie, Keynote Address to the First IVMC, June 1988: Pathways to Vacuum Microelectronics, *IEEE Trans. Electron Devices* **36**(11), 2637–2640 (1989).
506. H. M. Bizek, et al., Gigatron, *Trans. Plasma Sci.* **16**(2), 258–263 (1988).
507. V. A. Isaev, D. V. Sokolov, and D. I. Trubezkov, Lecture on microwave devices with emission modulation, *8th Winter School/Workshop of Engineers* (1989), Book 2, pp. 3–36 (in Russian).
508. A. V. Goldetsky, The devices of vacuum super high frequency microelectronics, *Microelectronika* **26**(2), 123–129 (1997) (in Russian).
509. N. I. Tatarenko and A. S. Petrov, Vacuum microelectronics: reality and prospects, *Microelectronics worldwide. Advances the Contemporary Radioelectronics* No. 7, pp. 10–31 (1998) (in Russian).
510. E. S. Snow and P. M. Campbell, Fabrication of Si nanostructures with an atomic force microscope, *Appl. Phys. Lett.* **64**(15), 1932–1934 (1994).
511. Kazuhiko Matsumoto, Masami Ishii and Kazuhito Segawa, *J. Vac. Sci. Tech. B*, **14**(2), 1331 (1996).
512. Y. Sugiyama, Recent progress on magnetic sensors with nanostructures and applications, *J. Vac. Sci. Tech. B* **13**, 1075–1083 (1995).
513. K. Uemura, S. Kanemaru, and J. Itoh, Fabrication of a vacuum-sealed magnetic sensor with a Si field emitter tip, *J. Micromech, Microeng.*, **11**, 81–83 (2001).
514. J. M. Orloff and L. M. Swanson, Study of field-ionization source for microprobe applications, *J. Vac. Sci. Technol.* **12**(6), 1209–1213 (1975).
515. M. Faubel, W. M. Hebber, and J. R. Toennies, Low background cold cathode ion source for molecular beam detection, *Rev. Sci. Instr.* **49**(4), 449–451 (1978).
516. B. S. Satyanarayana, A. Hart, W. I. Milne, and J. Robertson, Field emission from tetrahedral amorphous carbon, *Appl. Phys. Lett.* **71**, 1430–1432 (1997).
517. D. Hong and M. Aslam, Field emission from p-type polycrystalline diamond films, *J. Vac. Sci. Technol. B* **13**(3), 427–430, (1995).
518. L. M. Baskin, Dynamic of field emission processes in static and super high frequency fields. Thesis of doctor's dissert., Tomsk, 1990 (in Russian).
519. L. M. Baskin and G. N. Fursey, Decisive role of deep trap states in initiating vacuum breakdown in presence of dielectric insertions, *Proc. 13th ISDEIV, Paris* (1998), Part 1, pp. 31–33.
520. Yu. Zakharchenko, I. S. Nefedov, G. V. Torgashov, N. I. Sinitsyn, and Yu. V. Gulyaev, Calculation of the electrophysical parameters of electron gas in a carbon nanotube using the quantum well method, *12 IVMC, (7), Darmstadt*, pp. 88–89 (1999).
521. J.-M. Bonard, Field emission from carbon nanotubes: perspectives for applications and clues to the emission mechanism, *12 IVMC, (7), Darmstadt*, pp. 300–301 (1999).
522. Y. Chen, D. T. Shaw, and L. Quo, Field emission of different oriented carbon nanotubes, *Appl. Phys. Lett.*, **76**(17), 2468 (2000).
523. N. I. Sinitsyn, G. V. Torgashov, Yu. V. Gulyaev, et al., Electron field emission from nanocluster carbon-metal films, *12 IVMC, (7), Darmstadt*, 1999, p. 254.
524. W. B. Choi, N. S. Lee, W. K. Yi, et al., The First 9-inch carbon-nanotube based field-emission display for large area and color applications, *Digest SID 00*, pp. 324–327 (2000).
525. I. Brodie, Physical considerations in vacuum microelecronics devices, *IEEE Trans. Electron Devices* **36**(11), pp. 2641–2644, 1989.
526. D. Nicolaescu, Modeling of the field emitter triode (FET) as a displacement/pressure sensor, 41st IFES, 1994, Rouen, France (publ. In *Appl. Surf. Sci.* 87/88, 1995, p. 61).
527. J. C. Jiang, P. K. Allen, and R. C. White, Microvacuum diode sensor for measuring displacement and loads, *In 4th Int. Conf. On Vacuum Microelectronics* (Nagahama, Japan, 1991).
528. H. C. Lee and R. S. Huang, A novel field emission array pressure sensor, *In 6th Int. Conf. On Solid State Sensors and Actuators* (San Francisco, CA, 1991), p. 110.
529. H. H. Busta, J. E. Pogemiller, and B. J. Zimmerman, The field emitter triode as a pressure/displacement sensor, *In 6th Int. Conf. On Vacuum Microelectronics* (Newport, RI, 1993).

530. G. N. Fursey, A. V. Kotcheryzhenkov, D. V. Novikov, and V. M. Oichenko, Field emission from diamond-like films and tubelenes, *The International Topical Meeting on Field Electron Emission from Carbon Materials*, 2–4 July 2001, Moscow, pp. 50–51
531. J. M. Kim, N. S. Lee, J. H. You, J. E. Jung, C. G. Lee, J. W. Kim, K. S. Choi, et al., Development of carbon nanotube field emission displays with triode sructures, *The International Topical Meeting on Field Electron Emission from Carbon Materials*, 2–4 July 2001, Moscow, p. 29
532. G. N. Fursey, Field Emission in Vacuum Microelectronics, *Appl. Surf. Sci.* **215**, 113–134 (2003).
533. G. N. Fursey, D. V. Novicov, G. A. Dyuzhev, A. V. Kotcheryzhenkov, and P. O. Vassiliev, The field emission from carbon nanotubes, *Appl. Surf. Sci.* **215**, 135–140 (2003).

LIST OF MAIN NOTATION

$A(T_e, T_i)$	function describing nonlocalized transfer of energy from the electron subsystem to the lattice
$b(y)$	"effective" total density of mobile charges
B	current density per unit solid angle, or brightness limited by the Langmiur equation to a value B_{max}
c	velocity of light
$c(T)$	heat capacity
$C^+, C, (d^+, d)$	electron generation and annihilation operators
d_l	length of the most probable tunneling path
$D(E_x, F)$	barrier transparency
e_i	ion's charge
E_0	energy of the localized state
E_a	energy of the lowest state in the metal
E_F, μ	Fermi energy
$E(k), K(k)$	complete elliptic integrals of the first and second kinds
E_m	maximum energy at the normal to the direction of emission
$E(y)$	field strength inside the sample
$E_x = p_x^2/2m$	part of the electron kinetic energy carried by the momentum component p_x normal to the surface
$g(T)$	a temperature-dependent Tompson coefficient[242]
$G(r, t)$	volume density of heat evolution rate determined by the Joule and Tompson effects
$f(E_x, \vec{T})$	Fermi function
F	applied electric field
H_L	unperturbed cathode Hamiltonians
H_m	barrier height
H_R	unperturbed anode Hamiltonians
H_T	tunneling Hamiltonian
i, I	emission current

I_{couple}	current of paired electrons
j, J	field emission (FE) current density.
j_0	emission current density as defined by the Fowler-Nordheim equation (Eq. 1.8)
j_{em}	emission current density from the conduction band
$j(F_{\text{S}}, T_{\text{S}})$	emission current density depending on the emission surface temperature T_{S} and the electric field strength F_{S} at the surface
J_{tr}	Jakobian of the transition from coordinates p_x, p_y, p_z to ζ, ϑ, χ
k	Boltzmann constant
L	length of the tip
m	mass of electron
m_i	ion's mass
$m_x^* = m_x/m, m_y^* = m_y/m, m_z^* = m_z/m$	ratios of the effective mass components to the free electron mass.
$M = \varphi/F_0 a$	upon a dimensionless parameter, a—the micro-roughness height
$\vec{n}$	unit vector normal to the emitting surface
n_i	electron concentration in a pure semiconductor
n_∞	current density in the sample bulk
N_c	density of states at the bottom of a conduction band
$N_L = \sum_k C_k^+ C_k$	number of particles operator
$\vec{p}$	electron momentum
p_{F}	momentum of an electron at the Fermi surface
$p_x^2/m_x = \zeta^2$	
$p_y^2/m_y = \vartheta^2$	
$p_z^2/m_z = \chi^2$	
$P(\varepsilon)d\varepsilon$	total energy distribution of emitted electrons relative to the Fermi level.
r	radius of the tip endcap
r_{a}	anode radius
r_{k}	cathode radius
R	STM tip of radius
R_j	principal radii of curvature of the surface under investigation at their point of intersection with the most probable tunneling path
$T = T(\vec{r}, t)$	emitter temperature
$T^* = d/2K = 5.67 \cdot 10^{-5}/\sqrt{\phi} \cdot t(y)F$	inversion temperature
$T_{\text{e}}, c_{\text{e}}, \chi_{\text{e}}$	electron temperature, electron heat capacity, and thermal conductivity
T_0	temperature of the tip, far from the apex
T_{S}	a temperature of the emitter apex where $r = r_{\text{e}}$
U_0	potential in a spherical diode in the absence of space charge

$U(x)$	potential function
$U(x), \zeta(x)$	potential energies of the bottom of the conduction band and of the electrochemical potential level
v, k	velocity of incident particle and its wave vector, respectively
x_m	distance from the interface of the potential maximum
Γ_1 and Γ_2	localized state decay widths in the STM tip
$\Phi(x)$	a nonlinear potential function depending on the shape of the micro-roughness
Φ_p	polar angle in y, z plane
$\Phi(\vec{r}, t)$	electric potential
$\mathcal{F}_{1/2}(\eta)$	Fermi integral of the order 1/2 defined in the usual way
$\alpha(T)$	function related to $g(T)$[242]
β	geometric factor
γ	surface tension of the emitter tip material
δ	substance density
$\delta(E_k^r - E_q^r)$	spectral function of the electron gas in the metal
$\delta(k)$	phase shift between the incident and reflected waves
$\varphi_\perp$	angle in the plane perpendicular to the magnetic field
$\varphi_\parallel$	plane parallel to the field
φ	work function
$\psi(T)$	surface emissivity
$\rho_\perp$	density of states with the momentum normal to the surface barrier
ρ	resistivity constant
$\rho(T)$	resistivity as a function temperature
ρ_0	space charge density
$\lambda(T)$	thermal conductivity
σ	Stephan–Boltzmann constant
τ_{ee}	time of the electron–electron interaction
$\mu(E) = \mu_0(E_0/E)^\delta$	carrier mobility
μ_0	carrier mobility in field E_0
λ_{ep}	free path length of electron–phonon interaction
$\tau_{Wigner} = 2R/v + (2/v\partial/\partial k)[\delta(k)]$	time interval between when the centers of mass of the incident and scattered wave packets crossed the surface of a sphere of radius R
τ_{tun}	tunneling time is assumed to be the mean time of interaction between a tunneling electron and the potential barrier
$\tau_{Hartman}$	time was calculated as the time interval between when the maximum in the wave packet passes through the left side of barrier and the formation of the maximum in the transmitted wave packet at the right side of the boundary (used by Hartman[264])

τ_{ep}	characteristic time of the electron–phonon interaction
$\vartheta(x)$	degeneration parameter
ϑ_s	degeneration parameter near the surface
ϑ_∞	value $\vartheta(x)$ in the sample bulk
$\vartheta(Y), \Theta(Y)$	Nordheim function
Ω_j	a matrix representing the curvature of surface apex of the STM probe at the point of its intersection with the most probable tunneling path
ε_g	forbidden band width (for emission from the valence band)
ε_a	acceptor level
æ	high-frequency dielectric constant
ε_a and ε_d	energies of the acceptor and donor levels
$\bar{\varepsilon} = \bar{E}_{em} - \bar{E}_{cd}$	difference in average energies between the field emitted electrons ($\bar{E}_{em}$) and the conduction band *d* electrons ($\bar{E}_{cd}$)
ε_t	activation energy of traps
ω_L	Larmor precession frequency
υ_F	velocity of electrons on the Fermi surface
χ	electron affinity
χ_e	slopes depend on the electron affinity
χ_T	thermal conductivity coefficient

INDEX

Acceptor impurities 77
"accumulation" of surface micro-roughness 48
activation energy [migration a.e.] 47, 49, 95, 96
Activation energy of traps 95
activation energy 47, 49, 95, 96
Active element(s) 137, 138
adherent coating 73
adiabatic condition 58, 60
adiabatic conversions 57, 58, 60
Adsorbates (Al,Cr,Ge,Mb,Ti,Si,Zr,Hf) on W and Mo 125
adsorbed layers 37
Adsorbed Zr island(s) 124
Adsorption of zirconium (Zr) on W 54, 128
Allotropic forms of carbon 161, 162
Amorphous carbon 163
Amplifiers 69, 138, 139, 153, 154, 155, 157, 158, 169
Amray 117, 133, 135
Anisotropic adsorption of oxygen 101
Annealed emitter tip 124
Aperture 131
applied voltage 21, 23, 24, 48, 49, 78, 79, 82, 96, 128, 163
Array cathode 146, 148
array of field emitters 69
Arrhenius equation 38
atomic scale surface smoothness 6
atomic sharpness 6, 116
Atomic sharpness 6, 116
Atomically clean surface 107, 109
atomically clean 33, 49, 72, 98, 101, 103, 107, 109, 110, 144
"Atomically sharp" edges on the Mo 145
"Atomically sharp" edges on the W 145
atomically sharp emitters 6, XII
atomically sharp microtips 6
Atomically sized object(s) 131
atomically smooth emitters 6
attenuator 61
Auger spectrometers 115
Auger-electron spectrometers 136
"Auto-epitaxial" 119
Avalanche breakdown 96
average current 60, 61, 62, 128, 155, 168

Backward wave tubes (BWT) 157
Band bending changes 83
band bending 72, 83, 84
Bardeen's potential 27, 28
barrier width 7, 9, 10, 11, 27
Base electrode 146, 154
Baz's method 59
beam-bunching injectors 69
Bendgap semiconductors 163
Bias voltage 98
Bi-Sn-Pb alloy 67, 68
blade cathode(s) 24
Bloch function formalism 12
blunt emitter 30
Boguslavskii-Langmuire law 21
bombardment of emitter surface 57
breakdown event 67
breakdown, vacuum breakdown 32, 33, 39, 47, 64
bright ring 33
Broad-band 96
Broglie relation its wavelength 142

Build-up (effects) 48
build-up process 55, 118, 123, 128, 131
Built-up emitter tips 123
Built-up Zr/W 130
bunches 57

capillary emitters 67
Capture centers 92
Carbides 157
Carbon fibers 135, 162
Carbon Nanoclusters 161, 162, 163, 164, 167
carbon nanotubes 137, 161, 163, 165, 166, 167, 168
Carbon substrate 134
carrier concentration 71, 72, 81, 84, 88, 89, 91, 92, 98
Carrier generation 81, 89, 90, 98
"Cathode deposit" 162, 166, 167
cathode instabilities 36
Cathode-ray tube (CRT),(CRTs) 148, 150, 151, 152,
cavity 60, 61, 66
Ceramic(s) 111, 112, 113, 114
Clistrode 155
Close-packed crystal facets 119
Close-packed layer(s) 125, 126
Cluster(s) of atoms 122
Coherency 117
collector 36, 37, 53, 61, 65, 114, 138, 155, 156, 159, 160
Color rendition 150
commutator 61
conditioned electrodes 32
conducting liquid surface 68
conduction band 31, 32, 39, 40, 50, 75, 78, 81, 82, 83, 85, 88, 96, 97, 154, 163, XI
cone-shaped emitter 52
Contaminant(s) 77
contamination removal 73
Contrast 48, 73, 75, 125, 138, 151, 152
converters 69, 137, 138
Correlated electrons 105, 107
Correlation length 14, 111
Coverages of Zr 125
Crests 69, 117, 119
Critical temperature 47, 111, 113
CRT-monitors 152
Cryostat 53, 110, 111
Crystal facets 115, 119
Crystal grain 116
crystallographic orientation 116
crystallographic orientations 6, 116
Current converters 138
Current kinetics 98
current of a beam collector 61
cut-off potential 66

dc mode 50, 53, 54, 56
Debye temperature 50
deep cooling 56
Deep trap 92, 96, 153
Degeneration criterion 88
dence plasma 32
Density of pixels 146
Depletion region 88, 89, 92, 97
desorption in a strong electric field 72
desorption processes 72
destruction of the semiconductor tip 73
deviation from FN theory 9, 29
deviation from linearity of the current-voltage characteristics 29
deviation-point current 23
deviations from FN theory 20
Diamond-like films 97, 161, 163, 164, 166, 167, VI, XII
Diamond-like structures 140
Diffusion current 89, 90
Diffusion-drift equilibrium 98
directional coupler 61
dispersion law 12, 20
Display lifetime 148
Display panel(s) 138, 139, 144, 149, 152, 167
dissipation 39, 58, 62, 137
"Double" localization 131
Dwell time 148
Dyke's Experiments 33

Effective masses of electrons and holes 86
Efficiency of phosphors 150
"ejection effect" 64
Electron gun 117
Electric arc vaporization o graphite 161
electrical explosion 32, 34
Electroadsorption 101, 102
electrochemical etching 67, 111
Electrochemical potential 83, 84, 85
electrohydrodynamic instability 67

Electrolytic etching 77
electron affinity 71, 81, 82, 83, 85, 139, 140, 161, 163
Electron beam lithography 139, 158
electron energy spectra 64
Electron energy spread 115
electron holography 6, 20, 35, 39, 40, 139, V, XI
Electron lithography machines 115
electron lithography 115, 138, 140, 158, 161
electron microscopy 57, 101, 121, 122, 134, 135, 136, 139, 140, 142, 164, 166, 167, 169, V, XI
electron momentum 31, 143
Electron optical systems 57, 69, 136
electron pair tunneling 16
Electron pairs in FE 105
Electron sources 11, 69, 115, 117, 152
electron spectroscopy 6, 64
electron subsystem 51
electron tunneling time 15
electron-electron interaction 50, 51
Electron-electron interactions 105
electron-electron scattering 15
electron-electron tunneling time 15
electron-lattice interaction 51
electron-phonon free path 50, 54
electron-phonon interaction 50, 62
electrons in pairs 15
Electrostatic lens 140
electrostatic repulsion 25
"Elementary emission events" 105, 114
elliptic Fermi surfaces 32
elliptic integrals 3
Emission image 34, 37, 64, 71, 78, 79, 80, 81, 82, 142
emission images 33, 53, 78, 101, 104, 112, 119, VI
Emission localization to a small solid angle 125, 126
Emission localization to small solid angles 115
Emission spectra statistics 106
emission-included migration 49
emitter bulk 43, 46, 49, 71, 72, 95, 98
emitter cone angle 52
emitter destruction 33, 39, 43, 72, 89
emitter faces 6
emitter instability 72
emitter shape 78, 119
Emitting area 4, 12, 53, 78, IX
energy conversion devices 57
energy distribution 8, 9, 11, 12, 13, 40, 41, 57, 64, 65, 66, 69, 117, 155, 164, VI, IX
energy spectrum 4, 6, 12, XI
evaporation energy 71
Evaporation field 74, 102
exchange process (energy exchange mechanism) 39
excitation device 61
exitation 33, 60, 61, 67, 68, 106, 160, 161, IX
experimental chamber 60
"explosion" of the emitter tip 72
explosive breakdown 34
explosive emission 39, 67, XI
explosive-like evaporation 72
Extracting electrode 146
extremely high electric field 41, VI

facilitating addressing 148
Fast-acting high frequency devices 137
fatigue phenomena 72
FE display panels 144
FE electron sources 115, 117
(FE) Field Emission from semiconductors 71, 73, 75, 77, 79, 81, 83, 85, 87, 89, 91, 93, 95, 97, 99, 101, 103, VII
FE image for nanotubes 168
FE localization 116, 119, 124, 130
FE matrix 138
FE micrographs 120
FE microscope 38, 78, 122, 140, 142, 143, 144, 167
FE microscopy 33, 60, 101, 119, 138, 141, 143
FE multi-needle cathode 161
FE Miller microscope 48, 53
FE pattern(s) 35, 48, 73, 74, 102, 108, 111
FE projector 111, 113
FE statistics 107, 108, 109, 110, 111, 112, 113, 114
FE triode 139, 153, 154
FEAs 138, 139, 140, 149, 151, 152, 153, 156, 157, 158, 160, 166, 169
Feinman's procedure 59
Femitron 155
Fermi energy 4, 14, 32, 40, 41
Fermi function 4, 31
Fermi integral 86, 88
Fermi level 2, 4, 10, 15, 26, 31, 39, 40, 50, 82, 85, 87, 88, 89, 154, 163

Fermi surface structure 12, XI
Fermi surface 12, 31, 32, 51, XI
ferrite filter 61
Fiber-optic bundle 111
field (electron) emission from liquid metals 67
field desorption cleaning 73
field desorption 72, 73, 74, 100, 101, 103, 160
field electron image 60
(field emission) FE cathode 24, 32, 36, 47, 61, 64, 116, 131, 134, 136, 137, 139, 148, 149, 150, 154, 158, 159, 168
Field Emission (FE) pattern, images 74, 130, 164, 165, 167, VI
Field emission arrays (FEAs), FEA 153, 156
field emission current stability 65
Field emission displays 144, 146, 150, 169
Field emission localization by a local work function decrease 124
Field emission statistics from metals 108
Field emitter arrays 146, 149, 158
Field emitter explosion 32
Field enhancement 67, 116, 118, 125, 163, 164
Field evaporation 102, 119, 121, 131
Field geometric factor 79
Field ion imaging 74, 119
Field ion micrograph(s) 75, 123, 124
Field ion microscopy 75, 103, 119, 125, 142
Field ion picture 141
field outgrowths 121
field penetration 72, 80, 81, 85, 92, 94, 105
field stimulated reactions 72
"Field-surface melting" 131
Film of dielectric 139
Flat crystal plane 116
Flat-panel display 146, 152
FN coordinates 3, 9, 72, 75, 166
FN formula 3, 12, 23
FN theory 1, 6, 8, 9, 12, 13, 17, 19, 20, 26, 29, 58, 60, 69
Formula of Boguslavskii-Langmuire 20
Fowlel-Nordheim (FN) coordinates 3, 9, 72, 75, 166
Frame rate 148
free carriers 72
Free electron model 1, 6, 12, 31
Free-electron lasers 152
freezing the micro-protrusions 68
Frequency converters 137
Frequency multipliers 69, 138
frequency transformation 57
Full color displays 150, 151
Fullerene(s) 137, 140, 161, 162, 163, 164, 165, 166, VI

Gas lasers 160
Gas-filled proportional counter 106
Gate electrode 146, 153, 154, 158
Gate-to-cathode capacitance 148
Ge, Si, GaAs (atomically clean surface) 74, 144
gedanken experiment 59
Generation rate 81, 82, 95, 98
Generators for frequencies 139
Geometric factor 8, 79
geometric quotient 4
Glass faceplate 146
Graphite fibers 159
Green's function method 14
Grid electrode 154
grids 61, 157

half width of energy distribution 65
Hamiltonian anode 13
Hamiltonian cathode 13
Hamiltonian complete 13
Hamiltonian tunneling 13
Hartree-Fock approximation 27
heat capacity 45, 51
heat localization 47
Heat resistance of the emitters 137
Heterodynes 157
High beam coherence 115
high brightness 115, 131, 138, 139, 148, 151, 158, 167, 168, XI
high current densities 6, 16, 19, 24, 30, 31, 32, 34, 39, 47, 49, 56, 114, 131, X, XI
High energy tail 105
high energy tails 15, XI
High frequncies 137
high power 57, 69, 137, 140, 152
High radiation tolerance emitters 137
High resolution television 152
High-brightness field emission 115
High-definition television 148, 150
high-energy electron devices 69
High-power electron beams 152
high-power traveling wave tubes 69
High-resolution electron probe devices 115
Hitachi 117
Hole energies 85

Holograms 135
Honeycomb anode 146
"Honeycomb" control electrod (gate) 138
Horisontal microtriodes (vacuum transistors) 153
"hot" hole 15
Hydrogen field ion pattern 126
Hysteresis 96, 97, 163

image potential 9
Image shrinkage 78, 79
Image stability 152
initial properties 72, 77, 78, 104
initiation phase of breakdown 32
injecting carriers 71
input power 60, 61, 65, 67
Integrated circuits 138, 139, 140, 152
interelectrode capacitance 60
Interference fringes 131
Interferogram(s) 135
intrinsic magnetic field 25
Intrinsic noise (level) 155, 156
inversion temperature 39, 41, 43, 47
ion bombardment of emitter 57
ion emission microscopy 71
ion emission 67, 71
Ion microscope 121, 125, 142
ion motion 57, 63
Ion pattern(s) 74, 126
ion trajectories 63
Ionic image 142
Ionization potential 71, 163, 164
Iridium (Ir) FE cathode 159
irreversible changes 64, 72

Josephson tunneling 13
joule heating 36, 38, 39, 43, 45, 49, 50, X

Ka-Na alloy 67
Kernel 46
Kinks 82, 131
klystron(s) 69

Langmuir equation 115
Langmuir limit 115
large micro-protrusions 119
Large-area (LCD) displays 150
Large-pixel-count displays 148
Larmor precession frequency 59
LC-indicators 151
LC-materials 151
Lead 28, 43, 64, 71, 72, 125, 127, 163, 164
Lifetime 69, 82, 125, 148
Light probe 92, 98
limiting current densities 50, 89
liquid helium temperature 52, 154
liquid metals 61, 67, 69
Liquid nitrogen temperature 155
liquid surface(s), metals 67, 68
liquid-metal cathodes 67
liquid-metal ion sources (LMIS) 67
LMIS techniques 67
Local radii of curvature 116
localized emission 54, 122, 128, 135, X
localized Green functions 7
long current pulses 33
Lorentz force 25, 158
Low electron affinity 139, 140, 161
low energy tails 6, 15, XI
Low index planes 117
Low voltage displays 150, VI
low-energy electrons 11, 66, 136
low-energy tails 6, 11, XI
Low-voltage FED 150
Luminance 148, 150, 151

Magnetic field gauges 138
Magnetron and laser sputtering of graphite 161
magnetron(s) 57, 69, 161
Magnification 38, 71, 78, 133, 136, 140, 141, 142, IX
manybody problem 27
many-particle effects 13, 15, 108
Mass-spectrometers 140
Matrix field emission cathodes 148
maximum brightness 131, 133
maximum current densities 38, 49, 62, 125, 135, X
maximum electron energy at the resonator output 65
Memory cells 152
metal FE cathodes 159
Metal gate film 147
metal-vacuum boundary 19, 57, 59
Micro-channel plate intensifier 74
microcrystal field emitter 71
Micrometer size ...dimensions 137

Micro-protrusion growth 118, 119
Micro-ridges 117, 123, 124
microtip 8, 32, 63, 130
Microtransistor 138, 139
microwave apparatus 61
Microwave devices 69, 138, 155, 156, 157, 169, XI
microwave field power conversion 57
microwave power amplification 60
Microwave transit-time devices 157
migration process 37
Miller indices 142
Mo 19, 30, 43, 48, 50, 108, 109, 125, 131, 145, 157, 165
Mobility 90, 91, 93, 94, 128, 164
Modulation voltages 148
modulator 61
Monochrome CRT displays 150
Monochrome display, panels 150, 168
Muller microscope 48, 53
multidimensional tunneling of electrons 7
multidimensional 7, 8, 27
Multielectron field emission from high temperature superconducting ceramics 111
multielectron theory 13
Multi-emitter matrix 149
Multipeak spectra 108
Multiple-beam electron lithography 140
Multiple-beam electron sources 152
Multiple-beam lithography 158
Multiple-particle tunneling 114
Multiple-wall carbon nanotubes 166

nanoelectronics 6, 137, 165, V, XI
nanoemitters 11
Nano-FE(cathode(s)) 138
nanometer emitters 7
nanometer range 6, 140
nanometer-scale 17, 54, 137, 139, X
Nanometer-size protuberances 143
nanosecond duration 56, 62, 63, 113
Nanostructures 131, 140, 158
nanotip 54
Nanotubes micrograph 167
narrow capillaries 61
Near-surface region 13, 15, 30, 42, 56, 71, 81, 82, 83, 84, 85, 88, 92, 94, 163
Needle 61, 67, 68, 153, 161
Niobium (Nb) 12, 35, 108, 123, 158
Nokia 151
Nonlinear current-voltage characteristics 75, 76, 81, 83, 97
nonrelativistic approximation 22
"Nonzero current" 88, 92
Nordheim function 2, 3, 4, 80, 85, 143
Nottingham heating 39, 43, 50
Nottingham mechanism 50
Nucleation 122
numerical calculation 5, 8, 10

one dimensional (problem) 1, 6, 30
one-particle approximation 6, XI
operational lifetime 69
Optimum crystallographic orientation 116
Ordered array (of metal or semiconductor tip) 146
outgassing 73
Outgrowths 119, 121, 131
overheated core 46
oxygen 67, 101, 103, 128, 129, 149, 164

Paince-Bohm's formalism 28
pair electrons 16, 105, 108
Pair tunneling 16, 17, 105, 108
paired electrons 16, 17
paraxial approximation 65
Paraxial electron beam 124
Particle accelerators 152
phase [pre-breakdown phenomena] 32, 35, 36, 37, 38, 58, 65, 131, 138, 140, 143, 160, IX
phase shift 58
Phosphor anode 126
Photo-field emission 97
Pix Tech Company 150, 152
Pixel size 150, 152, 158
Pixel, individual pixel 148, 149
Planar (lateral) vacuum FE triode 153
Planar addressable matrix 148
planar FE cathodes 161
Planar FE microtriode 153
planar macro-cathode surface 8
Plasma panel displays 151
Point FE cathode 139
Point-cathode diode 20
Poisson equation 19, 21, 22, 92
polar angle 7, 40
power amplifier 61

Power consumption 146, 150, 151, 152
power measuring device 61
pre-breakdown phenomena 33, 34, 35
pre-breakdown situation 32
precession of tunneling electron spin 59
pre-explosion phase 38
Pressure gauges 138
prismatic resonator (H03 mode) 61, 67
Probe method 92
probe-hole 36, 37, 38
Proportional semiconductor detector 106
Protuberances 111, 119, 122, 123, 131, 143
p-Type Semiconductors 81, 83, 84, 88, 91, 92, 95
Pulse amplitude analyzer 106
pulse repetition 63
Pulsed illumination 97
Pyramidal nanocrystals 165
Pyramidal protrusion 121

quasi stationary conditions 33
quasi-classical (approach) 6
Quasi-Fermi level 87, 89
quasi-particle operators 13
quasistatic regime 61

radii of curvature 7, 61, 116
radio pulse 63, 64, 65
rearrangement of the emitter surface 47
Reconstruction (of the surface) 135, 143, 164
rectangular axially symmetrical barrier 7
rectangular barrier 7, 58, 59
relativistic effects 22, 23, 24, 25, 30
Relativistic electrons 30, 152
Repetition rate 156
repulsive potential 69, XI
residual gas pressure 53, 63, 100, 159
Resist-free electron lithography 158
resistivity 42, 45, 75, 76, 77, 78, 81, 99
resonator(s) 61, 64, 65, 66, 67, 157
Response time 151, 152
retarding potential curves 60, 64, 65, 66
"ring" 33, 34, 35, 36, 37, 38, 125, 126
Ring structure(s) of Zr 125

Saturation current 82, 90, 91, 92, 95, 101
saturation region 34, 76, 80, 81, 83, 90, 91, 96, 100, 101
Scanning electron microscope(s) 121, 123, 124, 135, 140, 147
scanning tunneling microscope 7
scattering theoretic technique 7
Schottky barrier 139
Schottky emission regime 128
Screen diagonal 150
Secondary emission coefficient 105
Seits-Vasiliev-Van Oostrum 28, 29
Selective adsorption 54, 124, 128
Selective chemical etching 139
Selective work function reduction 124
self heating 37, 44, 63
"Self organizing" systems 137
Semiconductor detector 106, 107, 110, 113
Semiconductor dielectric boundary 139
semiconductor emitters 78, 81
Semiconductor whiskers 75
Service life 117, 135, 150, 159
Shadow mask 150, 151
Shape of the potential barrier 9, 19
sharp edge-plane system 24
sharp microtips 6, 10
Sharp point(s) 116
Sharp protuberances 131
Sharp riges, edges 145
sharpening 72, 78
Silicon base 147
Silicon dioxide 139
Silicon pyramids 139
silicon substrate 153, 155, 157, 158
Single atom 7, 12, 116, 131, 132
Single crystal tip(s) 97, 143
Single-crystal grains 142
Single-electron, character 110, 111
SiO_2 dielectric 147
Small electron transport time 137
Small micro-protuberances 122
small-scale field emitters 10
small-scale objects 6, 10
small-size emitters 9
solid needle substrate 61
solid/vacuum interface 56
Sony 150, 151
Space charge region 78, 80, 82, 88, 92, 97
Space-charge limited 88
Space-charge region 78, 80, 97
spectral function 14
spherical diode model 25
spherically symmetric model 10

spin orientation of the incident wave 59
Spin rotation in a magnetic field 59
Spind-type field-emitter cathode array 147
Spot size 98, 139, 150
Stabilization by oxygen treatment 128
Stable semiconductor field emission cathode 99
static potential 58
static regime 65
Statistical distribution 105, 114
Statistical FE event(s) 106
Statistical processes in FE 106
Statistics of FE 105, 108
Sticking coefficient 101
STM tip 7
Stratton's theory 82, 83, 85
stream-liquid mixture 47
Strip lines 157
Submicron dimensions 138, 139, 156
submicron inhomogeneities 67
Subpicosecond range 139
super small emission spot 56
Superconductor(s) 110, 111, 112
surface cleaning 71, 72, 73, 101
surface diffusion in strong electric fields 71
surface impurities 72
surface migration 37, 48, 54, 63, 125, 128, 143
Surface of the dielectric 139
Surface self-diffusion 49, 115
surface tension forces 67
Surface tension 67, 71, 117
surface transport processes 71

Tantalum (Ta) in the (III) direction 123
Tektronix 151
Television modulation frequencies 148
thermal conductivity 42, 44, 45, 51
thermal destruction 36, 56, 62, X
thermal radiation 45
"Thermal-field build-up" 115, 116, 118
Thermal-field (T-F) emission 4
Thermionic-field emission 5
thermoelastic stresses 46
thin layers 61
Thorium atoms 123
three-dimensional (analysis) 44
Three-dimensional microtriodes 153
three-dimensional Schrodinger 7
threshold 64, 68, 71, 101, 149, 158, 161, 162, 163, 164, VI
"tight" 12
tight binding wavefunctions 12
time delay 37
time dependent problem 44, 59
time of tunneling, traversal time 59
Tin 146, 163, 164
Tip apex 78, 98, 134, 141, 142, 143, 148
Tip cathodes 117, 146, 149, 156, 157
tip heating 37, 42, 43, 44
Tip profile 53, 121
Titan (Ti) 158
Tl 157
Tonks-Frenkel field 68
transfer Hamiltonian methods 12
Transistors 138, 139, 153
transit effects 65, 66
transit phase of the emitted electron 65
transit space 66
transit time effects 64, 66, 69
Transition processes 97
transparency 1, 2, 5, 7, 9, 12, 31, 59, 81, 98, 100, 101, 105, 113
traps 72, 92, 95, 96, 98
Travelling wave tubes (TWT) 157
Tungsten (W) FE emitters 115
Tunnel microscopy 138
tunneling path 7
tunneling spectroscopy 6
turning points 26

Ultra short frame 151
ultra-high-resolution 6
Ultrasmall emission site at (100) plane 131
under-barrier friction 58

Vacancies 131
vacuum arc 32, 63
Vacuum condition(s) 63, 100, 106, 117, 124, 125, 127, 128
vacuum gap 4, 47, 65, 82, 98, 137, 149
Vacuum magnetic sensors (VMS) 158
Vapor-liquid-solid (VLS) method 75
Vertical FE triode 154
Vertical microtriodes 138, 153
very high current densities 19, 21, 23, 25, 27, 29, 30, 31, 131
very high electric fields 6, 19, 67
very high resolution electron microscopes 69
video pulse 63, 65

Virtual lack 137
VME, Vacuum microelectronics 6, 11, 17, 69, 71, 136, 137
Voltage drop 78, 79, 80, 81

W tip 116, 120, 123, 129, 141, 142
W tips oriented in the (100) direction 125
wave packet 58
wavelengths 66
weak magnetic field 59
Weissner effect 112
Wide band gap semiconductors 97
Wide band semiconductors 139
Wide-band amplifiers 157
Wigner model 7

X-band cavity 60
X-ray micro-analyzers 115

"Zero-field" approximation 83
Zero current approximation 82, 87, 89, 90
Zr layers on W 126
Zr layers 125, 126
Zr source 129
Zr/W layers 128
Zr/W system 125, 127, 128
ZrO on W 115, 125, 126
ZrO/W layers 128
β-factor 9